Introduction to the Physics of Electronics

PRENTICE-HALL SOLID STATE PHYSICAL ELECTRONICS SERIES
Nick Holonyak, editor

ANKRUM
  Semiconductor Electronics
BURGER & DONOVAN, eds.
  Fundamentals of Silicon Integrated Device Technology:
    Vol. I: Oxidation, Diffusion and Epitaxy
    Vol. II: Bipolar and Unipolar Transistors
GENTRY, GUTZWILLER, HOLONYAK & VAN ZASTROW
  Semiconductor Controlled Rectifiers: Principles and Applications of p-n-p-n Devices
LAUDISE
  The Growth of Single Crystals
NUSSBAUM
  Applied Group Theory for Chemists, Physicists, and Engineers
NUSSBAUM
  Electromagnetic and Quantum Properties of Materials
NUSSBAUM
  Semiconductor Device Physics
PANKOVE
  Optical Processes in Semiconductors
ROBERTS & VANDERSLICE
  Ultrahigh Vaccuum and Its Applications
STREETMAN
  Solid State Electronic Devices
UMAN
  Introduction to the Physics of Electronics
VAN DER ZIEL
  Solid State Physical Electronics, 2nd edition
WALLMARK & JOHNSON, eds.
  Field Effect Transistors: Physics, Technology, and Application
WESTINGHOUSE ELECTRIC CORPORATION
  Integrated Electronics Systems

Introduction to the Physics of Electronics

MYRON F. UMAN

Assistant Professor of Electrical Engineering
University of California, Davis

PRENTICE-HALL, INC., Englewood Cliffs, N.J.

Library of Congress Cataloging in Publication Data

UMAN, MYRON F
   Introduction to the physics of electronics.

   (Prentice-Hall solid state physical electronics series)
   Includes bibliographical references.
   1. Electronics. 2. Semiconductors. 3. Physics. 4. Solids. I. Title.
TK7815.U46   621.3815   73-9685
ISBN 0-13-492702-8

© 1974 by
PRENTICE-HALL, INC.
Englewood Cliffs, New Jersey

All rights reserved. No part of this
book may be reproduced in any way
or by any means without permission
in writing from the publisher.

10 9 8 7 6 5 4 3 2 1

Printed in the United States of America

PRENTICE-HALL INTERNATIONAL, INC., *London*
PRENTICE-HALL OF AUSTRALIA, PTY. LTD., *Sydney*
PRENTICE-HALL OF CANADA, LTD., *Toronto*
PRENTICE-HALL OF INDIA PRIVATE LIMITED, *New Delhi*
PRENTICE-HALL OF JAPAN, INC., *Tokyo*

1786878

To Sandy and Jennifer

Contents

	PREFACE	xi
1	INTRODUCTION	1

    1-1 Modeling Electronic Elements and Devices, 3
    1-2 Nonlinear Devices in d.c. Circuits, 7
    1-3 Nonlinear Devices in a.c. Circuits, 8
    1-4 Integrated Circuits, 11

2	ELEMENTS OF EQUILIBRIUM STATISTICAL MECHANICS	14

    2-1 Statistical Physics, 19
    2-2 Properties of the Distribution Function, 22
    2-3 Distributions for Classical Systems in Thermodynamic Equilibrium, 24

3	NONEQUILIBRIUM STATISTICAL PHYSICS	41

    3-1 Phenomenological Relations, 43
    3-2 The Kinetic Equation, 50
    3-3 Collisions, 57
    3-4 Conductivity and Mobility, 66
    3-5 Diffusion, 72
    3-6 Experimental Data, 77
    3-7 Fluctuations and Noise, 80

4 ELECTRON BEAMS AND PLASMA ELECTRONICS 87

4-1 Electron Beams, 88
4-2 Space-Charge Limited Current, 95
4-3 Plasma, Sheaths, and the Contact Potential, 101
4-4 Biased Sheaths, 107
4-5 Ambipolar Diffusion, 110
4-6 Plasma Oscillations, 113

5 ELEMENTS OF QUANTUM MECHANICS 119

5-1 Wave-Particle Duality, 120
5-2 Schrödinger's Equation, 126
5-3 Free States and Potential Barriers, 130
5-4 Wave Packets, 137
5-5 Bound Particles, 142
5-6 Quantum Statistics, 154

6 ELECTRONS IN SOLIDS 167

6-1 The Band Theory of Solids, 171
6-2 Electrons in a Periodic Potential, 176
6-3 The Pseudoclassical Description, 185
6-4 Real Crystals, 187

7 METALS 192

7-1 Electron Emission, 195
7-2 Contact Potentials, 204

8 SEMICONDUCTORS 208

8-1 Impurities and Extrinsic Semiconductors, 210
8-2 Electrons and Holes, 216
8-3 Current in Semiconductors, 229
8-4 Bulk Semiconductor Devices, 237
8-5 Metal-Semiconductor Contacts, 243

9 SEMICONDUCTOR JUNCTIONS 255

9-1 The p-n Junction, 256
9-2 The p-n Junction in Equilibrium, 261
9-3 The Biased p-n Junction, 267
9-4 The I-V Characteristic of the Junction, 276
9-5 Reverse Bias Breakdown, 282
9-6 Noise in p-n Junctions, 284
9-7 The p-n Junction Diode, 287

Contents   ix

      9-8   Microwave Diodes, 296
      9-9   Optical Diodes, 303

10   MULTIJUNCTION DEVICES   *311*

      10-1   Bipolar Junction Transistor, 311
      10-2   Four-Layer Devices, 332

11   FIELD-EFFECT DEVICES   *342*

      11-1   The Junction Field-Effect Transistor, 343
      11-2   MOS Capacitor, 353
      11-3   Metal-Oxide-Semiconductor Field-Effect Transistors, 359
      11-4   Charge-Coupled Devices, 367

12   INTEGRATED CIRCUITS   *371*

      12-1   Technology of Integrated Circuits, 373
      12-2   Monolithic Silicon Integrated Circuits, 383
      12-3   Thin-Film Integrated Circuits, 393
      12-4   Hybrid Integrated Circuits, 396

      APPENDICES   *399*

        A   Constants, 399
        B   Properties of Silicon and Germanium at Room Temperature, 400
        C   Answers to Selected Problems, 401

Preface

    This book deals with the physical process of electric conduction. Its purpose is to develop an understanding of the basic physical concepts that underlie this process.

    Why is such an understanding important for electrical engineers? Well, sooner or later most engineers have to deal with real devices: integrated circuits, Gunn-effect oscillators, field-effect transistors, tunnel diodes, and even vacuum tubes. Some invent them, but most just use them to solve an engineering problem. In either case an understanding of what is going on inside the device to make it work is almost essential to successful development or design.

    It goes without saying that those engineers who in their professional work will try to develop new devices must have a firm knowledge of the principles that have been so successful in describing the physical wonders of nature. To them, this text will serve as an introduction to the concepts that they will pursue in greater detail and with greater breadth as they continue their careers. But other electrical engineers need this fundamental understanding too. It is they, after all, who use the exotic devices that their fellows have invented to solve modern design problems. Using them to greatest advantage requires an understanding of the underlying mechanisms that make them work. It is no longer possible, as it was fifteen to twenty years ago, to study a limited number of devices for their own sake. The explosion of the solid-state technology has made the study of all of the current useful devices impractical for an introductory course. There are just too many of them.

Furthermore, another activity of circuits and systems design engineers is to specify what characteristics a new device, yet to be invented, should have so that it will be useful. Negative differential conductivity is an excellent example of such a desirable device property. To be able to specify practical characteristics and to communicate with reasonable efficiency with the device developers, these engineers need at least an introductory knowledge of the essential physical concepts.

We cannot say what new and even more exotic devices will be developed tomorrow, but we do know most of the basic physical principles by which the electronic properties of nature can be described. And these principles are the foundations for the understanding of the operation of all devices, past, present, and future. So it is these principles and their application to all manner of devices that we shall study.

As engineers who at one time or another are involved with problem-solving, that is, with design, we use all manner of electronic devices from vacuum and gas-filled tubes through p-n junction devices, such as p-n diodes, transistors, and junction field-effect transistors, to complex integrated circuits. Our understanding of the operation of these devices, their useful ranges, limitations, and other characteristics, permits us to make intelligent choices to solve our particular problems. Frequently our understanding is expressed in terms of models for the devices, each model for each device depending on its application. Two examples of such models, but by no means the only two, are the familiar d.c. current-voltage characteristic curves and a.c. incremental linear equivalent circuits. These models and others have their bases in a thorough understanding of the physical operation of the devices, an understanding that is required even to choose the proper model applicable to a given situation. While we shall not, because of limitations of space, develop all of the more common device models in our study, we shall explore the physical operations of some modern devices in order to lay a firm foundation for future study of the models and their use in problem solving.

The objective of this book, then, is to present the physical principles of electronic devices in such a manner as to serve both as an introduction to the study of device physics and as preparation for a further study of device modeling and applications.

This book was developed from course material presented to junior electrical engineering students who had previously undertaken introductory studies in electromagnetics, thermodynamics, and quantum mechanics. The usual freshman-sophomore sequence in physics is sufficient background for most students. My students also have taken an introductory course in linear electronic circuits and have studied analog and digital integrated circuits as idealized modular functionals, that is, as black boxes. The material presented in this book is considered in our curriculum to be prerequisite to a subsequent study of linear and nonlinear active electronic circuits. Because of the limitations of space and of scope, we do not deal with the dielectric and magnetic properties of materials. These are more naturally accommodated at an introductory level in a course in electromagnetics. Nor shall we be able to discuss quantum electronics, as that subject requires a more advanced understanding of the quantum mechanical description of nature than is usually evident at the introductory level.

It is a special privilege to acknowledge the inspirations and influences of my teachers, my many students, and my colleagues who have in one way or another contributed to this work. In particular, my original interest in this area was developed under the tutelage of Professor G. Warfield. I am also indebted to Professors M. A. Uman, S. K. Mitra, J. N. Churchill, H. J. Jensen, and V. R. Algazi, who have made helpful suggestions in the organization and presentation of this material, and to my students, who have responded to this effort with enthusiasm and constructive criticism.

Finally I owe debts of gratitude to Professor H. H. Loomis, Jr., for his encouragement and cooperation and to Miss Paula Buchignani for her tireless and dedicated typing of the manuscript.

MYRON F. UMAN

Introduction to the Physics of Electronics

1

Introduction

Electronic circuits utilize a multitude of components in a variety of interconnections to perform important and useful processes on electrical signals. A low-pass filter, for instance, removes the high-frequency components from a signal by an appropriate combination of passive elements, that is, resistors, capacitors, and inductors. An amplifier usually converts energy from a d.c. source into a large a.c. signal that reproduces a smaller a.c. reference signal. Amplifier circuits employ an appropriate combination of passive elements and at least one *active* element such as a transistor. In every case, circuits, whether they be discrete or integrated, are made up of passive and/or active components. Each element or device in an electronic circuit is characterized by its conductive property, a relationship that describes the current that flows through it in response to the voltage that is applied across it. The current-voltage relationship of a component or device can be used to develop models describing its characteristic behavior. Other models can be developed based on an understanding of the physical processes that underlie the conduction effects in the device. It is an understanding of the models of circuit elements that permits the engineer to analyze and to design circuits for specific applications. In many instances, as, for example, in integrated circuits, the terminals of a device may not be accessible for the measurement of a current-voltage characteristic. In these cases modeling of the device depends totally on a knowledge of the physics of the conduction processes. And competence in analysis or design in all cases depends on the accuracy and effectiveness of the models.

This book is an introduction to the physical processes in nature that have been and will continue to be utilized by engineers to create useful electronic devices. Our study will emphasize the relationships among the fundamental processes and the conduction properties of common devices so that the engineer may better understand not only the characteristic operation of devices but also their inherent capabilities and limitations.

Of the many components of electronic circuits, this book will not be very much concerned with the linear passive elements of resistors, capacitors, and inductors. These elements, characterized by their properties of resistance, capacitance, and inductance, respectively, are familiar to students who have studied an introductory course in circuits. The relationship between current and voltage for each of these components can be expressed analytically in the form of a linear differential equation. For a resistor, for example, the relationship is $I = (V/R)$, where I is the current through the resistor, V is the voltage drop across it, and R is the resistance. The behavior of networks of these elements can be analyzed using highly developed techniques that utilize linear mathematics. Application of Kirchhoff's current and voltage laws are two such techniques. The properties of capacitance and inductance are usually studied in an introductory course in electricity and magnetism. The property of resistance, or its inverse, conductance, is of importance not only in resistors but also in all other devices. Therefore we shall study this property in some detail.

The primary emphasis in this book will be on the operational characteristics of nonlinear devices that include passive diode switches and active amplifiers such as junction and field-effect transistors. It is the nonlinear devices that make possible the fascinating variety of signal-processing techniques that characterize modern electronics. For these elements the relationships between current and voltage are complex and depend critically on internal physical processes. Usually the current-voltage relationships are too complicated to express even in terms of a nonlinear equation. And if it were possible to write equations, their analytical solution would probably not be possible because nonlinear mathematics is not well developed. In lieu of nonlinear equations and analytical solutions, engineers have developed graphical displays and techniques of solution. A graphical display of a current-to-voltage relationship is called a *current-voltage characteristic curve* or *I-V* characteristic. Each element has its own *I-V* characteristic. In addition, computers can be employed to analyze and to design nonlinear networks by manipulating numerical data.

In this chapter we shall consider examples of ideal and real *I-V* characteristics as well as common methods of analysis of circuit problems with nonlinear elements. We shall also introduce the concept of an integrated circuit. The objective in this chapter is to place in perspective the work of the remainder of the book and to indicate why an understanding of the physical processes of electronics is important not only to those engineers who build devices but also to those who use them. The need for such an understanding becomes apparent when we compare idealized and realistic characteristics of devices and when we attempt to model their behavior in both discrete and integrated circuits.

1.1 MODELING ELECTRONIC ELEMENTS AND DEVICES

A nonlinear element in a simple circuit that otherwise would be linear is shown in Fig. 1-1. The relationship, either analytical or graphical, between

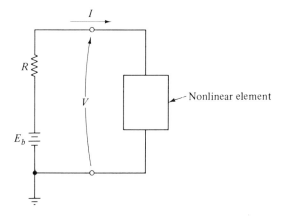

Fig. 1-1 Nonlinear element in a simple circuit.

the current flowing into the terminals of the device and the voltage difference between the terminals is called its *terminal characteristic*. Hypothetical devices that perform specialized functions in circuits can be described by idealized characteristics. How useful a real device is depends, after economic considerations, largely on how its *I-V* characteristic resembles or differs from the ideal and over what range of a parameter such as frequency or power level the device maintains that characteristic behavior.

A Switch

In an electronic circuit a switch is a device that is either opened or closed. When closed the ideal switch short-circuits its terminals; that is, any current can flow with no potential drop evident. When open, the ideal switch prohibits current flow for whatever voltage might be applied. An ideal diode is a switch that is open when the voltage across its terminals is negative in the sense of Fig. 1-1 and that is closed otherwise. The *I-V* characteristic of an idealized diode switch and its conventional circuit representation are shown in Fig. 1-2. A variety of real devices have terminal characteristics sufficiently like the ideal diode to be useful. Among these are vacuum diodes, gaseous diodes, *p-n* junction semiconductor diodes, Schottky barrier diodes, and silicon-controlled rectifiers (SCRs). The current-voltage characteristic of a typical *p-n* junction diode is sketched in Fig. 1-3. Notice how this character-

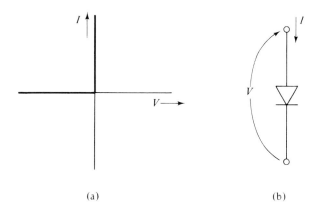

Fig. 1-2 Ideal diode: (a) *I-V* characteristic, (b) circuit representation.

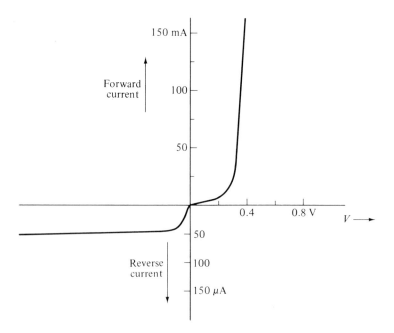

Fig. 1-3 Characteristic of a typical real *p-n* junction diode (note change in scale for reverse current).

Sec. 1-1 Modeling Electronic Elements and Devices

istic differs from the ideal. For negative voltages, called the *reverse direction*, some current does flow: The device is not a perfect open circuit. The device is not a perfect short circuit either since a finite potential drop appears between the terminals for a given forward current. Just what internal processes give rise to these characteristics, how they determine the nonideality of the device, and how they effect the range and limitations of the performance of the device are the questions we plan to answer in our study.

An Amplifier

Another useful functional operation in circuits is amplification. An ideal amplifier is a device the output of which is a magnified reproduction of some reference signal. Since amplifiers must have a reference or input signal and an output signal, they can be represented as a two-port device as in Fig. 1-4(a). In practice, two of the terminals of the two-port device are frequently common so that the three-terminal representation of Fig. 1-4(b) is also used.

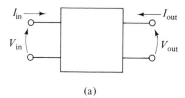

(a)

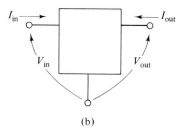

Fig. 1-4 Amplifier: (a) two-port representation, (b) three-terminal representation.

(b)

The ideal *I-V* characteristic of a current amplifier is shown in Fig. 1-5(a). The output current-voltage relationship is plotted for selected values of input current. By contrast, Fig. 1-5(b) indicates the output or collector characteristic of a typical transistor in what is called the *common-emitter configuration*. Notice that the characteristics of the real device do not extend to very low values of output voltage. Notice, also, that each curve has a nonzero slope everywhere, in contrast to the idealized curves. Last, notice that an output exists with no input. The terminal characteristics of this real device, as with the diode, arise as the result of internal properties and processes. We can

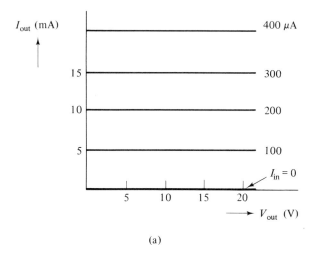

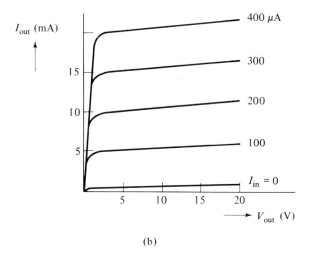

Fig. 1-5 Characteristics of a current amplifier: (a) ideal, (b) typical transistor in a common-emitter configuration.

understand the characteristic and the circuit models developed from them only by an examination of these properties and processes.

As an example of models of a familiar circuit element, consider the resistor. The terminal characteristic of an idealized resistor can be expressed analytically as $I = (V/R)$ or graphically as in Fig. 1-6(a). Three equivalent models of a practical resistor are shown in Fig. 1-6(b). One of these is that of the idealized circuit element and is valid at low frequencies. Depending on

Sec. 1-2 Nonlinear Devices in D.C. Circuits

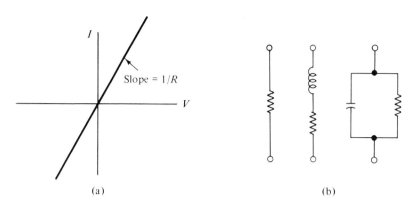

Fig. 1-6 Resistor: (a) low-frequency terminal characteristics, (b) circuit models.

the physical construction of the resistor, one of the other two models may be required if the signal frequencies are sufficiently high. For example, a resistor may be made from a long piece of resistive wire coiled into a small package in which case the inductance of the coil becomes important at high frequencies. Alternatively, a low-inductance resistor can be made, but it may evidence capacitive effects at high frequencies.

And so it is with every circuit element. The model used to represent its behavior in a circuit depends on its physical construction and the application to which it is put.

1.2 NONLINEAR DEVICES IN D.C. CIRCUITS

Analysis of circuits using nonlinear elements with d.c. currents and voltages can be accomplished by using a graphical technique known as *load line analysis*. The technique consists of determining the current-voltage characteristic of the linear part of the circuit and superimposing that curve on the *I-V* characteristic of the nonlinear element. The current-voltage curve of the linear part of the circuit will always be a straight line. Since the linear part represents a load on the nonlinear element, its characteristic is called the load line. The state of the circuit is determined by the point at which the two curves intersect, the point of intersection being the simultaneous solution of the equations that represent both parts of the circuit.

As an example of the technique, let us consider the simple circuit of Fig. 1-1 with a *p-n* junction diode as the nonlinear element. In such a series circuit, the current is of primary interest because the same current flows through each element.

The current-voltage relationship for the diode is given in Fig. 1-3. The current-voltage relationship for the rest of the circuit is a straight line since the resistor is a linear device. All that is required to sketch the load line are two points on the line. The easiest conditions to determine are the one for which no current flows through the circuit and the one for which no voltage appears across the terminals of the diode, that is, the one for which sufficient current flows so that the IR drop across the resistor equals the battery voltage. These two points are shown in Fig. 1-7. The linear characteristic is the

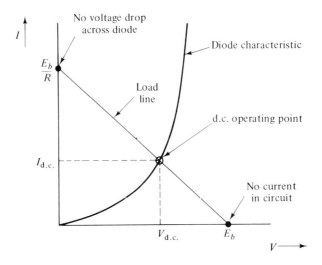

Fig. 1-7 Load line analysis for the circuit of Fig. 1-1 using the characteristics of the diode of Fig. 1-3.

line joining these two points and is shown superimposed on the forward characteristic of the diode. The load line represents all possible terminal current-voltage combinations for the battery in series with the resistor. It is the graphical representation of the relationship $V = E_b - IR$. But only one operating condition for the battery and resistor corresponds to a possible operating condition for the diode. The operating condition for the whole circuit is the intersection of the two curves and is called the *d.c. operating point* or the *quiescent operating point*. The current that flows in the circuit is $I_{d.c.}$ and the voltage drop across the diode is $V_{d.c.}$, as indicated in Fig. 1-7.

1.3 NONLINEAR DEVICES IN A.C. CIRCUITS

The currents and voltages in electronic circuits have both d.c. and a.c. components in general. A circuit, then, will have a quiescent operating condition corresponding to the average currents and voltages. But the currents and voltages will deviate from their averages as the a.c. components of the signals. Analysis of the response of a circuit with nonlinear elements to a.c. signals

Sec. 1-3 Nonlinear Devices in A.C. Circuits

depends on the *I-V* characteristics of the devices, the internal processes that give rise to the characteristics, and the magnitude and frequency of the a.c. signal.

The quiescent operating conditions of the output circuit of an amplifier of the type represented in Fig. 1-4 can be determined using the load line technique, where the load line in this instance is the current-voltage relationship for the linear circuit loading the output terminals of the amplifier. An example of such a load line analysis is shown in Fig. 1-8(a), where the nonlinear device is the common-emitter transistor of Fig. 1-5(b). Notice that in this case there are very many possible d.c. operating conditions for the output circuit, depending on the value of the d.c. component of the input current. Figure 1-8 indicates, for example, that if the input current is 250 μA, the d.c. output current and voltage are the coordinates of the point Q.

Low-Frequency Signals

When a low-frequency a.c. signal is added to the d.c. input current the operating condition of the output circuit changes in response to the instantaneous value of the input current. The instantaneous operating point is confined to the load line for a resistive output circuit. Figure 1-8(b) shows a magnified view of the region around the quiescent operating point Q. The shaded bars indicate the ranges of input current and output current and voltage for an a.c. input current of amplitude 20 μA.

If the amplitude of the a.c. input signal is sufficiently small, the instantaneous operating point for the output may sample only a small portion of the *I-V* characteristic, which can be approximated by equally spaced parallel straight lines. Under these conditions the a.c. output current will be linearly related to the a.c. input, and the device operates as a linear element for the a.c. signal. Just such a condition is represented in Fig. 1-8(b). By analyzing the *I-V* characteristic in the vicinity of Q, it is possible to devise a small-signal linear equivalent circuit composed of an ideal amplifier and resistors that simulates the behavior of the real circuit in response to the incremental a.c. signals. The validity of an incremental equivalent circuit analysis is restricted to small perturbations about the quiescent point, where *small* is determined by the linearity of the *I-V* characteristic in the vicinity of the quiescent operating point. For a larger signal the instantaneous operating point samples regions of the characteristic where a linear relationship no longer holds and the analysis is much more complicated.

High-Frequency Signals

The response of a circuit with a nonlinear element such as an amplifier to a.c. signals at high frequencies is complicated in two ways. First, the load line characteristic of the linear portion of the circuit will depend on frequency

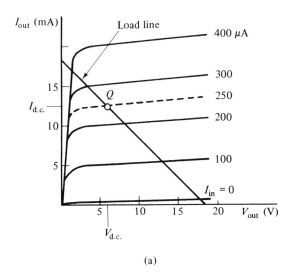

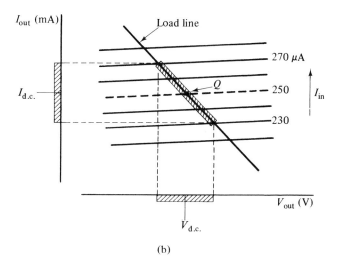

Fig. 1-8 Load line analysis for the output circuit of a common-emitter transistor: (a) determination of the quiescent operating point, (b) magnified view of the vicinity of Q.

if there are capacitors and inductors in the circuit. Second, the relationship between the current and voltage of the nonlinear device at high frequency differs from the d.c. or low-frequency I-V characteristic. In fact, what we mean by high frequencies are those frequencies at which deviations from low-frequency behavior occur. All is not lost, however, provided we understand what phenomena within the device itself cause the high-frequency deviations. If we understand the device physics, we shall still be able to devise a high-frequency incremental equivalent circuit by adding frequency-dependent elements to the low-frequency incremental circuit. An example of such a high-frequency circuit is the hybrid-π model, which will be introduced in Chapter 10. The added elements must be properly chosen to reflect the physical behavior of the device. An example of a high-frequency effect is the delay caused by the finite time required for electrons to traverse the device. This effect can be simulated in a circuit model by a capacitance called the *transit-time capacitance*.

Clearly, an understanding of the internal processes within a device is of great importance in the development and utilization of small-signal a.c. models for electronic devices. The internal physical processes determine not only the parameters of such an equivalent circuit model but also, and very importantly, the limitations of the device itself. The rest of this book, then, is devoted to the development of an understanding of the physical processes that determine the electrical properties of devices. While the emphasis will be on modern devices in widespread use, the underlying physical principles that we shall study are valid and are applicable to present-day devices that we shall not be able to study because of space limitations and to future conduction devices that have yet to be invented.

1.4 INTEGRATED CIRCUITS

A simple picture of an integrated circuit (IC) is a single piece of solid-state material such as a crystal of a semiconductor in which very many passive and active circuit elements have been formed. A detailed discussion of integrated circuits is presented in Chapter 12 based on the material on solid-state electronics of Chapters 6–11. Commonly, resistors, capacitors, diodes, and transistors occupy well-defined adjacent regions of the integrated device. The IC also includes as part of the structure interconnecting leads between the regions where the various circuit functions are performed on signals. Thus an IC is a complicated network constructed as a whole and is essentially indivisible. Because of the very small size of integrated circuits, only a limited number of connections can be made to an external circuit. In contrast, a discrete circuit is composed of individual elements manufactured separately and thereafter connected together in some fashion. The elements of a discrete circuit are individually available for testing and for measurement of their terminal characteristics. This is not true of integrated circuits.

Yet an integrated circuit is more than a discrete circuit transposed and manufactured in integrated form. Practical problems arise that are not confronted by the designer of discrete circuits. For example, inductors are difficult to make in integrated form so that generally they must be avoided in the design or their circuit function must be simulated by networks of active and passive elements. Additional examples of complications in integrated circuits are the problems of isolating neighboring regions of the device from one another electrically and of providing interconnecting conductive paths between the proper elements. In both cases the techniques used to accomplish isolation and interconnection introduce effects that complicate the behavior of the system as a whole. Modeling these effects depends on an understanding of physical processes since specific portions of the IC may not be available for measurement. It is important to recognize these complications and their effects on the performance of the IC both in design and in application. For example, an IC utilizing reversed-biased *p-n* junctions for isolation will not perform as well at high frequencies as one that employs a beam-lead isolation technique. A discussion of details of IC fabrication and techniques of isolation and interconnection is presented in Chapter 12.

In addition to restrictions and complications in design, integrated circuits offer new opportunities to the circuit designer. First, very large numbers of elements, active and passive, can be fabricated in a very small volume of the solid-state material. Furthermore, many identical integrated circuits can be produced at the same time; that is, they may be batch-processed. Thus the equivalent network of devices can be extremely complicated at very modest cost. For example, filters can be constructed without using inductors by simulating their function using active elements in conjunction with resistors and capacitors. Second, circuit elements exist in integrated form that have no practical counterparts in discrete form. Among these are multi-emitter, lateral, and thin-film devices.

It is clear from the foregoing discussion that integrated circuits present problems of understanding and modeling that are difficult if not impossible to surmount without a clear picture of what physical processes contribute to the conductive properties of such devices. On the other hand, the fundamental physical processes of conduction occur in all electronic systems including vacuum and gaseous devices as well as the solid-state forms. We shall therefore first develop the principal concepts that we shall need for an understanding of all electronic devices in simple classical systems before proceeding to the more complicated quantum mechanical problems of solid-state electronics.

BIBLIOGRAPHY

Introductory textbooks dealing with *I-V* characteristics, load lines, and incremental equivalent circuits:

DURLING, A. E., *An Introduction to Electrical Engineering*. New York: The Macmillan Company, 1969.

FITZGERALD, A. E., D. E. HIGGINBOTHAM, and A. GRABEL, *Basic Electrical Engineering* (3rd ed.). New York: McGraw-Hill Book Company, 1967.

SMITH, R. J., *Circuits, Devices, and Systems* (2nd ed.). New York: John Wiley & Sons, Inc., 1971.

An intermediate text on nonlinear circuits:

ANNER, G. E., *Elementary Nonlinear Electronic Circuits.* Englewood Cliffs, N.J.: Prentice-Hall, Inc., 1967.

PROBLEMS

1-1. (a) An approximation to the forward I-V characteristic of the p-n junction diode of Fig. 1-3 can be made by drawing segments of straight lines so that they resemble the real curve. Sketch such an approximate characteristic using two straight lines.
(b) Devise an equivalent circuit composed of a battery, a resistor, and an ideal diode that has this approximate I-V characteristic. Indicate the values of the battery voltage and the resistance.

1-2. A voltage-controlled current source is a two port device whose output current depends on its input voltage. Sketch an ideal I-V characteristic for such a device.

1-3. A voltage-controlled voltage source is a two port device whose output voltage depends on its input voltage. Sketch an ideal I-V characteristic for such a device.

1-4. The current gain is defined as the ratio of the change in output current to the change in input current. What is its value for the ideal amplifier of Fig. 1-5?

1-5. Write an equation for the voltage at the terminal of the nonlinear element in Fig. 1-1 in terms of the current flowing through the resistor. From this expression, plot the current as a function of voltage.

1-6. Modify the circuit of Fig. 1-1 to include a low-frequency sinusoidal voltage source $e(t)$ in series with the battery. Take the nonlinear element to be the diode of Fig. 1-3. Use graphical techniques to indicate the instantaneous diode current and voltage.

1-7. For the device of Fig. 1-8(a), find the maximum value of the output current that is possible for large values of the a.c. component of the input current. This effect is known as *saturation*. Assume a low-frequency sinusoidal a.c. component of the input current of sufficient amplitude to drive the device into saturation. Sketch the input current and what the corresponding output current looks like.

2

Elements of Equilibrium Statistical Mechanics

All electronic devices, just like all physical systems, are composed of incredibly large numbers of very small atoms. Just as the behavior of any system can be explained in terms of the behavior of all of its various parts, the properties of physical systems can be understood in terms of the aggregate behavior of its constituent atoms. In some cases the atoms or molecules may be in an ionized state, meaning that they are split up into electrically charged particles, negative electrons and positive ions. For instance, 1 cm^3 (one cubic centimeter) of air is actually comprised of about 2.5×10^{19} molecules of which about 78% are nitrogen and 20% oxygen. In unpolluted air at sea level and over land, between 500 and 600 of these molecules are ionized, the principal ionizing agent being terrestrial radioactivity. The semiconductor germanium contains about 5×10^{22} atoms/cm^3, and, at room temperature, about 10^{13} of these are ionized as a consequence of the nature of the material that will be discussed in a later chapter.* In any event, it is the behavior of these ions and electrons under the stimulus of forces, usually electromagnetic forces, that determines the ultimate electronic properties of the whole system. For example, the uniform motion of ions or

---

*In semiconductors the objects that carry positive charge and hence are analogous to positive ions are called *holes*. In this and the following two chapters we shall refer to positive charge carriers by the more general term *ions*. Since most of the ideas developed in these chapters are applicable to both gases and solids, the more general context emphasizes that the fundamental concepts are widely applicable.

Chap. 2 Elements of Equilibrium Statistical Mechanics

electrons through a medium under the influence of an external electric field is evidenced as an electric current flowing through the medium, and if the current is linearly proportional to the applied field, the medium is said to be *ohmic;* that is, Ohm's law is applicable.

The problem is that of relating the behavior of the microscopic electrons and ions to the properties of the whole system and to the terminal characteristics of the device. In classical physics, the behavior of a point-mass object is completely specified by knowledge of the position and momentum of the particle as functions of time for all time. The position (**r**) and momentum (**p**) can be calculated, in principle, by solving the equation known as Newton's second law,

$$\mathbf{F} = \frac{d\mathbf{p}}{dt} \qquad (2\text{-}1)$$

where **F** is the instantaneous force acting on the particle, and

$$\mathbf{p} = m\frac{d\mathbf{r}}{dt} \qquad (2\text{-}2)$$

where m is the mass. All that is required for a complete solution is knowledge of the forces acting in time and knowledge of **r** and **p** at some instant of time. The latter specification is usually called the *initial conditions*, although there is no reason some time other than the "beginning" would not suffice. The initial conditions, denoted by $\mathbf{r}_0$ and $\mathbf{p}_0$, are required in the mathematical sense to determine the two constants of integration for the second-order differential equation, Eq. (2-1). The difficulty, of course, is that it is impractical, even in the wildest dreams of the most dedicated builder of giant computers, to calculate simultaneously the position and momentum at each instant of time for all the real particles in any realistic system. The principal complication, over and above the comparatively small sizes of computers, is that the forces acting on any particular particle depend instantaneously on the behavior of all the other particles. This complication is particularly acute when the particles under consideration are charged particles interacting through the electric and magnetic fields, which are determined by the charge, position, and momentum of each particle. It is obvious that this is a very difficult approach.

Rather than throw our hands up in despair we shall take the more successful approach of statistical mechanics, or statistical physics, as it is sometimes called. The principal concept of statistical mechanics is the recognition that the measurement of the properties that we ascribe to a physical system, such as pressure, temperature, specific heat, electric conductivity, and so forth, does not depend on the measurement of detailed instantaneous behavior of each constituent particle, because we have no way, really, of sensing that behavior. Instead the gross properties of a large-scale system depend only on the average behavior of all the constituent particles taken as a whole. We shall try to illustrate this point by two interesting and familiar examples, though they be far afield from physics.

First, consider the interesting property of a system that consists of a pair of dice, that is, the sum of the two numbers facing up after the dice have been rolled. If, for the particular occasion, 7 is the desirable sum, then it may be some comfort to recognize that there is a variety of combinations of the numbers on the two dice that give this result. Just what combination gives the 7 — 1-6, 2-5, or 3-4 — is of no importance. In other words, the property of the system does not depend on the behavior of each individual die, but, rather, it depends on the two dice taken as a whole.

Furthermore, if we can assume that the dice are unweighted, that is, that each face of a die has equal probability of turning up, then we can calculate the probability of obtaining any result in the following way. Let us describe the state of the dice by the sum of the two numbers, which is the measurable property of importance and which is an integer number between 2 and 12. Such a state, described by a macroscopically measurable property, is usually called a *macroscopic state*, or *macrostate* for short. What we would like to know, presumably before we take the chance, is the probability that the system will be in a particular macrostate after we take the chance. To obtain this probability we recognize, as before, that there are a number of combinations that result in the same macrostate. For this particular example it is easy to enumerate all the combinations that make up a given macrostate, and we do this by describing the behavior of the individual dice. Such a complete description of the individual constituent parts of the system is called a *microscopic state*, or *microstate* for short. Let us assume that we are able to distinguish one die from the other by, for example, painting one red and the other blue. Color serves only to identify which die is which and does not effect the macroscopic description. We shall adopt a convention for the purpose of describing the microstate of always stating the number on the red die first. For example, one possible microstate is obtained when the red die reads 4 and the blue reads 1. This microstate, 4-1, is only one of several microstates that lead to the same macrostate, the sum of the numbers being 5. Another such microstate is 1-4. There are 36 possible microstates for our system corresponding to the possible permutations of two numbers, each ranging from 1 to 6. And there are 11 possible macrostates. If the dice are honest, each microstate will have an equal probability of occurring. In the statistical mechanical description of physics it is always assumed, a priori, that all microstates have equal probability of occurring. Then, the most probable macrostate of the system is that one that is obtained from the greatest number of microstates. It is the most probable macrostate that we are most likely to measure. For the pair of dice, the most probable macrostate is 7 since the

Chap. 2 Elements of Equilibrium Statistical Mechanics   17

greatest number of microstates, 6 in all, give this result. There are, then, 6 chances in 36 that 7 will be rolled. The least probable macrostates are 2 and 12, each of which can be obtained from only one microstate. These macrostates then have only 1 chance in 36 to be rolled. Furthermore, it is a wise man who learns from both his actual and his intellectual experiences.

A second illustration of the statistical method comes from demography, the study of populations. The study of some properties of a population, for example, birth rate, mortality rate, or geographical distribution, is actually more relevant to statistical physics than is the study of a pair of dice in that a population is more likely to be composed of a large number of constituent parts, as is a physical system. To demonstrate the essential features of a statistical description we shall consider just one simple parameter of a population, age. Each person, each fundamental part, has ascribed to him an age, and it is possible to write down the name of each person—or his tax-identifying number—and his age. In a population the size of that of the United States, for example, that is a formidable task, but a task that is nonetheless undertaken once a decade by the Census Bureau. For most common purposes, however, such a list represents too much information. Consider the television rating services, public opinion polls, or market research surveys. For these applications the behavior of a person who is 20 years and 10 days old is not much different from a similar one 20 years and 11 days old. Rather, more useful information may be the number of people between certain age limits, say 20 and 21 years, irrespective of their names or other identifying characteristics.

Let us pursue this point a bit further before we attempt to make some pertinent connections with statistical physics. In practice, demographers usually describe a population by the number of people with ages in certain well-defined ranges. A macroscopic description of the population system could be, then, a series of integers representing the number of people in each age group. Just what range constitutes an age group is a question that can be answered only when we know for what purpose the study is being made. Frequently a 5-year interval constitutes an age group, but smaller intervals could be used. However, so that we can make sense out of the system, the age range must be large enough so that there are very many people in it. In this way, as a few move into a particular group and as others move out, the total number in the group changes only by a small percentage. In fact, if a person whose tax-identifying number is 266-60-1647 moved out of the 20–24 age group just as 147-28-6141 moved in, the population would have assumed a new microstate but would be in the very same macrostate. When the distribution of numbers of people into age groups does not change with time, the population is said to be in a *steady state*. Thus the whole system may be in a steady state; that is, its macrostate does not change, even though the ages of all the individual people are continually changing—some are dying, and some are being born. Usually, however, the macroscopic description does vary in time. Even for a stable population there will be fluctuations in the macrostate. For instance, the aforementioned 266-60-1647 might have moved out of his age group a short time before 147-28-6141 moved in. If each age group contains a sufficiently large number of people, the fluctuations in the macrostate about its steady-state value will be small, and the equilibrium macrostate will provide the basis of a useful statistical description.

An important feature of the population that can be obtained from its macroscopic description is the average age of its citizens. Of course, this property can be obtained from the microscopic description harbored in the files of the Census Bureau simply by adding the ages of all the people and dividing by the total number of people. But we can accomplish the same thing, and hopefully get essentially the same results, with a lot less work using the macroscopic distribution of people into age groups. If we know the number of people in each age group *and* the average age of each group, we can obtain the average age of the population by multiplying these two numbers together, and then summing with similar products for each age group and dividing the sum by the total number of people. Try it! Although the latter technique appears to be more complicated than the former, it requires far fewer steps—only twice the number of age groups—and can be just as accurate provided we know the average age of the people within a single age group reasonably well. The smaller the range of ages in a group, the more accurately we can assign an average age to that group. For example, if the age group spanned 5 years, we might make a significant error in guessing the average age for that group, whereas if the group spanned only 1 year, our guess might be much better. This is especially true for the racehorse population since they all celebrate their birthdays on January 1 without regard to their actual dates of birth. On the other hand, the smaller the range of ages in a group, the larger the number of groups needed to describe the population and the fewer people in each group. The limit is reached, of course, when each age group contains only one person, in which case we are back to the microscopic approach to the problem. Furthermore, as we discovered above, in order that the macroscopic distribution be relatively stable in time, that is, to avoid significant fluctuations in the numbers in each group, it was necessary to insist that each age group contain large numbers of people. Clearly, we must strike some type of balance between large and small age groups—between stability in the macroscopic description and accuracy of calculation. Engineers are particularly well qualified to perform this balancing act in the statistical description by virtue of their experience in *trading-off* among incompatible design criteria.

All the concepts developed through these two convenient examples—macrostate, microstate, most probable macrostate, distributions, and averages—have their counterparts in the statistical description of physical systems.

2.1 STATISTICAL PHYSICS

The physical systems with which we shall be concerned in our study are composed of electrons, ions, and electrically neutral particles. Each of these particles has certain properties such as charge, mass, position in space, energy, linear momentum, and angular momentum, which may contribute to some property of the whole system. For instance, the mass of the system is the sum of the masses of its parts, neglecting relativistic effects. Similarly, the net momentum is the vector sum of the momenta of its parts. Just as in the demographic example in which the age of an individual was a property of interest, we can imagine some situations in which almost any property of the individual particles might be of interest. Consider, for example, kinetic energy. We know from our studies of thermodynamics that the temperature of a body is just a macroscopic embodiment of the kinetic energy of the random motion of its constituent particles. By virtue of its velocity $\mathbf{v}$, a mass has kinetic energy:

$$\text{kinetic energy} = \tfrac{1}{2} m \mathbf{v} \cdot \mathbf{v}$$

One of the principal notions of statistical physics for classical particles is that any particle in a system may have any energy. That is, some may have large velocities, others small, and so forth. This idea, that the particles of a system may have different velocities just as people in a population may have different ages, was introduced by James Clerk Maxwell in 1859. Furthermore, the energy of any given particle may change in time as it interacts with all the other particles.

Energy Distribution Functions

A valid microscopic description of a system based on the energy of its particles might be the instantaneous specification of the energy of each particle. A valid macroscopic description might be the enumeration of the particles that have energies that lie within certain specified ranges of energy. As in the discussion of the description of a population according to age, the ranges of energy must be large enough so that many particles always have energies in the range but small enough so that an average value for the range is well defined.

Let us consider for the moment a hypothetical system and assume that we know the energy of each particle. Instead of listing each particle and its energy, we shall draw a one-dimensional coordinate system, the axis of which represents particle energy. Each particle is represented by a point located in the energy space according to its instantaneous energy. Figure 2-1 shows a

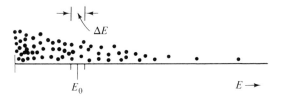

Fig. 2-1 Sketch indicating some representative points in energy space.

schematic representation of this idea, although there are too few particles indicated for this example to be representative of a statistical system. Notice, too, that the points have been displaced from the one-dimensional line so that they can be seen. Each point represents a particle and is therefore called a *representative point*. If we divide the energy space into regions such as ΔE and if we are able to specify the number of representative points in each range, then we shall have described a macrostate of the system. Remember, the range ΔE must be large enough so that it contains many points but small enough so that the points in the range have a well-defined average. If there are very many particles in the system, there will be very many representative points in each range, even if we make the ranges reasonably small.

The number of representative points ΔN in a small energy interval ΔE is related to the density of representative points in energy space by

$$\Delta N = (\text{density}) \cdot \Delta E \tag{2-3}$$

provided the density does not change over the interval. Since the designation of ΔN for each ΔE specifies the macrostate, the knowledge of the density of representative points also specifies the macrostate. This density is usually called the *energy distribution function* and is given the symbol $f(E)$, where $f(E)$ indicates that the value of the function f depends on the energy E. The distribution function may be described by a mathematical formula, a table of numbers, or a graphical plot. If the energy range ΔE around E_0 is allowed to get small in the sense usually implied in calculus, then the distribution function evaluated at E_0 is

$$f(E_0) = \lim_{\Delta E \to 0} \frac{\Delta N}{\Delta E}\bigg|_{E=E_0} = \frac{dN}{dE}\bigg|_{E=E_0} \tag{2-4}$$

The distribution function $f(E)$, known for all values of energy, specifies the macrostate.

But energy is only one example of a property of a particle that might be used as the basis for a statistical description. Others include the position of the particle in space or its velocity or momentum. To obtain a macroscopic description based on a given property we construct an appropriate parameter space and specify the density of representative points in that parameter space as the distribution function. When the property is position in space, the parameter space is just the ordinary three-dimensional real space, and the distribution function is just the real density of particles in space. This example is useful for demonstrating the concept of an elemental volume or range in the parameter space, one small enough for a well-defined average and yet large enough to contain many representative points, thus minimizing fluctuations. The density of particles in a gas at standard temperature and pressure is about $2.5 \times 10^{19}/\mathrm{cm}^3$. If our elemental volume were 10^{-19} cm³, we would expect on the average only two and one-half particles in it. And that number would probably fluctuate rather violently: zero, three, one, two, zero, etc. Obviously, this volume is too small for a stable macroscopic description. If our elemental volume were 1 cm³, there would be 2.5×10^{19} particles in it, but it might be too large for the practical use of calculus, especially if the size of the system is on the order of centimeters. If our elemental volume were, say, 10^{-12} cm³, we would expect that there would be 2.5×10^7 particles in it on the average and that fluctuations in this number would be small. In addition, a linear dimension for this volume would be 10^{-4} cm or 10^{-6} m; hence it is of a practical size for doing calculus with objects of ordinary size.

Phase-Space Distribution Functions

Because in classical physics the behavior of a particle is completely specified by its position and momentum as functions of time, these two properties are the ones most frequently used as bases for the statistical description of physical systems. Since position and momentum are each three-dimensional vectors, six components are required to specify the values of these properties. Therefore, we must construct a six-dimensional parameter space in which three coordinates are ordinary spatial coordinates and three are momentum coordinates. Such a space is called the six-dimensional *phase space*. Each particle is represented by a point in phase space, and the complete behavior of a particle is reflected in the motion of the representative point through phase space. We shall designate the phase-space distribution function by $f(\mathbf{r}, \mathbf{p})$. The phase-space distribution function is the keystone of classical statistical physics. Since the distribution function is actually the density of representative points in phase space, the number of particles with position in the range $d\mathbf{r}$ in the vicinity of $\mathbf{r}_0$ and with momenta in the range $d\mathbf{p}$ in the vicinity of $\mathbf{p}_0$ is just

$$dN(\mathbf{r}_0, \mathbf{p}_0) = f(\mathbf{r}_0, \mathbf{p}_0)\, d\mathbf{r}\, d\mathbf{p} \tag{2-5}$$

where $d\mathbf{r}\, d\mathbf{p}$ is a small, six-dimensional volume in phase space. Another representation we shall use for an elemental volume in phase space is $d^3r\, d^3p$, where the superscript indicates a three-dimensional volume.

2.2 PROPERTIES OF THE DISTRIBUTION FUNCTION

To facilitate our discussion of the properties of distribution functions and their uses, we shall return to the description based on energy since the energy distribution function depends on only one parameter. The number of particles with energy in a differentially small range dE in the vicinity of E_0 is

$$dN(E_0) = f(E_0)\,dE \tag{2-6}$$

This equation comes from the interpretation of $f(E)$ as a density as expressed in Eqs. (2-3) and (2-4). The total number of particles in the system is the same as the total number of representative points. This total number can be obtained by adding the incremental numbers from each interval dE. When the intervals are differentially small, the summation process becomes an integration so that the total number N of particles is

$$\boxed{N = \int_0^\infty f(E)\,dE} \tag{2-7}$$

where the integration is over all possible energies. Every distribution function when integrated over all possible values of its parameter space must yield the total number of particles.

The probability of finding a particle with energy in the range dE about E_0 is just the fraction of the total number that have energies in that range. This probability is designated by $P(E_0)\,dE$ and, using Eqs. (2-6) and (2-7), is

$$P(E_0)\,dE = \frac{f(E_0)\,dE}{\int_0^\infty f(E)\,dE} \tag{2-8}$$

The probability density $P(E)$ is frequently called the *probability distribution function*. It is nothing more than the distribution function normalized to the total number of particles. The integral of the probability over all energies is 1, as it should be since that is the probability of finding a particle with any energy.

The distribution function can be used to calculate properties of the system. For instance, we can estimate the total energy in a system if we know the average energy of all the particles whose energies lie in each range dE. We merely multiply the number of particles in the range, Eq. (2-6), by their average energy and then sum over all ranges. If the range dE centered about E_0 is small enough, the average energy in the range will most likely be E_0. Then the contribution to the total energy U from the particles whose energies are in this particular range is

$$dU(E_0) = E_0\,dN(E_0) = E_0 f(E_0)\,dE \tag{2-9}$$

The total energy in the system is the sum of the contributions from each

Sec. 2-2 Properties of the Distribution Function

range:
$$U = \sum dU = \int_0^\infty Ef(E)\,dE \tag{2-10}$$

This process is the same as the one we often encounter when, for instance, totaling the weight of ten people, three of whom weigh 150 pounds, three 160 pounds, and four 175 pounds. We add (3 × 150) plus (3 × 160) plus (4 × 175), getting a hefty 1630 pounds. The average energy of a particle, denoted $\bar{E}$, is just the total energy of all the particles divided by the total number of particles:

$$\boxed{\bar{E} = \frac{\int_0^\infty Ef(E)\,dE}{\int_0^\infty f(E)\,dE}} \tag{2-11}$$

The average weight of our ten people is 163 pounds. This result is also the mathematical formulation of the averaging process described in connection with the demographic example at the beginning of this chapter with the summation over discrete intervals replaced by an integration. In an entirely analogous way, the average value of any property that is a function of energy, $g(E)$, can be obtained by taking

$$\bar{g} = \frac{\int_0^\infty g(E)f(E)\,dE}{\int_0^\infty f(E)\,dE} \tag{2-12}$$

Notice, also, that the averaging process can be done using the probability function or the normalized distribution defined in Eq. (2-8) in which case

$$\bar{g} = \int_0^\infty g(E)P(E)\,dE \tag{2-13}$$

The properties of the six-dimensional phase-space distribution function can be written down by analogy with those of the energy distribution function. The total number of particles is

$$\boxed{N = \iint f(\mathbf{r}, \mathbf{p})\,d\mathbf{r}\,d\mathbf{p}} \tag{2-14}$$

where $\int d\mathbf{r}$ indicates a three-dimensional integration over spatial coordinates the limits of which are the physical boundaries of the system and where $\int d\mathbf{p}$ indicates a three-dimensional integration over all possible momenta. The probability of finding a particle in the volume $d\mathbf{r}$ in the vicinity of $\mathbf{r}_0$ with momentum in the range $d\mathbf{p}$ about $\mathbf{p}_0$ is

$$P(\mathbf{r}_0, \mathbf{p}_0)\,d\mathbf{r}\,d\mathbf{p} = \frac{f(\mathbf{r}_0, \mathbf{p}_0)\,d\mathbf{r}\,d\mathbf{p}}{N} \tag{2-15}$$

The average value of any property that is a function of either position or momentum or both, $h(\mathbf{r}, \mathbf{p})$, is

$$\bar{h} = \frac{1}{N} \int \int h(\mathbf{r}, \mathbf{p}) f(\mathbf{r}, \mathbf{p}) \, d\mathbf{r} \, d\mathbf{p} \qquad (2\text{-}16)$$

The overbar indicates an average over the entire system. A local average, the average value of h over all momenta, is denoted by angular brackets $\langle h(\mathbf{r}) \rangle$ and in general is a function of position:

$$\langle h(\mathbf{r}) \rangle = \frac{\int h(\mathbf{r}, \mathbf{p}) f(\mathbf{r}, \mathbf{p}) \, d\mathbf{p}}{\int f(\mathbf{r}, \mathbf{p}) \, d\mathbf{p}} \qquad (2\text{-}17)$$

An additional property of $f(\mathbf{r}, \mathbf{p})$ arises because of its functional dependence on both position and momentum. When $f(\mathbf{r}, \mathbf{p})$ is integrated over all momenta, the result is a functional that is dependent only on position. The subsequent integration of this functional over the spatial volume of the system yields the total number of particles according to Eq. (2-14). This functional, therefore, must be the spatial density of particles $n(\mathbf{r})$:

$$n(\mathbf{r}) = \int f(\mathbf{r}, \mathbf{p}) \, d\mathbf{p} \qquad (2\text{-}18)$$

Now that we know some general properties of distribution functions, we shall turn to specific examples: the distribution functions which describe systems in thermodynamic equilibrium.

2.3 DISTRIBUTIONS FOR CLASSICAL SYSTEMS IN THERMODYNAMIC EQUILIBRIUM

Thermodynamic equilibrium is that state of a closed system upon which no external forces are acting, which is represented by the most probable macrostate. Any system left to itself eventually reaches thermodynamic equilibrium through irreversible processes. In such a state, the entropy, which is a measure of the random nature of the behavior of the particles, has its maximum value. Since each microstate has equal probability of occurring, the most probable macrostate is that macrostate that can be obtained from the greatest number of microstates. In fact, the entropy of a macrostate, S, and the number of microstates, W, that lead to it are related since both are maximized in thermodynamic equilibrium. The form of the relationship is $S = \kappa \ln W$, where $\ln W$ is the natural logarithm of W and κ is Boltzmann's constant, which appears frequently in thermodynamics and which has a value 1.38054×10^{-23} J/°K (J = joules). One of the consequences of thermodynamic equilibrium is that the macroscopic variables can have no spatial variations. Temperature, density, pressure, and so forth are all uniform in space. Another consequence of thermodynamic equilibrium is that there can be no ordered motion in the system. We shall show later in this section that the average momentum of particles described by the equilibrium macrostate is zero. Since electric current is just the macroscopic manifestation of the ordered motion of electrons or ions, there can be no electric current. For that matter, there can be, by definition, no electric forces to drive the current.

Therefore, it would appear as though thermodynamic equilibrium is an uninteresting state since we cannot do anything to a system in that state or get anything out of it. And, indeed, that is the case. No device operates in such a state. Devices do operate under the influence of forces, usually electric and magnetic forces, and hence are in some nonequilibrium state that may be a steady state, may vary periodically, or may be a transitory state. Nonetheless, because of its simplicity and because it is frequently useful to approximate a nonequilibrium system as one being almost, but not quite, in equilibrium, we shall devote the rest of this chapter to the condition of thermodynamic equilibrium.

Phase-Space Distribution for Thermodynamic Equilibrium

Before deriving the phase-space distribution function for thermodynamic equilibrium, let us consider the procedures involved in such a derivation in order to emphasize the assumptions and approximations. Conceptually, the determination of the equilibrium distribution for classical particles is straightforward. First we divide phase space into small six-dimensional volumes $d\mathbf{r}\,d\mathbf{p}$ and define a macrostate as the specification of the number of representative points that lie within each volume. We then count the number of microstates that lead to the same macrostate by performing permutations of the arrangements of the representative points in phase space.

An important assumption is made in the calculation of the permutations that restricts the applicability of the results. This assumption is that each particle is distinguishable from every other particle, as if we could paint a number on each one and follow it around. This supposed feature of the particles is a fundamental assumption of classical physics, the so-called deterministic approach to physics. The validity of the assumption can be shown only by comparing a measurement or the results of an experiment with theoretical predictions based on the theory. Suffice it to say that in a great many instances the assumption can be taken as valid, while in a great many other instances it cannot. In particular, when so-called quantum mechanical effects are important, the deterministic theory generally does not work. It is common practice, therefore, to divide statistical physics into two branches, classical statistics and quantum statistics. We shall have more to say about quantum statistics later, and we shall apply them to practical problems in the latter part of this book. For the moment we shall concentrate on classical statistics to develop our understanding on familiar grounds.

Having determined how to count the number of microstates that lead to the same macrostate, we must find the equilibrium macrostate, the one for which entropy and the number of microstates that lead to it are maximized. We perform the maximization conceptually by redistributing the representative points among the various elemental volumes until we get the proper distribution. But we are constrained in our rearrangement of the representative points by certain conservation properties of the system. For example, the total energy of a system in equilibrium is conserved, as is the total number of particles. As we move particles around, we must not violate these principles. After we determine the macroscopic state that is obtained from the maximum number of microstates, we find the distribution function from its definition as the number of representative points per unit volume in phase space, Eq. (2-5).

Derivation of the Distribution Function

We shall now perform these steps and derive the classical equilibrium distribution. The first step is to divide phase space into small volumes $d\mathbf{r}\,d\mathbf{p}$, remembering that each volume should have a large number of representative points in it in order that the macrostate not be subject to violent fluctuations. Let us assume, for the sake of argument, that there are m such volumes, where m is a very large number. The macrostate is specified by the number of points in each volume. For the ith volume this number is N_i, and the total number of particles in the system is N, where

$$N = \sum_{i=1}^{m} N_i \tag{2-19}$$

The next step is to count the number of microstates that lead to the same macrostate—to the same set of numbers $\{N_1, N_2, \ldots, N_m\}$. The problem is to determine how many ways the N particles can be rearranged in phase space without changing the N_i numbers. First, the number of permutations of N distinguishable things taken one at a time is $N!$, where ! is the standard notation for the factorial:

$$N! = N(N-1)(N-2)\cdots 1$$

If we think of ordering the particles and assigning each one to a small volume one at a time, the number of different arrangements of the particles among the boxes is $(N!)$. One such arrangement is shown in Fig. 2-2(a). But each

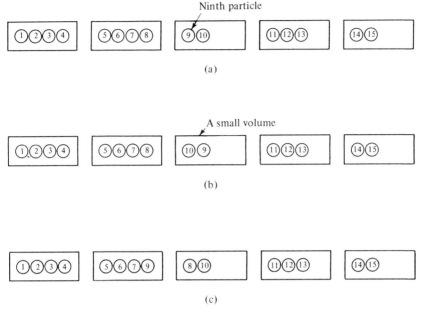

Fig. 2-2 Examples of the distribution of distinguishable particles into volumes. The microstates of (a) and (b) are the same, whereas (c) is a different microstate. Both microstates lead to the same macrostate.

arrangement does not necessarily represent a different microstate since some might just be a rearrangement of the N_i particles within the ith cell. The microstate of Fig. 2-2(b) is the same as that of Fig. 2-2(a) since only the particles of the third box were rearranged. The number of these rearrangements for the ith cell is $(N_i!)$. A different microstate leading to the same macrostate is shown in Fig. 2-2(c). The total number of distinct microstates leading to the same macrostate is, therefore,

$$W = \frac{N!}{(N_1!) \cdot (N_2!) \cdots (N_m!)} \tag{2-20}$$

or

$$W = \frac{N!}{\prod_{i=1}^{m} (N_i!)} \tag{2-21}$$

where $\prod$ indicates a repeated product.

From Eq. (2-21) it is evident that the number of microstates depends on the number of particles in each volume of phase space. We must now rearrange the particles among the volumes to find the equilibrium macrostate.

The equilibrium macrostate maximizes W as well as the entropy S since, as first formulated by Ludwig Boltzmann in 1872,

$$S = \kappa \ln W \tag{2-22}$$

It is useful to deal with the entropy because we shall need it later. The entropy can be expressed in terms of the number of representative points in each cell by combining Eqs. (2-21) and (2-22):

$$S = \kappa[\ln (N!) - \sum_{i=1}^{m} (\ln N_i!)] \tag{2-23}$$

The entropy is maximized in the mathematical sense when, for a small change in the parameters on which it depends, the resulting change in entropy, δS, is zero. In other words if a virtual change in the set of numbers $\{N_1, N_2, \ldots, N_m\}$ does not cause a corresponding change in S, then this set represents the equilibrium distribution. From Eq. (2-23)

$$\delta S = \kappa \delta(\ln N!) - \kappa \delta \left[\sum_{i=1}^{m} (\ln N_i!) \right] \tag{2-24}$$

Since the total number of particles N is a constant, Eq. (2-24) reduces to

$$\delta S = -\kappa \delta \left[\sum_{i=1}^{m} (\ln N_i!) \right] \tag{2-25}$$

Stirling's approximation for the logarithm of a factorial may be employed to simplify this result. The approximation, which is good for values of x greater than ten, is

$$\ln x! \simeq x[\ln(x) - 1]$$

Since the numbers N_i must be large for the statistics to be valid, we can use this approximation in Eq. (2-25) with the result

$$\delta S = -\kappa \delta \left\{ \sum_{i=1}^{m} N_i [\ln(N_i) - 1] \right\} \tag{2-26}$$

After expanding each term of the sum and taking the virtual change, we get

$$\delta S = -\kappa \sum_{i=1}^{m} \ln N_i \, \delta N_i \tag{2-27}$$

where in thermodynamic equilibrium

$$\delta S = 0 \tag{2-28}$$

As mentioned earlier, the changes δN_i in the number of points in each volume of phase space are not all independent because a system in equilibrium must have constant energy and a constant number of particles. Conservation of particles is expressed by the relationship

$$\sum_{i=1}^{m} \delta N_i = 0 \tag{2-29}$$

The net change in energy due to the change in the number of points in the ith volume is $E_i \, \delta N_i$, where $E_i = p_i^2/2m$ is the energy of each particle represented by those points. All the particles in the ith volume are assumed to have energy E_i because the volume is small. For the total energy of the system to be conserved, we must require

$$\sum_{i=1}^{m} E_i \, \delta N_i = 0 \qquad (2\text{-}30)$$

Equations (2-29) and (2-30) represent constraints on the solution of Eqs. (2-28) and (2-27).

To obtain the solution for the N_i that maximizes entropy under these constraints, we employ Lagrange's method of undetermined multipliers. We multiply each of the constraints, Eqs. (2-29) and (2-30), by an unknown factor, α and β, respectively. We then add these two constraint equations to the right-hand side of Eq. (2-27). Since each equation equals zero, their sum does also:

$$\sum_{i=1}^{m} (\ln N_i + \alpha + \beta E_i) \, \delta N_i = 0 \qquad (2\text{-}31)$$

This relation now represents the condition of maximum entropy along with the constraints.

Solution of Eq. (2-31) in that form is not easy since all the N_i are not independent. If we make arbitrary changes in all but two of the numbers N_i, the changes required in the remaining volumes will be completely determined by the conservation of energy and the number of particles. On the other hand, we have at our disposal the two multipliers α and β. If we choose α and β such that

$$\begin{aligned}\ln N_1 + \alpha + \beta E_1 &= 0 \\ \ln N_2 + \alpha + \beta E_2 &= 0\end{aligned} \qquad (2\text{-}32)$$

for the first and second volumes specifically, then we have left in Eq. (2-31) the terms

$$\sum_{i=3}^{m} (\ln N_i + \alpha + \beta E_i) \, \delta N_i = 0 \qquad (2\text{-}33)$$

In Eq. (2-33) all of the variations N_i are independent so that this equation can be satisfied if and only if

$$\ln N_i + \alpha + \beta E_i = 0 \qquad (2\text{-}34)$$

for each volume in phase space. Equation (2-34) has the solution

$$\boxed{N_i = \epsilon^{-\alpha} \epsilon^{-\beta E_i}} \qquad (2\text{-}35)$$

where the number in the ith volume depends on their energy. Notice that Eq. (2-35) is valid for all the volumes, including the first and second. We now must evaluate the multipliers α and β.

Thermodynamic Temperature

First let us return to Eq. (2-27), which expresses the virtual change in entropy due to a virtual change in the numbers N_i. If we insert into this expression our solution for N_i, Eq. (2-34), we obtain

$$\delta S = -\kappa \sum_{i=1}^{m} (-\alpha - \beta E_i) \, \delta N_i$$

$$\delta S = \kappa \alpha \sum_{i=1}^{m} \delta N_i + \beta \kappa \sum_{i=1}^{m} E_i \, \delta N_i$$

$$\delta S = \beta \kappa \, \delta E \tag{2-36}$$

where we have applied the constraint on the total number of particles in the system. Equation (2-36) expresses the virtual change in entropy in terms of the virtual change in energy. We can compare this expression with the combination of the first and second laws of thermodynamics,

$$dE = T \, dS - P \, dV \tag{2-37}$$

where T is the thermodynamic temperature, P is the pressure, and V is the volume of the system. From Eq. (2-37) we see that for a system at constant volume the change in energy due to a change in entropy is

$$\left(\frac{dE}{dS}\right)_V = T \tag{2-38}$$

Comparing Eqs. (2-38) and (2-36) we find that

$$\boxed{\beta = \frac{1}{\kappa T}} \tag{2-39}$$

Conservation of Particles

The constant α is related to the conservation of particles. The total number of particles N is the sum of the N_i. Using Eq. (2-35),

$$N = \sum_{i=1}^{m} \epsilon^{-\alpha} \epsilon^{-E_i/\kappa T}$$

or, rearranging, we find

$$\epsilon^{-\alpha} = \frac{N}{\sum_{i=1}^{m} \epsilon^{-E_i/\kappa T}} \tag{2-40}$$

The denominator of Eq. (2-40) is called the *partition function:*

$$Z = \sum_{i=1}^{m} \epsilon^{-E_i/\kappa T} \tag{2-41}$$

Sec. 2-3 Distributions for Classical Systems in Thermodynamic Equilibrium

The actual form of the macrostate depends on the partition function, while it in turn depends on what portion of phase space is available to the particles and on how we divide phase space up into small volumes. For classical particles, of course, all values of momentum and hence energy are allowed, while the particles are restricted in coordinate space by the finite volume of the system.

The results of our derivation, up to the determination of a specific Z, do not depend on how we divide phase space into small volumes $d\mathbf{r}\,d\mathbf{p}$. The partition function in Eq. (2-41) is represented by the sum of a finite but large number of terms, one for each phase-space volume. It is easier to evaluate Z when it can be represented by an integral. While it is not possible to obtain an integral representation for Z for all systems, we shall assume that we can do so for systems that are of interest to us. The final results, of course, are very useful; otherwise we would not have gone to all this trouble. In this case, at least, the usefulness of the result justifies our assumption.

We shall obtain an integral representation for Z in the following way. Suppose that we divide each of the small volumes $d\mathbf{r}\,d\mathbf{p}$ into smaller volumes σ, which we shall call *cells*. Further, let us suppose that there are η of these cells per small volume:

$$\eta = \frac{d\mathbf{r}\,d\mathbf{p}}{\sigma} \tag{2-42}$$

The partition function now has the same form as Eq. (2-41), but there are many more terms in the sum. In fact, for each term in the sum prior to the subdivision of $d\mathbf{r}\,d\mathbf{p}$ into cells there are η terms after the subdivision. Fortunately the η terms that replace, for example, the jth term in the original sum are all the same since the energy of the particles is the same in all of the cells made from the jth small volume. The sum of these η terms, then, is just

$$\eta \epsilon^{-E_j/\kappa T}$$

The partition function for the new subdivision of phase space is

$$Z = \frac{1}{\sigma} \sum_{i=1}^{m} \epsilon^{-E_i/\kappa T}\,d\mathbf{r}\,d\mathbf{p} \tag{2-43}$$

where we have substituted for η from Eq. (2-42). We can now transform the sum to an integral obtaining

$$Z = \frac{1}{\sigma} \int\int \epsilon^{-p^2/2m\kappa T}\,d\mathbf{r}\,d\mathbf{p} \tag{2-44}$$

where the integrals are over all allowed positions and all momenta and where we have expressed the energy of a particle in terms of the momentum.

Equation (2-44) represents a sixfold integral. The integrals over spatial coordinates are the easiest to perform because the integrand is not a function of position. The notation for the spatial integral can be expanded, for example, to

$$\int d\mathbf{r} = \iiint_{\text{all } x, y, z} dx\, dy\, dz$$

if we choose to use rectangular spatial coordinates. The result of the integration is the volume of the system V.

If we choose to use rectangular coordinates in momentum space as well, the momentum integrals are

$$\int_{-\infty}^{\infty}\int_{-\infty}^{\infty}\int_{-\infty}^{\infty} \epsilon^{-(p_x^2 + p_y^2 + p_z^2)/2m\kappa T}\, dp_x\, dp_y\, dp_z \qquad (2\text{-}45)$$

Equation (2-45) is the product of three integrals, each one of the form

$$\int_{-\infty}^{\infty} \epsilon^{-p_j^2/2m\kappa T}\, dp_j = (2\pi m\kappa T)^{1/2} \qquad (2\text{-}46)$$

The net result of the integration of Eq. (2-44), then, is

$$Z = \frac{V}{\sigma}(2\pi m\kappa T)^{3/2} \qquad (2\text{-}47)$$

We are at last able to find the number of particles in the ith cell. This number is given by Eq. (2-35), which is valid for the cells as well as for the small volumes. The Lagrange multiplier α is given by Eq. (2-40), where the partition function is that of Eq. (2-47). Combining these equations, we obtain

$$N_i = \frac{(N/V)\sigma}{(2\pi m\kappa T)^{3/2}} \epsilon^{-p^2/2m\kappa T} \qquad (2\text{-}48)$$

The phase-space distribution function is defined by Eq. (2-5) as the number of representative points in a small portion of phase space divided by the volume of that portion of phase space. Or, in other words, it is the density of points in phase space. Since Eq. (2-48) gives the number of points in a cell of volume σ, we obtain the phase-space density by dividing N_i by σ to get

$$\boxed{f_{\text{MB}}(\mathbf{r}, \mathbf{p}) = \frac{n}{(2\pi m\kappa T)^{3/2}} \epsilon^{-p^2/2m\kappa T}} \qquad (2\text{-}49)$$

where $n = N/V$ is the uniform spatial density of the particles. Equation (2-49) is called the *Maxwell-Boltzmann phase-space distribution function* and describes the macroscopic state of a system of classical particles in thermodynamic equilibrium. The notation $f_{\text{MB}}(\mathbf{r}, \mathbf{p})$ indicates a distribution function defined in six-dimensional phase space, where the subscripts refer to Maxwell and Boltzmann, both of whom were instrumental in the early formulation of statistical physics.

Maxwell-Boltzmann Distribution in Phase Space

There are a number of interesting features about the Maxwell-Boltzmann distribution. Although it is a density function defined in six-dimensional phase space, it is independent of spatial coordinates, which indicates that the system it describes is spatially uniform. Furthermore, it depends only on the magnitude of the momentum, not on its vector direction, so that it is *spherically symmetric* in momentum-space. A plot of the Maxwell-Boltzmann distribution vs. the magnitude of the momentum of a particle is shown in Fig. 2-3. This symmetry property also implies that there is no net momentum

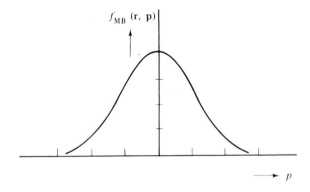

Fig. 2-3 Maxwell-Boltzmann distribution function $f_{MB}(\mathbf{r}, \mathbf{p})$ as a function of the magnitude of the momentum p.

for the system. For any specific value of momentum p_1, there are as many particles with momenta in the range dp about p_1 as there are with momenta in the range dp about $p_2 = -p_1$ so that the net momentum of these two groups of particles taken together is zero. Since this statement is true for any specific momentum, the average momentum of the whole system is zero.

It is also easy to prove mathematically that the average vector momentum of the system is zero. Applying the definition of average value, Eq. (2-16), we find

$$\bar{\mathbf{p}} = \frac{1}{N} \int\int \mathbf{p} f_{MB}(\mathbf{r}, \mathbf{p}) \, d\mathbf{r} \, d\mathbf{p}$$

This relation is a vector equation that has, for instance, an x-component

$$\bar{p}_x = \frac{1}{N} \int\int p_x f_{MB}(\mathbf{r}, \mathbf{p}) \, d\mathbf{r} \, d\mathbf{p}$$

Inserting the Maxwell-Boltzmann distribution, we find an integrand of the form

$$\int p_x \epsilon^{-p^2/2m\kappa T} \, d\mathbf{p}$$

If we choose to use rectangular coordinates for the integration in momentum space, we can write

$$\int_{-\infty}^{\infty}\int_{-\infty}^{\infty}\int_{-\infty}^{\infty} p_x \epsilon^{-(p_x^2+p_y^2+p_z^2)/2m\kappa T}\, dp_x\, dp_y\, dp_z$$

Integration over the x-component of the momentum is proportional to

$$\int_{-\infty}^{\infty} p_x \epsilon^{-p_x^2/2m\kappa T}\, dp_x$$

The integrand of this last relation is a product of two functions, one an odd function of p_x and the other an even function, each of which is sketched in Fig. 2-4. The product of the two functions is odd, as indicated in the figure, so that contributions to the integral for negative values of p_x are equal to but

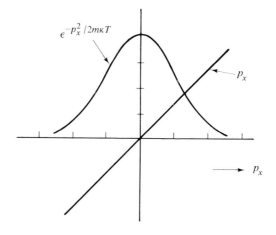

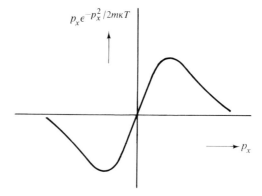

Fig. 2-4 Even and odd functions of p_x.

Sec. 2-3 Distributions for Classical Systems in Thermodynamic Equilibrium 35

of opposite sign from those for positive values of p_x and the integral is zero. Similar arguments hold for the y- and z-components so that the average vector momentum is zero.

There is an additional interesting feature about the Maxwellian distribution that we should like to point out. The probability distribution for momentum only, obtained by integrating $f_{MB}(\mathbf{r}, \mathbf{p})$ over the volume and dividing by N, is

$$P_{MB}(\mathbf{p}) = \frac{1}{(2\pi m\kappa T)^{3/2}} \epsilon^{-p^2/2m\kappa T}$$

If we write p^2 in terms of its cartesian components, we can rewrite the expression for $P_{MB}(\mathbf{p})$ as

$$P_{MB}(\mathbf{p}) = P_{MB}(p_x) P_{MB}(p_y) P_{MB}(p_z)$$

that is, as the product of three independent probabilities, each one of the form

$$P_{MB}(p_j) = \frac{1}{(2\pi m\kappa T)^{1/2}} \epsilon^{-p_j^2/2m\kappa T}, \quad j = x, y, z$$

The student who is familiar with mathematical theories of probability will recognize this probability density for one component of momentum as being in the same form as the so-called *normal distribution* in a single variable:

$$\psi(x) = \frac{1}{\sigma\sqrt{2\pi}} \epsilon^{-(x-\bar{x})^2/2\sigma^2}$$

where σ is the standard deviation about the mean value $\bar{x}$. Of course, we have just shown that in thermodynamic equilibrium the mean or average value of each component of momentum is zero. The important observation, insofar as statistical physics is concerned, is that the probability densities for the various components of momentum are independent of each other but that each has the same standard deviation, $(m\kappa T)^{1/2}$. In other words, the probability that a particle has a certain x-component of momentum is independent of its y- or z-component, but all the particles taken together have identical distributions for each component. The latter feature is an expression of the principle of equipartition of energy.

Equilibrium Distribution in Energy Space

While the phase-space distribution function is the cornerstone of statistical physics, another useful form of the equilibrium distribution that is frequently encountered is the energy-space distribution, which we shall denote by $f_{MB}(E)$. The energy distribution is an alternative macroscopic description of the same system of classical particles in equilibrium that is described by the phase-space distribution. It stands to reason, therefore, that it is possible to derive $f_{MB}(E)$ directly from $f_{MB}(\mathbf{r}, \mathbf{p})$. What we have to know to accomplish the derivation is the relationship between energy and momentum, which is

$$E = \frac{p^2}{2m} \tag{2-50}$$

What we need to ensure is that the integral of $f_{MB}(E)$ over all allowed energies yields the total number of particles as indicated in Eq. (2-7), which is repeated here for convenience,

$$N = \int_0^\infty f_{MB}(E)\,dE \tag{2-7}$$

The corresponding equation for the phase-space distribution is Eq. (2-14),

$$N = \iint f_{MB}(\mathbf{r}, \mathbf{p})\,d\mathbf{r}\,d\mathbf{p} \tag{2-14}$$

To derive $f_{MB}(E)$ we start with Eq. (2-14) using Eq. (2-49) for $f_{MB}(\mathbf{r}, \mathbf{p})$ and work toward an equation of the form of Eq. (2-7) using Eq. (2-50). We start with

$$N = \iint \frac{n}{(2\pi m\kappa T)^{3/2}} \epsilon^{-p^2/2m\kappa T}\,d\mathbf{r}\,d\mathbf{p} \tag{2-51}$$

which is a sixfold integral. Our objective is a single integral, so somewhere along the way we must perform five integrations. The three spatial integrations of Eq. (2-51) are straightforward, yielding the volume V since the integrand is independent of position. The product $nV = N$ so that we are left with

$$N = \int \frac{N}{(2\pi m\kappa T)^{3/2}} \epsilon^{-p^2/2m\kappa T}\,d\mathbf{p} \tag{2-52}$$

which is a threefold integral over momentum coordinates. Notice that the integrand is an even function of momentum depending only on the magnitude of the momentum, which, through Eq. (2-50), is proportional to the energy. Since we are not restricted in our choice of coördinate systems in momentum space, we shall choose to express $d\mathbf{p}$ in coordinates that are also easily related to the energy. The appropriate coordinate system is a spherical one, shown in Fig. 2-5. The coordinates of a point in spherical momentum space are the magnitude of the momentum vector from the origin to the point p, the angle of the vector from the polar axis θ_p, and the angle from the aximuthal axis ϕ_p. The system is strictly analogous to the spherical spatial coordinates (r, θ, ϕ). The differential volume of momentum space is

$$d\mathbf{p} = dp(p\,d\theta_p)(p\sin\theta_p\,d\phi_p)$$

and the limits of integration are

$$p:\ 0 \longrightarrow \infty$$
$$\theta_p:\ 0 \longrightarrow \pi$$
$$\phi_p:\ 0 \longrightarrow 2\pi$$

The integration of Eq. (2-52) over the angles is straightforward. The integral

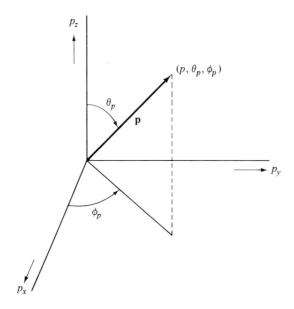

Fig. 2-5 Spherical coordinates in momentum space.

over ϕ_p yields 2π, and the integral of $\sin\theta_p$ over θ_p yields a factor of 2. Equation (2-52) thus reduces to

$$N = \int_0^\infty \frac{4\pi N}{(2\pi m\kappa T)^{3/2}} \epsilon^{-p^2/2m\kappa T} p^2 \, dp \tag{2-53}$$

Substituting for p^2 from Eq. (2-50),

$$N = \int_0^\infty \frac{8\pi m N E}{(2\pi m\kappa T)^{3/2}} \epsilon^{-E/\kappa T} \, dp \tag{2-54}$$

To change from integration over momenta to integration over energy we recognize from Eq. (2-50) that $p = (2mE)^{1/2}$ so that

$$dp = \left(\frac{m}{2E}\right)^{1/2} dE$$

As p varies from zero to infinity so does E, and so the limits of integration are the same. Equation (2-54) becomes

$$N = \int_0^\infty \frac{2N}{\pi^{1/2}(\kappa T)^{3/2}} E^{1/2} \epsilon^{-E/\kappa T} \, dE \tag{2-55}$$

Comparison of Eq. (2-55) with Eq. (2-7) yields the Maxwell-Boltzmann energy distribution:

$$\boxed{f_{\text{MB}}(E) = \frac{2N}{\pi^{1/2}\kappa T}\left(\frac{E}{\kappa T}\right)^{1/2} \epsilon^{-E/\kappa T}} \tag{2-56}$$

Notice that the energy enters not only in the exponential term but also as a square root in the multiplicative factor.

Steady-State Distributions for Particles in a Potential Field

One final comment on equilibrium statistics before we turn to more practical topics. Frequently we deal with systems that are in a steady state under the influence of an external force field that can be expressed in terms of a scalar potential energy U. While these conditions of steady state are not equilibrium conditions in the thermodynamic sense since an external force is present, they are frequently called equilibrium conditions because the phase-space distribution function resembles the Maxwell-Boltzmann distribution.

Examples of such external forces are the gravitational and electrostatic forces. The potential energies of a particle in these fields are mgh and qV, respectively, where g is the gravitational constant, h the height above a reference plane, q the electric charge of the particle, and V the electrostatic potential function. By the methods of the next chapter it can be shown that the distribution function that describes the steady state of such a system has the same form as the Maxwellian except that the kinetic energy is replaced by the sum of kinetic and potential energies. That is,

$$f(\mathbf{r}, \mathbf{p}) = \frac{n_0}{(2\pi m \kappa T)^{3/2}} \exp\left\{-\frac{p^2}{2m\kappa T} - \frac{U(\mathbf{r})}{\kappa T}\right\} \qquad (2\text{-}57)$$

where we have indicated that the potential energy $U(\mathbf{r})$ may be a function of position. The spatial density, according to Eq. (2-18), is the integral of $f(\mathbf{r}, \mathbf{p})$ over all momenta. For the current case, this procedure yields a nonuniform spatial density:

$$\boxed{n(\mathbf{r}) = n_0 \epsilon^{-U(\mathbf{r})/\kappa T}} \qquad (2\text{-}58)$$

According to this result, n_0 is the density where the particles have zero potential energy.

BIBLIOGRAPHY

Introductory textbooks discussing probability, distribution functions, and statistical thermodynamics:

REIF, E., *Statistical Physics*. New York: McGraw-Hill Book Company, 1967.

GIEDT, W. H., *Thermophysics*. Princeton, N.J.: Van Nostrand Rinehold Company, 1967.

MORSE, P. M., *Thermal Physics*. Reading, Mass.: W. A. Benjamin, Inc., 1962.

KITTEL, C., *Elementary Statistical Physics*. New York: John Wiley & Sons, Inc., 1958.

For advanced study:

CALLEN, H. B., *Thermodynamics*. New York: John Wiley & Sons, Inc., 1960.

LANDAU, L. D., and E. M. LIFSHITZ, *Statistical Physics*. Reading, Mass.: Addison-Wesley Publishing Company, Inc., 1958.

PROBLEMS

2-1. Find the average kinetic energy of a system of particles characterized by the Maxwell-Boltzmann distribution, $f_{MB}(\mathbf{r}, \mathbf{p})$. Obtain an expression for the square root of the average of the squared velocity [root mean square (r.m.s.) velocity] in terms of the temperature and mass. Calculate the r.m.s. velocity of electrons at room temperature ($T \simeq 290°K$).

2-2. Derive Eq. (2-27) from Eq. (2-26).

2-3. Demonstrate the principle of equipartition of energy by showing that $\overline{p_x^2} = \frac{1}{3}\overline{p^2}$ for classical particles in equilibrium.

2-4. The classical equilibrium distribution in speed space satisfies

$$N = \int_0^\infty f_{MB}(v)\, dv$$

(a) Find the appropriate speed-space distribution function, and
(b) Show that the average speed of these particles is

$$\bar{v} = \left(\frac{8\kappa T}{\pi m}\right)^{1/2}$$

(c) Sketch $f_{MB}(v)$ as a function of v, and
(d) Find the most probable speed for a particle.
(e) Compare the average speed, the most probable speed, and the r.m.s. velocity from Problem 2-1.

2-5. Using Eq. (2-56), verify that

$$\int_0^\infty f_{MB}(E)\, dE = N$$

2-6. Sketch the variation of $f_{MB}(E)$ with E and find the most probable energy for a particle. Compare the most probable energy with the average kinetic energy obtained in Problem 2-1.

2-7. Assume that a system of electrons in a uniform static electric field can be treated as classical particles with the steady-state distribution of Eq. (2-57). Take the electric potential $V(x)$ to be Ex. Sketch the variation of the electron density as a function of x. Calculate the scale length of the density variation for room temperature and an electric field strength of 10^4 V/cm (V = volts).

The scale length L is defined by

$$\frac{1}{L} = \left|\frac{1}{n}\frac{dn}{dx}\right|$$

2-8. What fraction of the particles of a Maxwell-Boltzmann distribution has magnitude of the x-component of momentum greater than $(m\kappa T)^{1/2}$, $2(m\kappa T)^{1/2}$, and $3(m\kappa T)^{1/2}$? *Hint:* Use tables of the normal curve of error or of the error function.

3

Nonequilibrium Statistical Physics

As we remarked in Chapter 2, a system in true thermodynamic equilibrium is a very dull one. All of its properties are completely homogeneous; its density, temperature, chemical composition—everything—all are constants, independent of position and time. Furthermore, it can be shown that all closed systems evolve toward such an equilibrium in a nonreversible way with increasing time.

It is the nonequilibrium condition that proves useful and interesting, be it a transient or a steady state. In the nonequilibrium case, because of the peculiar nature of a particular medium or material, electric currents may flow when electric fields or other forces are applied. The characteristics of the currents responsive to applied forces may lend themselves to device applications, the principal objectives of our study. After all, the utilization of real physical phenomena in the performance of some task is what engineering is all about.

The macroscopic state of a system that is not in thermodynamic equilibrium is still described by a distribution function. If transient conditions exist, the phase-space distribution function, for instance, will be an explicit function of time: $f(\mathbf{r}, \mathbf{p}, t)$. The mathematical description of the temporal development of the distribution is obtained from an equation, one term of which is the time derivative of $f(\mathbf{r}, \mathbf{p}, t)$. We shall develop this equation, called the *kinetic equation*, from physical arguments in Section 3.2. However, we shall not, for the most part, be concerned with transitory systems in this book because the statistical description of irreversible thermodynamic systems is an advanced and difficult subject. Suffice it to say that any real system always develops irreversibly with increasing time in such a way that the entropy increases on the average. *On the average* means the following: It is possible for a system to fluctuate from a macrostate that can be formed from many microstates to one formed from slightly fewer microstates, but this is a relatively rare occurrence. Usually the system develops from a less probable macrostate to a more probable one until it eventually reaches the most probable macrostate. If steady forces continue to act, this most probable macrostate will not be descriptive of thermodynamic equilibrium because, strictly speaking, the system is not a closed one. It will, however, be descriptive of a nonequilibrium steady state. Because every microstate is equally probable, the system may still fluctuate from its most probable state. Fluctuations give rise to noise signals in electronic devices. Noise is the subject of Section 3.7.

In a nonequilibrium steady state, externally applied forces maintain some kind of imbalance that otherwise would not exist. For instance, a temperature gradient can be imposed on an iron bar by placing one end in a bucket of ice and the other in an oven. To maintain the gradient in a steady state it is necessary to supply energy in the form of heat to the hot end and to extract it from the cool end. Inside the material a steady transport of energy from the hot to the cool end occurs. This transport of energy is in the form of a flux or flow of heat. The facility with which heat flows in a material is measured by a macroscopic property called *thermal conductivity*. If we did not maintain the gradient by some external means, then eventually the bar would assume a uniform temperature.

Thermal conductivity is just one example of a large number of so-called transport coefficients that relate the steady-state flux of some property to the steady-state forces that create the flux. Another example of a force and the resulting flux is a gradient in particle density that drives a current of particles or a particle flux from regions of higher density to regions of lower density. This effect is known as diffusion, and the transport coefficient is called the *diffusion coefficient*. Yet another example is a gradient in electric potential that drives an electric current of charged particles. The transport coefficient relating this force-flux pair is the *electrical conductivity*, the relation being a form of Ohm's law. Electric current can also flow in response to forces other than an electric field force. One such force is the gradient in the density of charged particles. The particles that diffuse because of the density gradient carry their charge with them, resulting in a flow of electrical current. The transport coefficient for this case turns out to be the product of the charge per particle and the diffusion coefficient. Another force that can cause electric current to flow under certain conditions is a temperature gradient. This effect is known as the Seebeck effect and is responsible for the operation of thermocouples. The coefficient is called, not surprisingly, the *Seebeck coefficient*.

Sec. 3-1 Phenomenological Relations

All of the transport coefficients, and the listing above does not begin to exhaust their number, are macroscopically measurable properties of a system. Historically, the coefficients arose in the descriptions of currents and forces observed experimentally so that their values were determined empirically. One purpose of nonequilibrium statistical physics is to predict the values of these coefficients theoretically on the basis of our understanding of the physical processes that underlie these effects. Comparison of theory with experiment indicates the range of validity of our ideas. If our physical models are indeed useful, we shall be able to use them to devise and to understand the behavior of other systems to which they might also be applicable. The purpose of this chapter is to develop physical ideas about electrical current so that we shall be able to model practical electronic devices and systems.

In this chapter we shall study the transport effects most commonly encountered in simple electronic devices, electrical conductivity, mobility, and diffusion. To do this, we shall have to study the relationships between the forces and the appropriate fluxes. This is a simple order if we know the nonequilibrium macroscopic steady state of the system, that is, if we know the distribution function. In Section 3.2 we shall develop the mathematical relationship between the distribution function and the forces. In the latter part of the chapter we shall use approximate solutions for the distribution function to estimate the important coefficients. But first in Section 3.1 we shall be more specific about the phenomenological relationships between the forces and fluxes and indicate the general problems to be solved in the remainder of the chapter.

3.1 PHENOMENOLOGICAL RELATIONS

Electrical current is the exhibition of the motion of net electrical charge past an observer per unit time. Similarly, particle current is the manifestation of the motion of particles past an observer per unit time. Obviously, particle current and electrical current must be related by the amount of electrical charge carried per particle. While it is currents that are measured in the laboratory, it is customary instead to study current densities. Current density is more fundamental since it is a function only of the forces and the intensive properties of the medium such as density, temperature, or chemical composition and not of its extensive properties such as length, cross-sectional area, or volume. Electrical current density is the net charge crossing a unit area in unit time and is related to the current by the familiar relationship

$$I = \int \mathbf{J} \cdot \hat{n} \, dA \qquad (3\text{-}1)$$

where I is the electrical current, $\mathbf{J}$ is the vector electrical current density, and the integral is taken over an appropriate cross section of area whose elemental area is dA with unit outward normal $\hat{n}$. In S.I. units, electrical current density is measured in amperes per square meter (A/m²). If 1 A flows uniformly through a #16 AWG wire, the current density is 765,000 A/m². Particle current density will be denoted by a capital Greek gamma, $\boldsymbol{\Gamma}$. If the particles under consideration each carry charge q, the electrical current density is just

$$\mathbf{J} = q\boldsymbol{\Gamma} \qquad (3\text{-}2)$$

In the case of electrons, $q = -e$, where

$$e = 1.602 \times 10^{-19} \text{ C (coulombs)}$$

Notice that the electrical current density is antiparallel to an electron particle current density.

The particle current density at a point in space, being the rate at which particles cross a unit area, is given by the product of the particle density and the average velocity of the particles at that point:

$$\boldsymbol{\Gamma}(\mathbf{r}) = n(\mathbf{r})\langle\mathbf{v}(\mathbf{r})\rangle \qquad (3\text{-}3)$$

where $\langle\mathbf{v}(\mathbf{r})\rangle$ represents the local average velocity, which, according to Eq. (2-17), is

$$\langle\mathbf{v}(\mathbf{r})\rangle = \frac{\int \mathbf{v} f(\mathbf{r},\mathbf{p})\,d\mathbf{p}}{\int f(\mathbf{r},\mathbf{p})\,d\mathbf{p}} \qquad (3\text{-}4)$$

and where $n(\mathbf{r})$ is the local density, which, according to Eq. (2-18), is

$$n(\mathbf{r}) = \int f(\mathbf{r},\mathbf{p})\,d\mathbf{p} \qquad (3\text{-}5)$$

Notice that the denominator of Eq. (3-4) is just the local density $n(\mathbf{r})$. Combining the three equations above, we obtain the relationship between the particle current density and the distribution function:

$$\boldsymbol{\Gamma}(\mathbf{r}) = \int \mathbf{v} f(\mathbf{r},\mathbf{p})\,d\mathbf{p} \qquad (3\text{-}6)$$

We can see from either Eq. (3-3) or Eq. (3-6) that if the system is in thermodynamic equilibrium there will be no net particle current because the average velocity of an equilibrium distribution, which is proportional to the average momentum, is zero, as shown in Chapter 2.

The electrical current density, according to Eqs. (3-2) and (3-3), is

$$\mathbf{J}(\mathbf{r}) = q n(\mathbf{r})\langle\mathbf{v}(\mathbf{r})\rangle \qquad (3\text{-}7)$$

or

$$\mathbf{J}(\mathbf{r}) = \int q\mathbf{v} f(\mathbf{r}, \mathbf{p})\, d\mathbf{p} \qquad (3\text{-}8)$$

Equation (3-8) is one example of the flux of a particle property, in this case charge. For a general property $h(\mathbf{r}, \mathbf{p})$, the flux or current density is given by

$$\mathbf{\Gamma}_h(\mathbf{r}) = \int h(\mathbf{r}, \mathbf{p})\mathbf{v} f(\mathbf{r}, \mathbf{p})\, d\mathbf{p} \qquad (3\text{-}9)$$

Some other specific properties that may be transported are mass, momentum, and energy. The transport of momentum is related to the kinetic pressure, and the transport of energy is a heat flux.

Conductivity

Now that we understand how the current densities can be expressed in terms of the statistical description of the system, we shall turn to the phenomenological relations between these fluxes and the forces that drive them. These relations, which define the transport coefficients, were developed historically from experimental observation and are applicable to linear, isotropic, homogeneous materials. Perhaps the most familiar of these to electrical engineers is the intensive form of Ohm's law,

$$\mathbf{J} = \sigma\mathbf{\mathcal{E}} \qquad (3\text{-}10)$$

which defines the electrical conductivity σ as the coefficient relating the electric force field and the electric current density. Ohm's law in terms of extensive variables is the terminal characteristic $I = (V/R)$, where R is the resistance of the object that has length L and cross-sectional area A, $R = (L/\sigma A)$, and where V is the potential drop along its length, $V = -\int \mathbf{\mathcal{E}} \cdot d\mathbf{L}$. Electrical conductivity has the dimensions in S.I. units of (ohm-meter)$^{-1}$ or mhos/meter, where mho $=$ (ohm)$^{-1}$. Typical values range from 10^{-17} for nonconductors to 10^7 for good conductors. Table 3-1 indicates the measured values of d.c. electrical conductivity for a variety of materials at room temperature. When the electric field is a static one, it can be derived from the scalar electrostatic potential function V,

$$\mathbf{\mathcal{E}} = -\nabla V \qquad (3\text{-}11)$$

in which case Eq. (3-10) takes the form

$$\mathbf{J} = -\sigma\nabla V \qquad (3\text{-}12)$$

The latter form indicates that for positive conductivity the current flows in the direction opposite to that of the vector gradient. Since the vector gradient

Table 3-1. Electrical Conductivity and Electron Mobility at Room Temperature for Intrinsic Materials

Materials*	Conductivity (mhos/m)	Mobility (m²/volt-sec)
Silver	6.14×10^7	1.0×10^{-3}
Aluminum	3.54×10^7	5.6×10^{-3}
Indium antimonide	1.67×10^4	7.7
Germanium	10	0.39
Quartz	$< 2 \times 10^{-17}$	—

*Silver and aluminum are classed as conductors, indium antimonide and germanium as semiconductors, and quartz as an insulator or nonconductor.

points in the direction of greatest *increase* in the value of the potential, or, in simple terms, up the potential hill, the current flows in the direction of greatest decrease, from regions of high potential to regions of low potential.

Electric Mobility

Another transport coefficient, the electric mobility μ, is frequently encountered, especially in discussion of the properties of solids. The mobility is directly related to the conductivity and arises from the empirical fact that a d.c. current flows in response to an applied d.c. electric field. According to Eq. (3-7), a d.c. current means that the charge carriers must have an average velocity. For linear materials, that average velocity is proportional to the applied field, the constant of proportionality being defined as the electric mobility,

$$\langle \mathbf{v} \rangle = \pm \mu \mathcal{E} \qquad (3\text{-}13)$$

where the sign depends on the sign of the charge. The average velocity represented by Eq. (3-13) is frequently called the *drift velocity*. If we write the product qn as ρ, the charge density in coulombs per cubic meter, and combine Eqs. (3-7), (3-10), and (3-13), we find the relation between the conductivity and mobility:

$$\sigma = \pm \rho \mu \qquad (3\text{-}14)$$

The mobility is a useful parameter because it focuses on one of the two effects, average velocity and charge density, that relate the current to the field. This feature makes the description of mobility particularly useful in semiconductors in which it is possible to alter the charge density by artificial means, such as by exposing the material to an appropriate light source. In most circumstances such a change in the charge carrier density does not affect the mobility. In S.I. units, mobility should be measured in square meters per volt-second;

however it is common practice to use dimensions of square centimeters per volt-second. Typical values of the mobility of electrons at room temperature in some materials are indicated in Table 3-1. Conductivity and mobility are the subjects of Section 3.4.

Diffusion

Another important transport effect relates a density gradient to the resulting particle flux and is described by Fick's law for diffusion,

$$\boxed{\mathbf{\Gamma} = -D\,\nabla n} \qquad (3\text{-}15)$$

where D is called the diffusion coefficient. Notice that Fick's law is in the same form as Ohm's law when the latter is written as in Eq. (3-12). Therefore, a similar macroscopic interpretation applies: For a positive diffusion coefficient, net particle current density flows in the direction of greatest *decrease* in density. As we shall see later, the diffusion coefficient is always positive, so the net particle flux is always from regions of high to regions of low density. Although Eqs. (3-12) and (3-15) are of similar form, the physical processes that underlie them are quite different. We shall study the diffusion process in Section 3.5.

Conservation of Particles

When a spatially nonuniform particle flux occurs, the density of particles in space must change, assuming that particles can neither be created nor destroyed. For example, if a net particle current flows to a given point, the density at that point must increase with time because the current represents the net transport of particles. The mathematical expression that relates the time change of the density to the spatial nonuniformity of the current is called the *equation of continuity*.

A simple heuristic argument can be used to derive the continuity equation. Consider the small test volume in space $\Delta V = \Delta x\,\Delta y\,\Delta z$ sketched in Fig. 3-1. The volume ΔV contains ΔN particles, where $\Delta N = n\,\Delta V$. Since the density may in general be a function of time, ΔN will also be a function of time. We shall derive the continuity equation by calculating how ΔN changes with time due to the currents of particles flowing across the surfaces of the volume.

Suppose that the particles have an average velocity $\langle \mathbf{v} \rangle = \mathbf{v}_0$ and hence a flux $\mathbf{\Gamma}_0$ at one corner of the volume (x_0, y_0, z_0), as indicated in the figure. Let us concentrate on the x-component of the particle motion, which carries particles into or out of the volume across the surfaces that are normal to the x-direction. These surfaces, each with area $\Delta y\,\Delta z$, are numbered I and II in

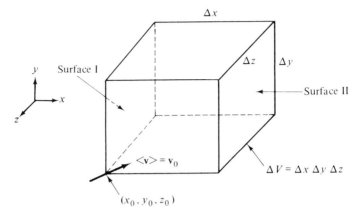

Fig. 3-1 Test volume in space used to demonstrate the continuity equation.

Fig. 3-1. The x-component of particle current density at surface I is Γ_{xI} and the corresponding current into the volume is $\Gamma_{xI}\,\Delta y\,\Delta z$ assuming that Γ_{xI} is constant over the surface. Therefore the number of particles that enter the volume across this surface in time δt is just $\Gamma_{xI}\,\Delta y\,\Delta z\,\delta t$. Another way of calculating the particles entering across surface I is to recognize that all of the particles within a small volume whose base is surface I and whose height is $v_{0x}\,\delta t$ above the base will manage to pass through the surface in time δt. This small volume is shown in Fig. 3-2. The number of these particles is just the density n_0 multiplied by the volume $\Delta y\,\Delta z v_{0x}\,\delta t$ provided that the volume is sufficiently small. The number of particles that enter across surface I, then, is $n_0 v_{0x}\,\Delta y\,\Delta z\,\delta t$, which is also $\Gamma_{xI}\,\Delta y\,\Delta z\,\delta t$. This second technique, incidentally, is usually used to derive the definition of particle current density, Eq. (3-3).

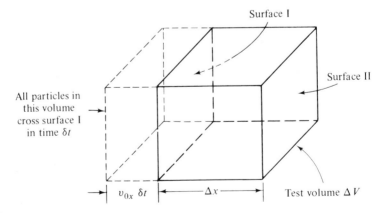

Fig. 3-2 Location of the small volume that contains all the particles that cross surface I in time δt.

Sec. 3-1 Phenomenological Relations

The number of particles that leave the test volume because of their x-directed motion do so at surface II according to our assumptions. By reasoning similar to either approach above, the number leaving in such a manner will be $\Gamma_{xII} \Delta y \Delta z \delta t$. Because Δx is assumed to be small, we can relate the particle current density Γ_x at surface II to that at surface I by performing a Taylor expansion about the point $x = x_0$ keeping only the terms to first order,

$$\Gamma_x(x) = \Gamma_x(x_0) + \frac{\partial \Gamma_x}{\partial x}(x - x_0) + \cdots$$

where $\Gamma_x(x_0) = \Gamma_{xI}$. If we evaluate this expression at the position x such that $(x - x_0) = \Delta x$, we obtain Γ_{xII}:

$$\Gamma_{xII} \simeq \Gamma_{xI} + \frac{\partial \Gamma_x}{\partial x} \Delta x$$

We are now in a position to obtain the change in ΔN due to the x-directed motion of the particles by subtracting the number of particles leaving from those entering. Using the Taylor expansion, the result is

$$\delta(\Delta N)_{\text{due to } V_{0x}} = \Gamma_{xI} \Delta y \Delta z \delta t - \Gamma_{xII} \Delta y \Delta z \delta t$$

$$= -\frac{\partial \Gamma_x}{\partial x} \Delta x \Delta y \Delta z \delta t$$

By exactly the same arguments we can calculate the changes in ΔN due to the y- and z-components of v_0 with the results

$$\delta(\Delta N)_{\text{due to } V_{0y}} = -\frac{\partial \Gamma_y}{\partial y} \Delta y \Delta x \Delta z \delta t$$

and

$$\delta(\Delta N)_{\text{due to } V_{0z}} = -\frac{\partial \Gamma_z}{\partial z} \Delta z \Delta x \Delta y \delta t$$

If we combine these three contributions and take the limit as δt becomes infinitesimally small, we obtain the time rate of change of ΔN, which is, using the notation of vector calculus,

$$\frac{\partial}{\partial t}(\Delta N) = -\nabla \cdot \Gamma \Delta x \Delta y \Delta z,$$

where

$$\nabla \cdot \Gamma = \frac{\partial \Gamma_x}{\partial x} + \frac{\partial \Gamma_y}{\partial y} + \frac{\partial \Gamma_z}{\partial z}$$

is the divergence of Γ. Since $\Delta N = n \Delta V$ and $\Delta V = \Delta x \Delta y \Delta z$, this reduces to the equation of continuity,

$$\boxed{\frac{\partial n}{\partial t} + \nabla \cdot \Gamma = 0} \qquad (3\text{-}16)$$

where n and Γ are functions of position and time.

In words, the time rate of increase of particle density is the negative of the divergence of the particle flux. If particles are being created and/or annihilated, which is a distinct possibility especially in semiconductors, the appropriate rates should be added to the right-hand side of the equation. In semiconductor physics the rate at which particles are being generated is frequently denoted by g and the annihilation rate by r. The equation of continuity in the presence of generation and destruction of charge is therefore

$$\frac{\partial n}{\partial t} + \mathbf{\nabla} \cdot \mathbf{\Gamma} = g - r \qquad (3\text{-}17)$$

By multiplying Eq. (3-16) by the charge carried by each particle, we obtain the equation of charge continuity, which, in the absence of the creation or annihilation of charge, has the form

$$\frac{\partial \rho}{\partial t} + \mathbf{\nabla} \cdot \mathbf{J} = 0 \qquad (3\text{-}18)$$

3.2 THE KINETIC EQUATION

In the previous section we discussed how macroscopic currents are related to the distribution function and how the transport coefficients relate these currents to the applied forces. In this section we shall investigate the relationship between the forces and the nonequilibrium distribution function itself. If we can determine the distribution function given some disturbance of the system from equilibrium, then we can calculate the transport coefficients using the definitions of the previous section. The mathematical expression for the distribution function in terms of applied forces is called the *kinetic equation*, which we shall now develop.

Since the distribution function, $f(\mathbf{r}, \mathbf{p}, t)$, is in general a function of position, momentum, and time, we should expect partial derivatives of the distribution function with respect to all three of these parameters to appear in the equation. The principal clue to understanding the kinetic equation is the recognition that the distribution is the density of representative points in phase space so that the number of representative points $\Delta^6 N$ in a small six-dimensional volume $\Delta^3 r \, \Delta^3 p$ is

$$\Delta^6 N = f(\mathbf{r}, \mathbf{p}, t) \, \Delta^3 r \, \Delta^3 p \qquad (3\text{-}19)$$

If the distribution changes in time, then this number also changes proportionally. To determine how it changes, we need only calculate the rates at which particles enter and leave the small volume in phase space. The problem is analogous to that of the continuity of particles in real space discussed in Section 3.1. In fact the kinetic equation is sometimes called the equation of continuity of representative points in phase space.

Sec. 3-2 The Kinetic Equation   51

There are, however, three ways that a representative point can move about in phase space, whereas there is only one way that a particle can move about in real space. A particle in real space moves about by virtue of its velocity, which is its rate of change of position in real space. A representative point may change its phase-space coordinates both by virtue of a change in its spatial coordinate and by virtue of a change in its momentum coordinate. A particle's velocity **v** is just the time rate of change of its position coordinate **r**; the time rate of change of its momentum, $\mathbf{p} = m\mathbf{v}$, is the force **F** acting on the particle. In addition to forces, another mechanism, collision, exists for changing the momenta of particles. The momenta of two particles may change almost instantaneously as a result of a collision between them. Think of two billiard balls whose "before" and "after" pictures are displayed in Fig. 3-3. While most particle collisions are not exactly like the collisions of billiard balls, the example is useful for a visual picture of what is going on. The

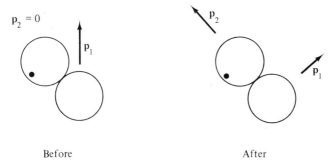

Fig. 3-3 Billiard balls before and after collisions.

actual forms of typical interactions will be discussed in the next section. Naturally, what causes the change in momenta of the colliding particles is a mutual interaction force of some nature. However, for reasons we shall see later it is customary to distinguish between general forces acting on the particles and collisional effects by considering the two separately. One way of distinguishing between the two types of forces is by the time interval during which they act. General forces such as gravity or the Lorentz force in electric and magnetic fields act on a particle for long periods of time as the particle moves about in a continuous field. Collisional forces, however, are very short range forces and act only for the very short intervals of time during which the particles are close to one another.

We shall now turn to obtaining mathematical expressions for the change in the number of points $\Delta^6 N$ in the small volume $\Delta^3 r\, \Delta^3 p$ due to these three mechanisms. First we shall consider the effect of motion through real-space coordinates. For simplicity we shall concentrate on one component of this motion. Because a six-dimensional volume is hard enough to visualize, much less to draw, we shall represent a view of the volume in Fig. 3-4(a) by showing schematically the edges of the two surfaces normal to the x-direction, labeled I and II. Each of these surfaces has an area $\Delta y\, \Delta z\, \Delta p_x\, \Delta p_y\, \Delta p_z$. The number

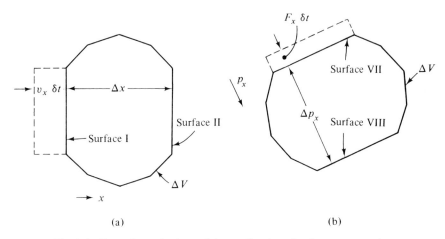

Fig. 3-4 Two schematic views of the small volume in phase space used to derive the kinetic equation.

of points entering $\Delta^3 r\, \Delta^3 p$ through surface I in time δt is just the number contained in the small volume whose base is surface I and whose height is $v_x\, \delta t$, evaluated at surface I. We have assumed that v_x is positive. The number of points entering through surface I by virtue of their drift through the x-component of spatial coordinates is then $(f\, \Delta y\, \Delta z\, \Delta p_x\, \Delta p_y\, \Delta p_z v_x\, \delta t)_\text{I}$ since f is the density of points. The number of points leaving through surface II by the analogous effect will be $(f\, \Delta y\, \Delta z\, \Delta p_x\, \Delta p_y\, \Delta p_z v_x\, \delta t)_\text{II}$. Notice that fv_x plays the part of the x-component of the current density of representative points through real-space coordinates and that fv_x is also $f(dx/dt)$. The value of this current density at surface II can be related to that at surface I by a Taylor expansion of the type used in Section 3.1 provided Δx is small:

$$(fv_x)_\text{II} \simeq (fv_x)_\text{I} + \frac{\partial}{\partial x}(fv_x)\, \Delta x \tag{3-20}$$

If we subtract the number leaving from the number entering in time δt due to v_x, we obtain

$$\delta(\Delta^6 N)_{v_x} = -\frac{\partial}{\partial x}(fv_x)\, \Delta x\, \Delta y\, \Delta z\, \Delta p_x\, \Delta p_y\, \Delta p_z\, \delta t$$

Furthermore, if we consider the analogous contributions due to v_y and v_z, we obtain, using the notation of vector calculus,

$$\delta(\Delta^6 N)_\mathbf{v} = -\boldsymbol{\nabla} \cdot (f\mathbf{v})\, \Delta^3 r\, \Delta^3 p\, \delta t \tag{3-21}$$

Sec. 3-2 The Kinetic Equation

where the dot product in the divergence is between the components of the del operator and those of the velocity. The term $\mathbf{V} \cdot (f\mathbf{v})$ represents the divergence of the current density of the representative points through real-space coordinates.

Next we shall consider the contribution to the change in $\Delta^6 N$ due to the drift of representative points through momentum space. Again we shall concentrate on one component of this motion and represent the volume in Fig. 3-4(b), which shows schematically the two surfaces normal to the p_x coordinate axis. Each of these surfaces, labeled VII and VIII, has an area $\Delta x \, \Delta y \, \Delta z \, \Delta p_y \, \Delta p_z$. The number of representative points that enters $\Delta^3 r \, \Delta^3 p$ through surface VII in time δt is just the number contained in a small volume whose base is surface VII and whose height is $(dp_x/dt) \, \delta t$ evaluated at surface VII. The rate at which particles drift through momentum space in the p_x-direction, dp_x/dt, is the force F_x. We shall consider *instantaneous* jumps through momentum space due explicitly to collisions later. The number entering is then $(f \Delta x \, \Delta y \, \Delta z \, \Delta p_y \, \Delta p_z F_x \, \delta t)_{\text{VII}}$ since f is the density of points. Similarly, the number leaving through surface VIII is $(f \Delta x \, \Delta y \, \Delta z \, \Delta p_y \, \Delta p_z F_x \, \delta t)_{\text{VIII}}$. Notice that the role of the p_x-component of the current density of representative points through momentum space is played by fF_x by analogy to the current density of particles through real space. The current density at surface VIII can also be related to that at surface VII by the familiar Taylor expansion

$$(fF_x)_{\text{VIII}} \simeq (fF_x)_{\text{VII}} + \frac{\partial}{\partial p_x}(fF_x) \Delta p_x \qquad (3\text{-}22)$$

since the surfaces are separated by a small range Δp_x in the x-component of momentum. Subtracting the number leaving from the number entering in time δt due to F_x, we find that the change resulting in $\Delta^6 N$ is

$$\delta(\Delta^6 N)_{F_x} = -\frac{\partial}{\partial p_x}(fF_x) \Delta x \, \Delta y \, \Delta z \, \Delta p_x \, \Delta p_y \, \Delta p_z \, \delta t$$

Furthermore, if we consider the contributions due to F_y and F_z, we shall obtain analogous results. To express the three contributions in a compact vector notation, we shall define a new operator called the del-p operator $\mathbf{V}_p$,

$$\mathbf{V}_p \equiv \hat{x} \frac{\partial}{\partial p_x} + \hat{y} \frac{\partial}{\partial p_y} + \hat{z} \frac{\partial}{\partial p_z} \qquad (3\text{-}23)$$

where $\mathbf{V}_p$ operates in momentum space the same way $\mathbf{V}$ operates in real space. Using this notation, the contributions to the change in $\Delta^6 N$ due to the three components of a force can be written compactly as

$$\delta(\Delta^6 N)_F = -\mathbf{V}_p \cdot (f\mathbf{F}) \, \Delta^3 r \, \Delta^3 p \, \delta t \qquad (3\text{-}24)$$

where the dot product is between the components of the $\mathbf{V}_p$ operator and those of the force $\mathbf{F}$. The term $\mathbf{V}_p \cdot (f\mathbf{F})$ represents the divergence of the current density of the representative points through momentum space.

Calculation of the contributions to the change in $\Delta^6 N$ due to collisions is a difficult task that depends specifically on the details of the forces of interaction between the particles. We shall leave detailed calculations to a more advanced course in kinetic theory and, for the present purposes, take a phenomenological approach, writing

$$\delta(\Delta^6 N)_{\text{coll.}} = \left(\frac{\delta f}{\delta t}\right)_{\text{coll.}} \Delta^3 r \, \Delta^3 p \, \delta t \tag{3-25}$$

where the δ is used to indicate that this is not an ordinary time derivative. We shall discuss later in this section and in the next section just what $(\delta f/\delta t)_{\text{coll}}$ represents. Suffice it to say for now that the collision term is generally represented by a complicated integral expression. Consequently, the term $(\delta f/\delta t)_{\text{coll}}$ is frequently called the *collision integral*.

We have now obtained expressions for the three mechanisms for changing the number of representative points in a small volume of phase space in a short interval δt. If we collect Eqs. (3-21), (3-24), and (3-25), divide through by δt, and take the limit as δt goes to zero, we get

$$\frac{\partial}{\partial t}(\Delta^6 N) = -\nabla \cdot (f\mathbf{v})\Delta^3 r \, \Delta^3 p - \nabla_p \cdot (f\mathbf{F}) \, \Delta^3 r \, \Delta^3 p + \left(\frac{\delta f}{\delta t}\right)_{\text{coll.}} \Delta^3 r \, \Delta^3 p$$

Each term in this expression contains the factor $\Delta^3 r \, \Delta^3 p$, which cancels out, leaving an equation that is independent of location in phase space and is therefore generally valid. This equation is

$$\frac{\partial f}{\partial t} + \nabla \cdot (f\mathbf{v}) + \nabla_p \cdot (f\mathbf{F}) = \left(\frac{\delta f}{\delta t}\right)_{\text{coll.}} \tag{3-26}$$

Notice that the second and third terms on the left-hand side represent the divergences of current densities of the representation points in the real- and momentum-space coordinates, respectively, and that f is the density of these points in the six-dimensional space. Therefore, Eq. (3-26) appears to be analogous to the continuity equation for particles in the three-dimensional real space, Eq. (3-16), except for the collision term. In fact it is analogous because we have effected an artificial distinction between general forces and the particle interaction forces that lead to collisions. In principle if we were sufficiently clever and diligent, we could incorporate the collision integral into the third term on the lefthand side of Eq. (3-26). Fortunately, it is not the custom to do so.

Equation (3-26) is not the kinetic equation in its usual form. The reader will be pleasantly surprised to learn that the kinetic equation that is usually written is a simplified version of this relation. The simplification arises if we expand the two terms that involve the divergence of a product of a scalar and a vector. We may use the vector identity

$$\nabla \cdot (\phi \mathbf{A}) = \mathbf{A} \cdot \nabla \phi + \phi (\nabla \cdot \mathbf{A}) \tag{3-27}$$

Sec. 3-2 The Kinetic Equation

Application of Eq. (3-27) to the second term in Eq. (3-26) yields
$$\nabla \cdot (f\mathbf{v}) = \mathbf{v} \cdot \nabla f + f(\nabla \cdot \mathbf{v})$$
The second term on the right is zero because the velocity (or momentum) and position of a particle are treated as independent parameters in phase space so that the derivative of one with respect to the other is zero. Therefore,
$$\nabla \cdot (f\mathbf{v}) = \mathbf{v} \cdot \nabla f \tag{3-28}$$
Application of the identity to the third term in Eq. (3-26) yields
$$\nabla_p \cdot (f\mathbf{F}) = \mathbf{F} \cdot \nabla_p f + f(\nabla_p \cdot \mathbf{F})$$
It is customary to restrict the class of forces to exclude those for which $\nabla_p \cdot \mathbf{F}$ is not zero so that $(\nabla_p \cdot \mathbf{F}) = 0$ and
$$\nabla_p \cdot (f\mathbf{F}) = \mathbf{F} \cdot \nabla_p f \tag{3-29}$$
This restriction is not as severe as it appears because most ordinary forces satisfy it since the x-component of the force is usually independent of the x-component of the momentum, and so forth. Specifically it is true for electric and magnetic forces. Even though the magnetic force $\mathbf{F} = q(\mathbf{v} \times \mathbf{B})$ does depend on momentum, the x-component of the force depends on the y- and z-components of momentum, not on p_x. Therefore $\nabla_p \cdot \mathbf{F} = 0$. One class of forces that can, however, depend on the momentum in such a way as to violate the restriction is the forces acting between colliding particles. Fortunately, collisional effects have already been segregated from the term in question and Eq. (3-29) stands. In fact, this is one reason we want to handle collisional interactions separately from other forces.

After making these simplifications, Eq. (3-26) takes on the usual form of the kinetic equation:

$$\boxed{\frac{\partial f}{\partial t} + \mathbf{v} \cdot \nabla f + \mathbf{F} \cdot \nabla_p f = \left(\frac{\delta f}{\delta t}\right)_{\text{coll.}}} \tag{3-30}$$

This expression relates the derivatives of the distribution function with respect to time, space, and momentum to the applied forces and to collisional effects. If we know what the forces are and how to handle the particle interactions, we can in principle solve for the distribution function that describes completely the state of the system. Then from the distribution function we can calculate the measurable averages, as in Chapter 2, and the transport coefficients, as in Section 3.1. Clearly, the only conceptual hurdle left is how to handle the collision term in the kinetic equation.

The Role of Collisions

We have already indicated that it would be prudent, for the purposes of this book, not to concern ourselves with detailed analysis of the many types of particle interactions that can occur but instead to take a phenomenological approach. What phenomenological role do collisions play in determining the distribution function? To answer this question, consider first a system in thermodynamic equilibrium. No forces are acting, and the steady-state distribution function is spatially uniform. Therefore, every term on the lefthand side of Eq. (3-30) is zero, and hence the collision term is also zero. This fact does not mean that collisions have ceased to occur just because the system is in equilibrium; it does mean that the collisions occur in such a way as not to disturb on the average the configuration of the representative points in phase space from their equilibrium distribution. The system continues to go from microstate to microstate by virtue of the collisions but remains on the average in the same macrostate.

Next let us consider a system that is not in thermodynamic equilibrium, but let us assume that it is, however, spatially uniform. Now suppose we suddenly turn off all the forces acting on the system and watch what happens. We know from thermodynamics that the system will develop toward the state of thermodynamic equilibrium as time develops. The only nonzero terms in the kinetic equation are

$$\frac{\partial f}{\partial t} = \left(\frac{\delta f}{\delta t}\right)_{\text{coll.}} \qquad (3\text{-}31)$$

In other words, the mechanism that drives the system toward equilibrium is the interparticle collisions. Furthermore, our intuition tells us that a system far from equilibrium will develop toward equilibrium at a fast rate, whereas one closer to equilibrium will relax more slowly since at equilibrium the collisions no longer affect the distribution function. In other words, the effect of collisions depends on the deviation from equilibrium. If we denote the time-independent equilibrium distribution by f_0, the deviation from equilibrium is $(f - f_0)$. The collision integral, therefore, can be represented phenomenologically, as

$$\boxed{\left(\frac{\partial f}{\partial t}\right)_{\text{coll.}} = -\frac{(f - f_0)}{\tau}} \qquad (3\text{-}32)$$

where the minus sign ensures that if we use Eq. (3-32) in Eq. (3-31) we shall eventually relax to equilibrium and where τ is a phenomenological parameter called the *relaxation time*. Inserting Eq. (3-32) into Eq. (3-30) yields a version of the kinetic equation known as the relaxation approximation:

$$\boxed{\frac{\partial f}{\partial t} + \mathbf{v} \cdot \nabla f + \mathbf{F} \cdot \nabla_p f = -\frac{(f - f_0)}{\tau}} \qquad (3\text{-}33)$$

The relaxation approximation is valid in many applications provided the distribution f does not deviate too greatly from its equilibrium value f_0.

Of course, we have just replaced one can of worms, the collision integral, by another, the relaxation time. However, the second can tends to catch more fish. The next section describes the phenomena of collisions in more detail and presents some alternative descriptions of the relaxation time in terms of measurable parameters.

3.3 COLLISIONS

The interactions between the particles that make up a system are very important for determining the characteristics of that system. There are very few systems, such as very hot, rarefied gases, where the effects of collisions can be neglected for practical purpose. In irreversible thermodynamics, collisions are the essential feature of a system that drives it toward equilibrium. In electrical systems, collisions provide the mechanism for the irreversible transfer of energy from its electrical form into heat; that is, collisions are responsible for resistivity. In fact, the particle interaction mechanisms determine all of the transport coefficients. Therefore, it behooves us to develop some general models of collision processes in order that we might better understand their effect on the behavior of the objects of our study.

We have already decided, as mentioned in the previous section, to ignore the detailed kinematics of even the most common collision processes and instead to take a more phenomenological approach. Taking this approach means that we shall have to deal with at least one measurable parameter that will characterize some behavior of the system rather than with fundamental physical principles. While a theoretical physicist may not approve, most engineers will be satisfied with a phenomenological approach provided they have some understanding of the meaning of the parameters invoked—where they come from and what affects their value. The simple model of collisions that we shall study uses several related parameters: mean collision time τ_c, mean free path λ_c, and effective collision cross section σ_c.

Phenomenological Description of Collisions

The *mean collision time* τ_c is the average time between collisions for any given particle. Any particle, such as a gas molecule or an electron in a solid, is in continuous motion by virtue of its energy. The macroscopic manifestation of the individual particle energies is just the thermodynamic temperature. For classical particles the average kinetic energy is $(3/2)\kappa T$. As a particle moves, it exhibits so-called *Brownian motion*; that is, in the absence of external forces its path is made up of very many straight-line segments, each segment randomly oriented with respect to the previous segment. Such a path is sketched in Fig. 3-5. At the end points of each straight segment a collision

Fig. 3-5 Path of a particle undergoing Brownian motion.

occurs that changes the momentum of the particle and starts it off on a new straight-line segment. In fact, a collision may be defined as any particle interaction in the absence of general forces that causes a deviation from a straight-line path.

Interactions exist for which it is not necessary for the particles to touch physically, as is the case for hard spheres. A very important example of such a distant interaction is the electrostatic force between charged particles. Electrons, therefore, do not act like hard spheres when they interact with each other or with other particles. However, we shall study the hard sphere or billiard-ball type of interaction for the sake of developing physical insight into collisional effects and an understanding of the roles of the various phenomenological parameters, τ_c, λ_c, and σ_c, that are customarily invoked to describe these effects. The hard sphere model has surprisingly wide applicability despite its simplicity. If discrepancies do arise between our models and empirical fact, we may be able to trace them to the billiard-ball approximation. Later in this section we shall compare the hard sphere case with more realistic interactions and indicate how the use of the more realistic models affects our results.

In any event, the mean collision time is just the average of the time elapsed between successive collisions for a single particle as it undergoes Brownian motion, the average being taken over a large number of intervals. Alternatively, the mean collision time can be defined as the average over all the particles of the time elapsed for each particle between two successive collisions with other particles. A detailed investigation would show that these two definitions of the mean collision time are essentially the same. For our purposes, we shall accept the equivalence of the two definitions without proof.

Furthermore, we can argue that the relaxation time discussed in Section 3.2 is, for all intents and purposes, the mean collision time. In the relaxation approximation the system is assumed to be almost, but not quite, in its equilibrium state; that is, the deviation from equilibrium is small. Under such conditions the motions of the particles are almost, but not quite, randomized since in equilibrium their motions would be completely randomized. One mean collision time later, each particle, on the average, will have suffered a collision. The momenta of the particles after they collide will be randomized

Sec. 3-3 Collisions

with respect to their previous momenta. After all the particles on the average suffer a collision, the system will be essentially completely randomized and therefore will be in thermodynamic equilibrium. Of course, this argument is an approximate one, particularly since Eqs. (3-31) and (3-32) predict only an exponential relaxation toward equilibrium. Nonetheless, that relaxation time constant, by the above argument, should essentially be the mean collision time. At the very least it should be some simple function of the collision time. We shall, therefore, take it to be the mean collision time because by doing so we shall certainly not make too large an error and, as we shall see, the collision time is easily obtained from empirical observations.

The *mean free path* λ_c may also be defined in one of two ways with equivalent results. The mean free path is the average of the distances traveled between successive collisions for a single particle undergoing Brownian motion, the average taken over a large number of free paths. In other words, it is the average length of the straight-line segments along which the particle travels free from all forces. Alternatively, the mean free path is the average over all particles of the distance traveled by each particle between two successive collisions with other particles. The mean free path must obviously be related to the mean collision time, the connection being the average speed of the particles $\bar{v}$:

$$\lambda_c = \bar{v}\tau_c \qquad (3\text{-}34)$$

The average speed is both the average speed of a single particle observed for a long time and the average of the speeds of all of the particles observed at the same instant. For gas molecules at atmospheric pressure (760 mm Hg or torr = 1.033×10^4 kg/m^2) and room temperature (290°K $\rightarrow$ 0.025 eV = 4×10^{-21} J), the mean free path is about 10^{-5} cm and the mean collision time about 10^{-10} sec.

The effective *collision cross section* σ_c is related to the mean free path and the particle density. To illustrate this relation and to define the cross section, we shall calculate the mean free path of a group of particles whose interactions are those of hard elastic spheres; that is, we shall calculate the mean free path for billiard balls. To do this, we must explore what it means for a particle to suffer a collision and what the probability is that a collision will occur.

For the purposes of illustration only we shall artificially divide a system of particles of radius R into two groups, one composed of particles that act as stationary targets and the other composed of particles that we shall designate as projectiles. A collision, illustrated in Fig. 3-6(a), occurs when a projectile comes close enough to a target so that they touch. At the time of collision the centers of the two particles are separated by a distance equal to the sum of their radii, in this case $2R$. When the centers are separated by more

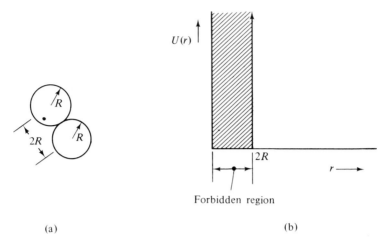

(a)　　　　　　　　　　　　　　　(b)

Fig. 3-6 Collisions between hard spheres: (a) the physical configuration, (b) the interaction potential as a function of the separation of centers.

than the sum of their radii, no interaction occurs. Because we are assuming a hard sphere interaction, the center of a projectile cannot approach to a distance less than $2R$ from a target. The potential function that describes such an interaction is sketched in Fig. 3-6(b). The interaction force is proportional to the gradient of the potential, so an *infinite* repulsive force acts when the separation is $2R$.

What is the probability that a specific projectile will collide with a specific target? If the center of the projectile, as illustrated in Fig. 3-7, lies within a

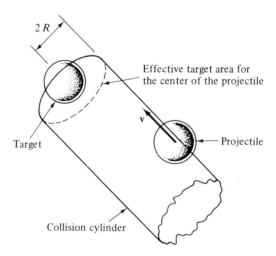

Fig. 3-7 A collision is going to occur.

right circular cylinder the base of which has radius $2R$ and is perpendicular to the direction in which the projectile is moving, then a collision will certainly occur. If not, then not. The base of this cylinder defines an effective target area called the *effective collision cross section* σ_c. Notice that this effective cross section, usually just called the cross section, depends on the size of both the target and the projectile. This is the case because the location of particles is ordinarily indicated by the location of their centers and their separation by the separation of their centers, especially in the description of the interaction potential. In the present case the effective cross section $\sigma_c = \pi(2R)^2 = 4\pi R^2$, which is four times the actual physical cross section of a single particle. This result can be obtained by examining either the collision cylinder in Fig. 3-7 or the plot of the interaction potential in Fig. 3-6(b).

Collision Probability

The probability that a single projectile will collide with any one of a number of targets is just the ratio of the total effective target area to the total shooting area. The probability of hitting a milk bottle with a baseball at the carnival, assuming that no aim is taken, depends on the ratio of the total collision cross section of the bottles to the area of the back wall of the tent. Figure 3-8 indicates that if the bottles are placed more than a ball's width

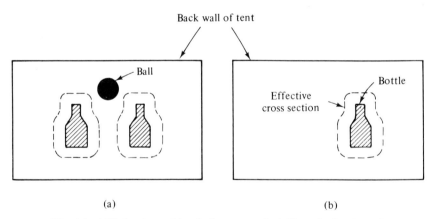

Fig. 3-8 Milk bottles and baseballs at a carnival. If no aim is taken, the probability of hitting a bottle is twice as great in (a) as in (b).

apart, the probability of hitting one of two bottles is twice as great as that of hitting one when only one is standing. (Of course, the likelihood of knocking a bottle over when it is hit depends on how heavily it has been weighted.)

Let us consider now the problem of shooting a projectile into a three-dimensional volume occupied by targets with density n targets per unit volume as illustrated in Fig. 3-9. For simplicity we shall assume that the targets are stationary. The probability of hitting a target in a thin layer dx of cross-sectional area A will depend on the number of targets in the volume $A\,dx$, which is

$$dN = nA\,dx$$

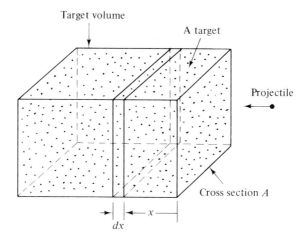

Fig. 3-9 Shooting a projectile into a volume of targets.

If we assume that the density is low enough and that dx is small enough so that there is no overlapping of targets in dx, then the total target area in dx is $\sigma_c\,dN = \sigma_c nA\,dx$. The total shooting area, of course, is just A, so that the probability of collision in the range dx is

$$P\,dx = \frac{n\sigma_c A\,dx}{A} = n\sigma_c\,dx \tag{3-35}$$

This probability is denoted $P\,dx$ because it depends directly on the size of the interval dx.

Mean Free Path

We are now in a position to calculate the mean free path, which is the average distance a particle travels into the volume of Fig. 3-9 before it suffers a collision. Instead of one projectile, let us assume that a beam of N_0 projectiles is incident upon the target volume. Furthermore, we shall assume that a projectile is removed from the beam when it collides with a target. At each interval dx some projectiles collide and are removed, while others penetrate farther into the volume. The average depth of penetration for the N_0 particles is also the mean free path. In practice the vector velocity of a beam projectile after collision with a target is in a direction other than that of the original

Sec. 3-3 Collisions

beam so that projectiles are removed from the beam upon collision. As we shall see, the mean free path can be obtained from a measurement of the number of projectiles left in the beam as a function of the distance traveled through the target volume.

The number of projectiles remaining in the beam at a depth of penetration x will be denoted by $N(x)$. The number lost from the beam in an interval dx at a depth x is equal to the product of the number in the beam at that depth and the probability of collision in the interval,

$$dN = -N(x)P\, dx = -Nn\sigma_c\, dx \tag{3-36}$$

using Eq. (3-35). Equation (3-36) can be solved for the beam number as a function of the depth of penetration by rearranging it in the form

$$\frac{dN}{N} = -n\sigma_c\, dx$$

which has the solution

$$N(x) = N(x=0)\epsilon^{-n\sigma_c x}$$

The value of $N(x=0)$ is just N_0 so that the number of projectiles left in the beam as a function of the depth of penetration is

$$N(x) = N_0 \epsilon^{-n\sigma_c x} \tag{3-37}$$

Equation (3-37) represents an exponential decay of the beam as it penetrates the target volume. Since dN particles travel a distance x and then are lost in the interval dx, the average depth of penetration for a projectile is

$$\lambda_c = \frac{\sum x\, dN(x)}{\sum dN(x)} \tag{3-38}$$

where the sums are taken over all intervals in the target volume. Since the intervals are small, it is convenient to use integral calculus to perform the summation. The integral form of Eq. (3-38), using Eqs. (3-36) and (3-37), is

$$\lambda_c = \frac{\int_0^\infty xn\sigma_c N_0 \epsilon^{-n\sigma_c x}\, dx}{\int_0^\infty n\sigma_c N_0 \epsilon^{-n\sigma_c x}\, dx}$$

which has the value

$$\boxed{\lambda_c = \frac{1}{n\sigma_c}} \tag{3-39}$$

Note that the mean free path in this simple model depends only on the density of targets and the effective collision cross section, which in turn depends only on the types of particles interacting.

According to Eqs. (3-37) and (3-38), the number of beam particles is reduced by a factor equal to ϵ, the base of the natural logarithms, when the beam has penetrated to a depth of one mean free path. The mean free path for collisions between any two types of particles or between two particles of the same type can be measured by shooting a beam through a target volume and measuring its decay length. If the density of targets is known or can be measured independently, the collision cross section can be calculated using Eq. (3-39). The mean collision time may be expressed in terms of the cross section by combining Eqs. (3-34) and (3-35), obtaining

$$\tau_c = \frac{1}{n\bar{v}\sigma_c} \qquad (3\text{-}40)$$

More Realistic Collisional Models

Two important refinements need to be made to the hard sphere model as expressed in Eqs. (3-39) and (3-40). The first comes with the recognition that both the "projectiles" and the "targets" are always moving and that their velocities are distributed statistically according to some distribution function. It turns out, however, that if the particles are colliding among themselves and if they are described by a Maxwell-Boltzmann distribution, the mean free path is

$$\lambda_c = \left(\frac{1}{2}\right)^{1/2} \frac{1}{n\sigma_c} \qquad (3\text{-}41)$$

a result that differs from Eq. (3-39) by a factor of 0.707. Notice that the mean free path is independent of the temperature of the particles in this model.

Frequently in electronics we must deal with collisions between two different species of particles such as, for example, electrons and gas molecules in a gas-filled diode. The two species may even have different distributions. For the case in which both types of particles are described by Maxwellian distributions but at different temperatures, the mean free path for collisions of type 1 with type 2 particles is

$$\lambda_{12} = \frac{1}{n_2\sigma_c[1 + \bar{v}_2^2/\bar{v}_1^2]^{1/2}} \qquad (3\text{-}42)$$

where n_2 is the density of type 2 particles and $\bar{v}_1^2$ and $\bar{v}_2^2$ are the respective mean square speeds. For collisions of light electrons with heavy atoms, $\bar{v}_a^2$ is usually much less than $\bar{v}_e^2$ so that Eq. (3-39) is a good approximation and

$$\lambda_{ea} = \frac{1}{n_a\sigma_{ea}}$$

where the subscript a refers to the atoms and e to the electrons.

The second important refinement comes with the recognition that most collisions are not of the hard sphere variety. A more reasonable and typical

Sec. 3-3 Collisions

interaction potential is shown in Fig. 3-10. At large distances of separation, the forces between particles may be attractive, such as the electrostatic interaction between positive and negative charges, the van der Waals effect between polarized molecules, and induced polarization of a neutral by a charged particle. At close range, forces become repulsive due, for instance, to the electrostatic repulsion of atomic electrons and so forth. The practical result of such an interaction potential is that the actual collisional effects depend on the energy of the projectile particle. The collision cross section and the mean free path are functions of projectile energy. Because of the existence of attractive forces between particles, we should expect the cross section to have a maximum value at a particular energy, corresponding to the energy at which particles are most influenced by the attraction. More energetic particles would come closer to the target before colliding and hence see a small effective cross section. The cross section, then, should be a decreasing function of energy for large energies. Figure 3-11 shows the experimental collision cross section for electrons with the molecules of noble gases as functions of the square root of the electron energy measured in electron volts. For purposes of comparison, the "diameter" of a gas molecule is on the order of 10^{-8} cm.

The concept of effective collision cross section is utilized in gaseous electronics, solid-state electronics, and almost every branch of physics. It is common to refer to cross sections for an ionization interaction or for excitation or recombination as well as for collision. We shall make use of these cross sections in later chapters. Nuclear physics also utilizes nuclear cross sections for fission and fusion reactions, slow neutron capture, and the like. These are measured in units of a barn (b), which is 10^{-24} cm^2, the broad side of which is indeed difficult to hit.

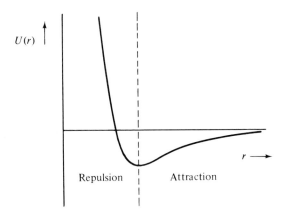

Fig. 3-10 Typical realistic collisional interaction potential as a function of the separation of the centers of the colliding particles.

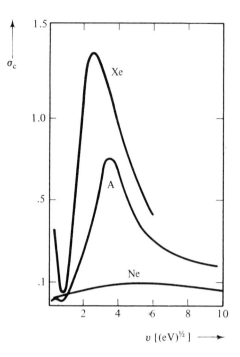

Fig. 3-11 Collision cross sections for electrons with noble gas molecules as functions of electron speed measured in (electron volts)$^{1/2}$. The unit of measure for the cross sections is $\pi \times 10^{-16}$ cm^2.

3.4 CONDUCTIVITY AND MOBILITY

Ohm's law states that a d.c. current flows in a linear, isotropic, homogeneous medium in response to a d.c. electric field. According to Eq. (3-7), a steady current implies that the charged particles carrying the current have a d.c. average velocity. When this average velocity is due to an imposed electric field, it is called the *drift velocity*. The electrical conductivity and mobility relate the current density and average velocity, respectively, to the electric field according to Eqs. (3-10) and (3-13). On the other hand, the equation of motion for a single charged particle in an electric field is

$$m\frac{d\mathbf{v}}{dt} = q\mathcal{E} \tag{3-43}$$

which, for constant $\mathcal{E}$, predicts a velocity that increases uniformly in time without limit. If the motion of all the electrons of, say, a carbon resistor obeyed such a relation as Eq. (3-43), then the current represented by the motion of these electrons would increase without limit for a constant $\mathcal{E}$. Of course we know this does not happen: Ohm's law is generally valid. The effect that is neglected in Eq. (3-43) and that is responsible for the d.c. current response to a d.c. field is collisions between the charge-carrying particles and other particles in the system.

Mean Particle Model for Conductivity and Mobility

Let us consider the motion of electrons in a system that is composed of negatively and positively charged particles and neutrals. We shall assume that the density of charged particles is low compared with that of the neutrals and that only the electrons are able to move in response to an electric field. In other words, we shall study the current carried by the electrons assuming that the principal collision interaction is between electrons and neutrals. Our objective is to develop a physical model for electron conductivity and mobility based on the behavior of an average electron in the system.

The restrictions we have imposed on our model are not as severe as they might appear at first glance. The assumption that the positive charge carriers contribute negligibly to the current is valid in many instances. Examples are heavily doped n-type semiconductors, where the density of the mobile electrons is much higher than that of the mobile holes, and partially ionized gases, where the ions are so much more massive than the electrons as to be essentially immobile. For situations where the positive carrier current cannot be neglected it is a simple matter to calculate the positive carrier conductivity and to combine the two effects. The assumption that the principal collisional interaction is between electrons and neutrals is also valid provided the charged particle density is low. Electrons suffer collisions with other electrons, with positive charge carriers, and with neutrals. Electron-electron collisions cannot effect the average velocity of the electrons because momentum is conserved in each collision. The momentum of two colliding particles taken together is the same before and after the collision so that if the particles are of the same species, the net momentum of this species and its average value are unchanged by the collision. On the other hand, the net electron momentum or velocity can change during collisions with positive carriers or with neutrals since momentum may be transferred from one species to another during interspecies collisions. Thus the average electron velocity, and hence the electron current density, is affected by collisions of electrons with positive carriers and with neutrals but not by collisions of electrons with other electrons. The cross sections for collisions between electrons and positive carriers are generally larger than those for electron-neutral collisions because of the long-range attractive Coulomb force between charge particles. However, if the density of positive carrier targets is sufficiently low, the mean free path for negative-positive interactions can be much longer than that for negative-neutral interactions since the neutral density is high. A gas that is less than 0.1% ionized satisfies these criteria, as do many semiconductors. In some situations in semiconductors, however, it is necessary to consider the interactions between negative and positive charge carriers, but the extension of the model to include these effects is not difficult.

If we ignore the random or thermal motion of the electrons and consider only their response to the electric field and to collisions with neutrals, the velocity of a single electron as a function of time might be that sketched in Fig. 3-12. Collisions occur at times $t_1, t_2, \ldots, t_7$, and so forth. The average of these time intervals is, of course, the mean collision time for electron-neutral interactions τ_{en}. If the electron were a *mean* one, its equivalent velocity vs. time behavior would be as sketched in Fig. 3-13 since the average of the velocities of all the electrons after each has suffered a collision is zero. The average velocity of the mean electron is just one-half its maximum value or

$$\bar{v} = -\frac{1}{2}\left(\frac{e\mathcal{E}_z}{m_e}\right)\tau_{en} \qquad (3\text{-}44)$$

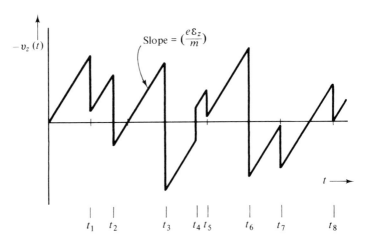

Fig. 3-12 Velocity as a function of time of a single electron in an electric field $\mathcal{E}_z$ with collisions.

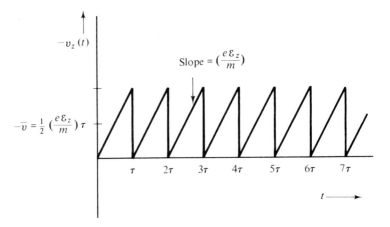

Fig. 3-13 Velocity of the mean electron vs. time.

Sec. 3-4 Conductivity and Mobility

which also is the average velocity of all the electrons. The maximum velocity of the mean electron can be obtained by the time integration of Eq. (3-43) between $t = 0$ and $t = \tau_{en}$ assuming zero initial velocity. The mobility of these electrons, using the definition in Eq. (3-13), is

$$\boxed{\mu_e = \frac{e\tau_{en}}{2m_e}} \quad \text{mean electron model} \quad (3\text{-}45)$$

and the conductivity, using Eq. (3-14), is

$$\sigma_e = \frac{e^2 n_e \tau_{en}}{2m_e} \quad \text{mean electron model} \quad (3\text{-}46)$$

To avoid possible confusion between the conductivity and cross section, both of which are customarily denoted by sigma, we shall in the future concentrate on the mobility and the mean collision time.

Although this *mean particle* model for mobility and conductivity is a much simplified one, it does serve to point out some important features. First, the mobility and conductivity depend inversely on the mass of the charge carrier, thus validating our earlier argument that the mobility of massive ions might be neglected compared with that of electrons and similarly for the respective conductivities if the charge densities are about the same. Second, the mobility and conductivity are related directly to the mean collision time, or, using the results of Section 3.3, inversely to the density of targets and the collision cross section. The larger the cross section, the lower the mobility. Furthermore, the mobility will be a function of temperature even if the cross section is not a function of particle energy or momentum.

Mobility as a Transport Coefficient

The mean particle model for mobility and conductivity is a useful one, but it fails to consider the statistical nature of the system. An alternative, and more fundamentally sound, model is based on the properties of the nonequilibrium distribution function for the charge carriers as determined by the kinetic equation, Eq. (3-30). The mobility, after all, is defined by the relation

$$\langle \mathbf{v} \rangle = \pm \mu \mathbf{\mathcal{E}}(\mathbf{r}, t) \quad (3\text{-}47)$$

where the electric field has its local value and the average velocity is a local average defined, according to Eq. (2-17), by

$$\langle \mathbf{v} \rangle = \frac{\int \mathbf{v} f(\mathbf{r}, \mathbf{p}, t)\, d\mathbf{p}}{\int f(\mathbf{r}, \mathbf{p}, t)\, d\mathbf{p}} \quad (3\text{-}48)$$

If the distribution function $f(\mathbf{r}, \mathbf{p}, t)$ that results when the system is subjected to an electric field $\mathbf{\mathcal{E}}(\mathbf{r}, t)$ can be obtained by solution of the kinetic equation, then the average velocity can be calculated using Eq. (3-48). The result will be in the form of Eq. (3-47), hence determining the mobility.

To simplify the calculation we shall assume a constant uniform electric field directed along the z-axis, $\boldsymbol{\mathcal{E}} = \mathcal{E}_z \hat{z}$, imposed on a uniform infinite system. In the steady state the electron distribution function will depend on the electric field and will therefore also be spatially uniform. The kinetic equation in the relaxation approximation reduces to

$$q\boldsymbol{\mathcal{E}} \cdot \nabla_p f = -\frac{(f - f_0)}{\tau} \tag{3-49}$$

So that the relaxation approximation will be valid, the deviation from equilibrium must be small, and hence the electric field must be small. What it means for the electric field to be small is that the average velocity due to the field, the drift velocity, must be small when compared with the root mean square speed or thermal speed of the particles as described by the temperature of the equilibrium distribution. That is, the relaxation approximation is valid provided

$$\frac{\langle \mathbf{v} \rangle \cdot \langle \mathbf{v} \rangle}{\overline{v^2}} \ll 1 \tag{3-50}$$

where

$$\tfrac{1}{2}m\overline{v^2} = \tfrac{3}{2}\kappa T$$

for the equilibrium distribution.

The technique commonly used to obtain an approximate solution to Eq. (3-49) is an iterative one. First we rewrite the equation in the form

$$f = f_0 - \tau q \mathcal{E}_z \frac{\partial f}{\partial p_z} \tag{3-51}$$

Since the real distribution is almost the equilibrium value f_0, the second term on the right-hand side can be approximated by using f_0 instead of f. A first approximation f_1 to f would then be

$$f_1 = f_0 - \tau q \mathcal{E}_z \frac{\partial f_0}{\partial p_z} \tag{3-52}$$

A second approximation f_2 to the real distribution f may be obtained by using f_1 from Eq. (3-52) in the second term on the right of Eq. (3-51):

$$f_2 = f_0 - \tau q \mathcal{E}_z \frac{\partial f_1}{\partial p_z} \tag{3-53}$$

Successive iterations can be used to obtain an approximate solution for f that is as close to being the correct value as diligence and energy allow. For the present purposes it is sufficient to stop with the first approximation and take

$$f \simeq f_0 - \tau q \mathcal{E}_z \frac{\partial f_0}{\partial p_z} \tag{3-54}$$

Sec. 3-4 Conductivity and Mobility

If we assume that the particles can be described classically, the equilibrium distribution is the Maxwell-Boltzmann distribution

$$f_0 = f_{\text{MB}}(\mathbf{r}, \mathbf{p}) = \frac{n}{(2\pi m\kappa T)^{3/2}} \epsilon^{-p^2/2m\kappa T} \qquad (3\text{-}55)$$

where

$$p^2 = p_x^2 + p_y^2 + p_z^2$$

It is easy to show that

$$\frac{\partial f_0}{\partial p_z} = -\frac{p_z}{m\kappa T} f_0 \qquad (3\text{-}56)$$

so that

$$f \simeq f_0 \left[1 + \frac{q\tau p_z \mathcal{E}_z}{m\kappa T} \right] \qquad (3\text{-}57)$$

Notice that the deviation from the equilibrium distribution is directly proportional to the applied electric field.

Now that we have obtained a solution for the distribution of the particles in the presence of an electric field, it is a straightforward matter to calculate their average velocity using Eq. (3-48). Since the electric field has only a z-component, the drift velocity will also have only a z-component,

$$\langle v_z \rangle = \frac{\int (p_z/m) f \, d\mathbf{p}}{\int f \, d\mathbf{p}} \qquad (3\text{-}58)$$

where the denominator is the local density $n(\mathbf{r})$. Inserting Eq. (3-57) into Eq. (3-58), we obtain two terms,

$$\langle v_z \rangle = \frac{1}{nm} \int p_z f_0 \, d\mathbf{p} + \frac{1}{n} \int \frac{q\tau}{\kappa T} \left(\frac{p_z}{m}\right)^2 f_0 \mathcal{E}_z \, d\mathbf{p} \qquad (3\text{-}59)$$

the first of which involves the average value of the z-component of momentum for a Maxwellian distribution, which is zero. The remaining term is

$$\langle v_z \rangle = \left[\frac{q}{m^2 \kappa T} \langle p_z^2 \tau \rangle \right] \mathcal{E}_z \qquad (3\text{-}60)$$

where $\langle p_z^2 \tau \rangle$ is the average over a Maxwellian distribution of the product $p_z^2 \tau$, assuming that the relaxation time may depend on the particle momentum through Eq. (3-40). Comparison of this result with the definition of mobility yields an expression for the mobility:

$$\mu = \frac{e}{m^2 \kappa T} \langle p_z^2 \tau \rangle \qquad \text{limited statistical model} \qquad (3\text{-}61)$$

For the oversimplified hard sphere model with τ independent of momentum and since $\langle p_z^2 \rangle = m\kappa T$,

$$\boxed{\mu_{\text{hard sphere}} = \frac{e\tau}{m}} \quad \text{limited statistical model} \qquad (3\text{-}62)$$

Interestingly, this result of the statistical calculation of the transport coefficient for hard spheres differs from that of the mean particle model, Eq. (3-45), by only a factor of 2. More importantly, the dependencies of the two models on charge, mass, and temperature are the same.

The fact remains, however, that the billiard-ball model is also a simplification of the true facts. It is frequently necessary to refine the model by taking into account the dependence of the collision time on momentum. In an isotropic medium, τ generally is a function of the magnitude of the momentum, $\tau = \tau(p^2)$, or, equivalently, of the particle energy. In anisotropic media, as most semiconductors are, the collision cross sections and hence τ may depend on the direction of momentum as well, $\tau = \tau(\mathbf{p})$. In either case, τ is usually known in graphical form since the cross section is a measured function and may have no simple analytic mathematical form. Therefore, the integration indicated in the average of Eq. (3-61) is usually done numerically with the aid of a computer.

Finally, a word or two about the validity of the statistical calculation is in order. If the calculation is valid, the result is a current density that depends linearly on the electric field; that is, Ohm's law is valid. The restriction on the validity of the calculation is that the deviation from equilibrium be small. Mathematically this restriction takes the form of Eq. (3-50), which relates the magnitudes of the drift velocity and the thermal speed. For a given material at a specified temperature, this restriction can be translated into one on the magnitude of the electric field since the mobility is known. Suppose that we take Eq. (3-50) to imply

$$\frac{\mu\mathcal{E}}{v_{\text{thermal}}} \leq 0.1 \qquad (3\text{-}63)$$

For electrons at room temperature, v_{thermal}, which is the root mean square velocity, is about 10^5 m/sec. If we take typical values of mobility from Table 3-1 as being 5×10^{-3} m²/V-sec for metals and 0.5 m²/V-sec for semiconductors, we find that Eq. (3-63) restricts the electric fields to values less than 2×10^6 and 2×10^4 V/m, respectively. The value for metals is not particularly restrictive, but such fields are not uncommon in semiconductor junctions. Therefore, we should expect that Ohm's law may not be valid in semiconductor junctions for high fields.

3.5 DIFFUSION

Fick's law states that a particle current flows in the presence of a nonuniform particle density. The physical process that underlies macroscopic diffusion is

Sec. 3-5 Diffusion

the random motion of each individual particle. Each particle continually undergoes Brownian motion, its orbit between successive collisions consisting of randomly oriented free paths. At any one instant, as many particles are moving in one direction as in another provided that the system is at least locally in equilibrium and that no external forces are acting. However, since there are more diffusing particles in regions of high density than in regions of low density, there will be more particles moving away from the region of high density than moving toward it. Thus, as illustrated in Fig. 3-14, there is a net

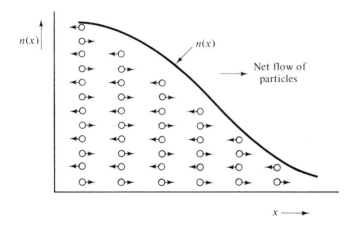

Fig. 3-14 Schematic representation of a density gradient and the instantaneous direction of the motion of representative particles.

flow of particles away from regions of high density, a process that, in the absence of an externally imposed gradient, continues until the density is uniform throughout the system. The diffusion process occurs independently for every species of particle in the system. In Fig. 3-14 each circle represents many diffusing particles and the arrows indicate the directions of these instantaneous velocities. Because there are more of these particles to the left of any particular point x than to the right, there will be a net motion of particles at that point from left to right. The rate at which the particle flow occurs will depend on the collisions that the particles suffer. If only the diffusing species is present, collisions among themselves determine the flow rate and the process is called *self-diffusion*. If the nonuniform density exists in a matrix of other species, then interspecies collisions may dominate the effective diffusion of one type through another.

Diffusion as a Transport Coefficient

The diffusion coefficient relating the particle flux to the density gradient is defined by Eq. (3-15), which we repeat as

$$\mathbf{\Gamma}(\mathbf{r}, t) = -D \, \nabla n(\mathbf{r}, t) \tag{3-64}$$

where, from Eq. (3-6),

$$\mathbf{\Gamma}(\mathbf{r}, t) = \int \mathbf{v} f(\mathbf{r}, \mathbf{p}, t) \, d\mathbf{p} \tag{3-65}$$

Thus, knowledge of the distribution function should permit us to obtain an expression for the diffusion coefficient by performing the calculation indicated in Eq. (3-65) and comparing the result with Eq. (3-64).

The distribution function for a system of particles under the influence of a steady, externally imposed gradient but in the absence of other external forces can be obtained by solving the kinetic equation in the relaxation approximation, Eq. (3-33), which takes the form

$$\mathbf{v} \cdot \nabla f = \frac{-(f - f_0)}{\tau} \tag{3-66}$$

since the distribution function will be dependent on the spatial coordinate if the density is. Equation (3-66) can be rewritten as

$$f = f_0 - \tau \mathbf{v} \cdot \nabla f \tag{3-67}$$

Let us suppose for simplicity that the gradient is along the z-direction only so that $f(\mathbf{r}, \mathbf{p}, t)$ will depend spatially only on the z-coordinate. Then Eq. (3-67) becomes

$$f = f_0 - \tau v_z \frac{\partial f}{\partial z} \tag{3-68}$$

and the particle flux will have only a z-component,

$$\Gamma_z(z) = -D \frac{\partial n}{\partial z} \tag{3-69}$$

Since the relaxation approximation requires that the system be close to equilibrium, the solution to Eq. (3-68) can be approximated by an iterative technique similar to the one used in the previous section.

Unfortunately, if we were to substitute f_0 into the second term of the right-hand side of Eq. (3-68) to obtain a first approximation to f, we would find trouble immediately since the true equilibrium distribution f_0 is independent of position. On the other hand, the objective of the calculation is to determine $\Gamma_z(z)$ at the local point z given the density as a function of position. Therefore, we shall take for the "equilibrium" distribution at a local point z a Maxwellian distribution that has a density appropriate to that local point. That is, for f_0 we use

$$\boxed{f_0(\text{L.T.E.}) = \frac{n(z)}{(2\pi m \kappa T)^{3/2}} \epsilon^{-p^2/2m\kappa T}} \tag{3-70}$$

This approximation is known as *local thermodynamic equilibrium* (L.T.E) and is an appropriate one to make provided that the distance over which the

Sec. 3-5 Diffusion

density varies appreciably is much greater than the collisional mean free path λ_c:

$$\left|\frac{\lambda_c}{n}\frac{\partial n}{\partial z}\right| \ll 1 \tag{3-71}$$

In other words, if we look at the system on a local scale, the density essentially does not change over a distance that may be many mean free paths wide. These local particles, since they interact with each other, will be distributed according to an equilibrium distribution using the local density. For systems in which there is a thermal gradient as well as a density gradient, Eq. (3-70) also serves to describe local thermodynamic equilibrium if the temperature is taken as spatially variable provided a restriction on the temperature gradient similar to Eq. (3-71) is satisfied. These restrictions ordinarily are easily met because mean free paths are much shorter than scale lengths of density and temperature gradients. Even in semiconductors where a charge density may drop by as much as 10^{15} carriers/cm^3 over a distance of 10^{-4} cm, the mean free path is on the order of 10^{-6} cm so that

$$\frac{\lambda_c}{n}\frac{\partial n}{\partial z} \simeq 10^{-2}$$

In fact, only in shock waves does any intensive property such as density or temperature change on a scale length comparable to a mean free path. Therefore, we shall assume that the steady-state system is in local thermodynamic equilibrium and that Eq. (3-70) is valid for the equilibrium distribution.

The first approximation f_1 to the solution of Eq. (3-68) for the actual distribution function f will then be

$$f_1 = f_0 - \tau v_z \frac{\partial f_0}{\partial z} \tag{3-72}$$

where f_0 is the L.T.E. distribution given in Eq. (3-70). For present purposes this is as far as we need to go in the iterative procedure. It is easy to show for uniform temperature that

$$\frac{\partial f_0}{\partial z} = \frac{\partial f_0}{\partial n}\frac{\partial n}{\partial z} = \frac{f_0}{n}\frac{\partial n}{\partial z} \tag{3-73}$$

Then, using f_1 to approximate f, we have

$$f \simeq f_0\left[1 - \frac{\tau v_z}{n}\frac{\partial n}{\partial z}\right] \tag{3-74}$$

Notice that the deviation from local thermodynamic equilibrium is directly proportional to the density gradient.

We are now able to calculate the particle current using the z-component of Eq. (3-65). By inserting Eq. (3-74) for f and Eq. (3-70) for f_0 into Eq. (3-65), we obtain two terms:

$$\Gamma_z = \int v_z f_0 \, d\mathbf{p} - \frac{1}{n} \int v_z^2 \tau \frac{\partial n}{\partial z} f_0 \, d\mathbf{p} \qquad (3\text{-}75)$$

The first of these terms is proportional to the average z-component of velocity of a Maxwellian distribution, which is zero. Therefore, only the second term remains, which we rewrite in terms of momentum as

$$\Gamma_z = -\left[\frac{1}{nm^2} \int p_z^2 \tau f_0 \, d\mathbf{p}\right] \frac{\partial n}{\partial z} \qquad (3\text{-}76)$$

Applying the definition of a local average value, Eq. (2-17), this becomes

$$\Gamma_z = -\frac{1}{m^2} \langle p_z^2 \tau \rangle \frac{\partial n}{\partial z} \qquad (3\text{-}77)$$

where the average is understood to be taken over the local Maxwellian distribution. Again we have taken the precaution of permitting τ to be a function of particle momentum so that this result can be applied to any suitable practical system of classical particles for which τ is a known function, as was the case with the expression for the mobility derived in the previous section.

Comparing Eq. (3-77) with Eq. (3-69), we can extract the diffusion coefficient:

$$D = \frac{1}{m^2} \langle p_z^2 \tau \rangle \qquad (3\text{-}78)$$

For the oversimplified hard sphere model with τ independent of momentum and since $\langle p_z^2 \rangle = m\kappa T$,

$$\boxed{D_{\text{hard sphere}} = \frac{\kappa T}{m} \tau} \qquad \text{limited statistical model} \qquad (3\text{-}79)$$

Some additional observations about diffusion are appropriate here. The mechanisms controlling the diffusion process are the random motions of the particles, reflected in their temperature, and the collision process. Furthermore, massive particles diffuse more slowly, all other things being equal. The same observations can be made about the electric mobility. In fact, if we compare the expressions for mobility and the diffusion coefficient obtained from the model based on Maxwell-Boltzmann statistics, Eqs. (3-61) and (3-78), respectively, we see that the ratio is independent of the collision mechanism and depends only on the charge and the temperature:

$$\boxed{\frac{D}{\mu} = \frac{\kappa T}{e}} \qquad (3\text{-}80)$$

Equation (3-80) is known as Einstein's relation. Thus, we need to know but one of these coefficients and the temperature to know the other.

Einstein's Relation

To explore Einstein's relation further, consider a system of charged particles under the influence of a static electric field but through which no net current flows. As indicated in Fig. 3-15 this can be accomplished by putting the

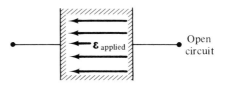

Fig. 3-15 System to demonstrate Einstein's relation.

→ Increasing potential

system in an open electric circuit. Both conductive and diffusive effects will take place inside this system. First, by virtue of the electric field a conduction current will flow related to the field by the charge density and mobility. Second, the presence of the field creates a nonuniform density that drives a diffusion current. The potential energy of a particle of charge q in an electric field is denoted by $U = qV$ where V is the electrostatic potential and $\mathcal{E} = -\nabla V$. In Chapter 2 we noted that the density of particles in a potential field was given by Eq. (2-58):

$$n(\mathbf{r}) = n_0 \epsilon^{-U(\mathbf{r})/\kappa T} \tag{2-58}$$

The electrical current density due to diffusion is

$$\mathbf{J}_{\text{diff.}} = -qD\,\nabla n \tag{3-81}$$

which becomes after some substitutions using Eq. (2-58),

$$\mathbf{J}_{\text{diff.}} = n\frac{q^2}{\kappa T}D\,\nabla V \tag{3-82}$$

The electrical current density due to conduction effects is

$$\mathbf{J}_{\text{cond.}} = -qn\mu\,\nabla V \tag{3-83}$$

Since no net current flows because of the open circuit, the sum of these two currents must vanish:

$$n\frac{q^2}{\kappa T}D\,\nabla V - qn\mu\,\nabla V = 0$$

from which we obtain Einstein's relation, Eq. (3-80).

3.6 EXPERIMENTAL DATA

The crucial test that must be applied to any theory is the comparison of its predictions with actual experimental results. If the data agree reasonably well with the calculations, then it is safe to assume that the simplified models on which the theory is based are sufficiently accurate representations of the physical truth to be useful. If, on the other hand, the theoretical predictions and experimental observations disagree, then fault may be found either in the experimental method or in the theoretical model. If the empirical results are conclusive, then the assumptions built into the theory must be examined, tested, and revised until a model is obtained that does reflect the actual physics with sufficient accuracy. The usefulness of an accurate theory lies in the application of the physical model on which it is based to the design and invention of practical devices and in the development of circuit models of actual devices.

The transport theory for gases is a highly developed and thoroughly understood theory, the beginnings of which go back to Maxwell and Boltzmann in the middle of the nineteenth century. It is an interesting fact that, however well developed the theory is, it is still unable to predict the results of some fundamental measurements. Pure gases are, conceptually, the simplest form of matter to understand, and very pure gases are available for experimentation, but some of their characteristics are the most difficult to model. Principally, the difficulties arise because of a wide variety of complicated inelastic collision mechanisms that occur between gas particles. Because momentum is not conserved during inelastic collisions, these effects make the theory difficult, although not always impossible, to formulate. Experiments frequently do deviate from the predictions of the simple theories we have been discussing. Nonetheless, the simple theoretical pictures are useful tools for basic understanding.

The theory for semiconductors, on the other hand, is in much better agreement with the data despite the fact that, conceptually, solids are more difficult to model because of certain inherent complexities. Among these are the fact that in semiconductors there may be two types of charge carriers, one carrying positive and the other negative charge, denoted by p and n, respectively. The same situation exists in gases, too, with electrons and ions, but in semiconductors the two species may have comparable mobilities. Hence both types of charge carrier must be taken into account. For the statistical description, this means that two distribution functions, one for each type of carrier, must be considered simultaneously. A second difficulty is the fact that these charge carriers in semiconductors are not classical particles but are instead more accurately described by the quantum mechanical model of physics. The statistical description of the particles, then, requires a quantum mechanical approach. Quantum mechanics, quantum statistics, and the pertinent theories of solids and semiconductors will be the subjects of study in Chapters 5, 6, 7, and 8.

Another difficulty that ostensibly should arise (and it does occasionally) when comparing theory and experiment in semiconductors is the not incon-

Sec. 3-6 Experimental Data   79

siderable difficulty in obtaining pure, regular samples of the material on which to do the experiments. The theories, expressed as they are in mathematical terms, require very regular materials. Even impurities, when they are introduced in the theory, are introduced in a mathematically regular way. Nature is seldom so kind to experimentalists. Yet semiconductor data are reasonably well understood in terms of the theories that have been developed.

Figure 3-16 shows the electric mobility in a particular sample of silicon

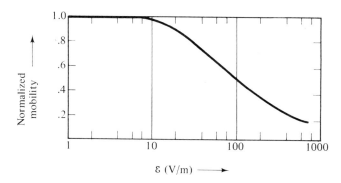

Fig. 3-16 Mobility of a silicon sample normalized to its low-field value as a function of applied field.

as a function of the applied field. The data indicate that the mobility is a function of the electric field for strong fields, which means that Ohm's law is not valid above some critical value of the field. The derivation of the mobility in Section 3.4 anticipated this difficulty due to the limitation of the applicability of the relaxation approximation to low fields. At high fields the electrons gain significant energy in each free path, and the mechanism of collision between the electrons and the crystalline lattice structure of the material changes, resulting in a change in mobility. The effects that come into play are well understood theoretically, and there is good agreement between theory and experiment.

The wide range of application of semiconductor materials in present-day devices is due largely to the strong dependence of the electrical properties of semiconductors on the type and amount of impurities present in the material. That is, the electrical properties can usually be tailored to meet the specifications of the particular device application. Modern integrated circuit technology depends largely on the ability of the fabricator to dope the material with a variety of impurities differentially. This doping with impurities can be done by performing a controlled preferential diffusion of the impurity material into the body of the sample. Hence, diffusion coefficients are of prime importance to the application of semiconductor materials to integrated circuits as well as to discrete devices. Diffusion coefficients for a wide variety of materials in silicon are displayed in Fig. 3-17. Notice that the values of these coefficients range over many orders of magnitude from one impurity to another. Measurements of the ratio of the diffusion coefficient to the mobility in

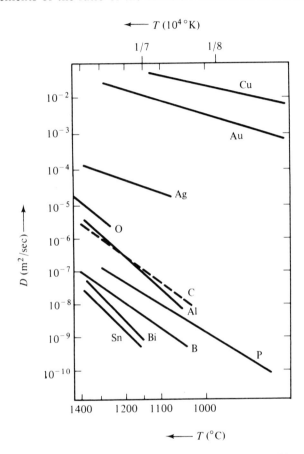

Fig. 3-17 Coefficients of diffusion for several impurities into silicon as functions of temperature.

solids have generally been restricted to low electric fields and moderate temperatures. The ratio given by Einstein's relation has been measured for charge carriers in semiconductors at room temperature with the result being 0.025 V, in excellent agreement with theory.

3.7 FLUCTUATIONS AND NOISE

So far we have been concerned with the behavior of an electronic system that is characterized by its most probable state whether that state is the equilib-

Sec. 3-7 Fluctuations and Noise

rium state of Chapter 2 or a nonequilibrium steady state of the first six sections of Chapter 3. Unfortunately a system is not always in its most probable state. The probabilistic picture predicts only that the system will be in that specific state on the average. Instantaneously, the macroscopic state may fluctuate from its average value. Fluctuations occur because every microstate has equal probability of occurring but not every microstate leads to the same macrostate.

As an example of the concept of fluctuations, consider the case discussed in Chapter 2 of a pair of dice as a probabilistic system. The most probable macrostate for the dice is 7, which means that if we were to roll the dice repeatedly many times the average number would be 7. However, we know from sad experience that the state of this system fluctuates from its average value. The fluctuations in this particular system are large and important because there are only two dice. For systems composed of very large numbers of constituent parts, fluctuations are much less severe and, depending on the circumstances, may not be important.

In electronic systems, on the other hand, fluctuations may be very important in circuits that are intended to amplify very small signals. As we shall see, fluctuations in the macrostate of a device lead to fluctuations in the current flowing through it or in the voltage drop across it. The fluctuation signal is called *noise* because when amplified and put through an audio speaker that is what it sounds like.

The purpose of electronic systems is to process in some way a desired signal that contains information. If at some stage in the processing an undesirable noise signal appears that is comparable in magnitude to the information signal, the possibility of losing the desired signal exists. Because of this concern, it has become customary to apply figures of merit such as noise temperature and the signal-to-noise ratio to electronic devices and systems to indicate their inherent limitations for processing small signals.

Theories of noise in electronic devices are based on the theories of mathematical probability. An excellent treatment of noise in electronic devices based on the mathematics can be found in the book by van der Ziel that is referenced in the Bibliography at the end of this chapter. For the most part, specification of noise figures and parameters and the development of idealized noise sources that can be used in equivalent circuits to represent noise effects are the result of experimental determinations involving measurement. In other words, except perhaps in the areas of research and development of devices, a phenomenological approach to noise effects is satisfactory. This is the approach we shall take in the latter chapters of this book, discussing noise properties and characteristics of the more common devices phenomenologically as we study their most probable behavior. In the rest of this section we shall develop the general concepts of thermal noise in a little more detail and introduce the concept of corpuscular noise.

Thermal Noise in a Resistive Material

Thermal noise is that noise due to the random motions of the charge carriers in an electronic system. Individual particles perform random or Brownian motion by virtue of their kinetic energy and the collisions that they suffer. The property of thermodynamic temperature is the macroscopic effect of the kinetic energies of the particles. Therefore the noise that arises because of fluctuations due to the random motion of particles is called *thermal noise*. Because electrons carrying current also always perform Brownian motion, there will always be thermal noise in any electronic system. To avoid the development of thermal noise from probability theory we shall consider two rough heuristic arguments for the existence of a noise signal due to thermal fluctuations.

The first of these arguments is physical. When a system fluctuates from its most probable state the densities of the charged particles may also deviate from their most probable values. The possibility arises that extraneous electric fields will result from the charge imbalances of the deviation. These electric fields and the resulting current densities that they drive appear at the terminals of the device as voltages and currents, respectively. The voltages and currents that arise from the fluctuations are called thermal noise signals. Because the fluctuations are purely random phenomena, the noise signals are also random and nonperiodic. Fourier analysis of such a random signal into sinusoidal components would reveal that all frequencies are present with equal magnitude.

An alternative argument is based on the mathematical definition of a macroscopic property such as the instantaneous current through a device. The local current density, according to Eq. (3-7), is proportional to the local average value of the particle velocity, which has instantaneous value:

$$\langle \mathbf{v}(\mathbf{r}, t) \rangle = \frac{1}{n} \int \mathbf{v} f(\mathbf{r}, \mathbf{p}, t) \, d\mathbf{p} \tag{3-84}$$

The fluctuating distribution function $f(\mathbf{r}, \mathbf{p}, t)$ can be written as

$$f(\mathbf{r}, \mathbf{p}, t) = f_{mpm} + f_{fl}(t)$$

where the subscript *mpm* identifies the distribution for the most probable macrostate and the subscript *fl* identifies the fluctuating part of the distribution. Although we have explicitly indicated that the fluctuating distribution depends on time, there is no reason the most probable macrostate cannot also depend on time, as would be the case for a.c. systems. When we use this distribution in Eq. (3-84) we obtain a result of the form

$$\langle \mathbf{v}(\mathbf{r}, t) \rangle = \langle \mathbf{v}(\mathbf{r}) \rangle + \mathbf{v}_{fl}(\mathbf{r}, t) \tag{3-85}$$

where the first term on the right-hand side is the most probable value and the second term represents the fluctuation of the instantaneous average about the

Sec. 3-7 Fluctuations and Noise

most probable value. Generally the configuration of the most probable macrostate contains the information that is being processed and is therefore the desired signal.

The time average value of the fluctuation velocity $\mathbf{v}_{fl}(\mathbf{r}, t)$ must be zero because it is a random process. However, the noise signal itself is not as important as its energy content. The same comment applies to desirable signals as well because it is the energy of a signal that is actually generated, transmitted, processed, and detected. The instantaneous energy is proportional to the square of the current and hence to the square of the instantaneous average velocity of the charge carriers. Therefore, from Eq. (3-85) the time-averaged energy is proportional to the square of the most probable value plus the time average of the *square* of the fluctuation velocity since the average of $\mathbf{v}_{fl}(t)$ itself is zero. The time average of the square of the fluctuation velocity is not zero since as a square it is always positive. For this reason noise currents and voltages are always expressed in terms of root mean square (r.m.s.) values.

The thermal noise power in a resistor of resistance R can be calculated using the theories of mathematical probability alluded to earlier. The result for the power in the noise signal in the range of frequencies Δf is

$$\boxed{I_n^2 R = 4\kappa T \Delta f} \qquad (3\text{-}86)$$

where I_n is the r.m.s. thermal noise current, κ is Boltmann's constant, and T is the thermodynamic temperature. Two observations are important. First, the power in the noise signal depends on the temperature because the temperature is the macroscopic measure of the kinetic energy of the random motion of the particles, which is responsible for it. Second, the noise power per frequency bandwidth is independent of the frequency around which the range is centered. The latter fact is a consequence of the presence of all frequencies in the noise signal. It is very important to recognize that finite noise power must be expressed in terms of the bandwidth Δf of a measuring instrument that can respond only to signals in a limited band of frequencies.

The comments of the previous paragraph also apply to the thermal noise voltage, which can be obtained from Eq. (3-86) by using Ohm's law:

$$V_n = (4\kappa T R \, \Delta f)^{1/2} \qquad (3\text{-}87)$$

The signal-to-noise voltage ratio is the ratio of the r.m.s. value of the signal voltage to the r.m.s. thermal noise voltage V_n. The signal-to-noise ratio at the input to an ordinary amplifier must be favorable so that the signal will not be lost.

Corpuscular Noise

Thermal noise arises from the collective effects of all of the charge carriers in a system. Another general class of noise phenomena exists because of the finite value of charge on a single carrier. For instance, when current levels are very low it may be possible to detect the arrival of individual charged particles at a collecting contact or electrode. The collected current in such a case would be a series of independent random current pulses, which has an average value. The instantaneous value of the current would fluctuate about its average value, the fluctuation being evidenced as noise. This type of noise arises because charge densities are not actually continuous functions of position as we usually picture them. In reality, of course, charges exist in discrete quanta. The continuous charge densities that we ordinarily use are valid approximations only on a macroscopic scale when there are very many charges in the system and when the distances between them are negligibly small. The general class of noise phenomena that results from the discrete nature of charged particles is called *corpuscular noise*.

The example of corpuscular noise used in the preceding paragraph is called *shot noise* since the charges appear in individual shots. It can be shown using statistical techniques that the r.m.s. shot noise current in the frequency range Δf accompanying any current I that consists of the independent motion of discrete particles each carrying charge of magnitude e is

$$I_{ns} = (2eI\,\Delta f)^{1/2} \tag{3-88}$$

Another type of corpuscular noise arises with the creation or annihilation of charge, which in a gas is called ionization noise and in a solid, generation-recombination noise.

When appropriate we shall discuss both thermal and corpuscular noise in connection with the electronic systems in which they may be important as we encounter them in the remainder of our study. One final comment about signals and noise. It is technically feasible to extract a small signal from unfavorably large noise using phase-sensitive detection instruments. The techniques employed, however, are very expensive and therefore are not practical for ordinary applications in electronics, usually being restricted to research applications.

BIBLIOGRAPHY

Introductory textbooks discussing kinetic theory and transport phenomena:

REIF, E., *Statistical Physics*. New York: McGraw-Hill Book Company, 1967.

MORSE, P. M., *Thermal Physics*. Reading, Mass.: W. A. Benjamin, Inc., 1962.

KITTEL, C., *Elementary Statistical Physics*. New York: John Wiley & Sons, Inc., 1958.

Uman, M. A., *Introduction to Plasma Physics*. New York: McGraw-Hill Book Company, 1964.

Chap. 3 Problems

For advanced study:

CALLEN, H. B., *Thermodynamics.* New York: John Wiley & Sons, Inc., 1960.

SMITH, A. C., J. F. JANAK, and R. B. ADLER, *Electronic Conduction in Solids.* New York: McGraw-Hill Book Company, 1967.

For discussions of noise, see the book by Reif, pp. 32–35, as well as

PIKE, A. L., *Fundamentals of Electronic Circuits.* Englewood Cliffs, N.J.: Prentice-Hall, Inc., 1971. See Chapter 14.

VAN DER ZIEL, A., *Noise: Sources, Characterization, Measurement.* Englewood Cliffs, N.J.: Prentice-Hall, Inc., 1970.

THORNTON, R. D., et al., *Characteristics and Limitations of Transistors*, SEEC Vol. 4. New York: John Wiley & Sons, Inc., 1966.

PROBLEMS

3-1. Derive the extensive form of Ohm's law, $I = (V/R)$, from its intensive form, Eq. (3-10), by integrating over a uniform resistor of conductivity σ, length L, and cross-sectional area A. *Hint:* Find V and I in terms of $\mathcal{E}$ and $\mathbf{J}$, respectively.

3-2. A gas tube contains 10^{12} singly charged positive ions and 10^{12} electrons/m³. The electron drift velocity due to an applied electric field is 6×10^6 m/sec, while the ion drift is five-hundredths of 1% of this value. Calculate the electric current through the device assuming a cross-sectional area of 10^{-2} m².

3-3. Calculate the thermal velocity (r.m.s. velocity) of electrons in copper at room temperature assuming their mass is that of classical free electrons. Then
  (a) Estimate the average drift velocity of the electrons for a current density of 100 A/cm² assuming 8.5×10^{22} electrons/cm³. This current density is usually the design limit in copper.
  (b) Suppose that all of the electrons in a copper conductor of cross-sectional area equal to 10^{-2} cm² moved simultaneously in the same direction at the thermal velocity. What current would result?

3-4. Calculate the charge density and the electron density n in aluminum assuming that electron motion is responsible for current flows. Use $\sigma = 3.5 \times 10^7$ mhos/m and $\mu = 5.6 \times 10^{-3}$ m²/V-sec.

3-5. The average electron density in an emitter-base junction in a silicon *n-p-n* transistor may be 10^{10}/cm³. The electron mobility may be taken as 600 cm²/V-sec.
  (a) Find the conductivity.
  (b) Calculate $\mathbf{J}$ for an electric field of 10^5 V/cm.

3-6. The density gradient for electrons across the emitter-base junction of a silicon *n-p-n* transistor may be 10^{20} cm^{-4} if the density drops by a factor of 10^{16} electrons/cm^3 over a distance of 10^{-4} cm. For a diffusion coefficient of 40.0 cm^2/V-sec, find the electric current density driven by the gradient.

3-7. Solve Eq. (3-31) for $f(t)$ in the relaxation approximation using Eq. (3-32).

3-8. Calculate the mean free path and mean collision time in an atmosphere composed of nitrogen molecules at standard temperature and pressure assuming an average collision cross section of 6×10^{-15} cm^2.

3-9. The distribution function for a system in a small steady-state electric field is given by Eq. (3-57). Rewrite this equation to express the deviation of $f(\mathbf{r}, \mathbf{p})$ from equilibrium in terms of the velocity obtained by an electron in the field during one collision time and the r.m.s. velocity.

3-10. The distribution function for a system with a small steady-state density gradient imposed upon it is given by Eq. (3-74). Rewrite this equation to express the deviation of $f(\mathbf{r}, \mathbf{p})$ from local thermodynamic equilibrium in terms of the distance an electron moves in one collision time and the scale length of the gradient. The scale length L is defined by

$$\frac{1}{L} = \left| \frac{1}{n} \frac{\partial n}{\partial z} \right|$$

3-11. Use the continuity equation and Fick's law to obtain the diffusion equation for the temporal behavior of a nonuniform particle density assuming that the diffusion coefficient is independent of position. Then
(a) Show that the solution to this equation for particles originally concentrated at the origin of a one-dimensional space is

$$n(x, t) = \frac{N}{(4\pi Dt)^{1/2}} \epsilon^{-x^2/4Dt}$$

(b) Find the r.m.s. displacement from the origin $(\overline{x^2})^{1/2}$ at time t.
(c) Calculate the mean time for the molecules of a noxious gas to diffuse through the air a distance of 10 m if the diffusion coefficient is 10^{-5} m^2/sec.

3-12. Use the continuity equation, Ohm's law, and Fick's law to obtain an equation for the temporal behavior of a charge density in the presence of a density gradient and under the influence of an electric field.

3-13. Show that Eq. (2-57) is a steady-state equilibrium solution to the kinetic equation for a nonuniform system under the influence of an external force that can be expressed as the gradient of a scalar potential, $\mathbf{F} = -\nabla U(r)$.

3-14. Use Einstein's relation to calculate the diffusion coefficient for electrons in intrinsic germanium.

3-15. Calculate the thermal noise voltage that appears across a 100 kΩ resistor at room temperature as measured by an instrument with a 10,000 Hz bandwidth.

4

Electron Beams and Plasma Electronics

In this chapter we shall introduce several physical processes that have wide applicability in modern electronics, the most important of these being the formation of electron beams, space-charge limited current, the contact potential, and ambipolar diffusion. The concepts of the statistical physics of classical particles are all that are required to understand these phenomena, so therefore we shall consider them before proceeding to the more difficult concepts of quantum mechanics and the statistics of quantum mechanical particles. It should be noted, however, that most of the physical concepts treated in this chapter are applicable to quantum mechanical systems as well as to classical ones. For example, electrons in solids constitute a quantum mechanical system for reasons that will be described in Chapters 5 and 6. Space-charge limited current and ambipolar diffusion occur in solids. And the contact potential that occurs naturally between two dissimilar solids is of very great importance in modern electronics, being responsible for the nonlinear electrical characteristics of the class of devices known as junction devices. Among the devices in this class we shall study later are the Schottky barrier diode, *p-n* junction diode, bipolar junction transistor, and junction fiield-effect transistor.

Another topic introduced in this chapter that is of economic importance in modern electronics is the class of electron beam devices. In current technology the applicability of solid-state devices as electronic amplifiers and oscillators is limited to power levels less than a few kilowatts and to frequencies less than a few times 10^9 Hz. Power limitations are dictated by problems associated with dissipation of heat, and frequency limitations are related to physical processes such as diffusion, recombination, and the time required for the electrons to traverse a device of given physical dimensions, the so-called transit-time limitation. These problems will be defined and discussed in more detail in later chapters when we study the appropriate devices. On the other hand, electron beam devices can operate at high frequencies and at practical power levels up to tens of megawatts. They are of critical importance to a variety of industries, including radio and television communications, radar, and microwaves. Therefore no introduction to and survey of the concepts of physical electronics would be complete without their inclusion.

4.1 ELECTRON BEAMS

A directed stream of electrons is called an *electron beam*. In this section we shall discuss the creation of such beams and their application to a variety of practical devices.

A beam is created in a device called an *electron gun*, sketched schematically in Fig. 4-1. The gun consists of a cathode, a first anode, a drift space, and a

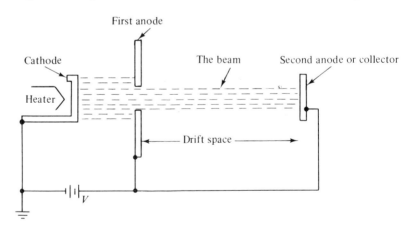

Fig. 4-1 Electron gun and beam.

second anode or collector, all inside an evacuated chamber. The cathode is made of a substance, usually a metal or a metal coated with an oxide, that when heated gives off electrons. The heater indicated in the sketch is a resistive material, sometimes the cathode itself, that dissipates 60 Hz power at 6.3 V furnished by an external power supply that is not shown in the illustration. The details of thermal emission from a heated metal will be discussed in

Chapter 7, which is concerned with the electronic properties of metals. Once emitted from the cathode the electrons are accelerated by the electric field in the region between the cathode and the first anode. The first anode has a hole in it, and the cathode and the first anode are mechanically designed so that the electric field between them and the presence of the anode hole combine to create a well-collimated beam. The electrons that pass through the first anode, traverse the drift space, and are collected by the second anode.

While in the region between the cathode and the first anode, the electrons are accelerated in the electric field maintained by an external power supply. Now the total energy of an electron in an electrostatic field is a constant

$$\tfrac{1}{2}mv^2 - eV = \text{constant} \tag{4-1}$$

where $-eV$ is the potential energy of an electron of charge $-e$ in an electric potential of V volts. If the electrons are emitted from a cathode at zero potential with no initial speed, when they reach the first anode biased at $+V$ volts their speed will be

$$v = \left(\frac{2eV}{m}\right)^{1/2} \tag{4-2}$$

The kinetic energy gained by the electrons in the acceleration process is supplied by the external source that establishes the potential field. In actuality electrons are emitted from a heated cathode with a distribution of random velocities. However, the random velocites are generally so much smaller than the velocity gained during acceleration that for all practical purposes the beam that enters the drift space is considered to be monoenergetic, that is, composed of electrons all with the same energy.

In practical devices the beam energy, supplied originally by the external source, is delivered to some other form either in the drift space or at the collector. The mechanisms of the transfer of energy from the beam to this other form are many and varied. In high-frequency amplifiers and oscillators the electrons interact with the electric fields in resonant cavities or in propagating electromagnetic waves. In the cathode-ray tube, the beam energy is delivered to a phosphorescent material that coats the collector. The energized phosphor subsequently emits visible light.

In most beam devices, the behavior of the beam in the drift region is critical to optimum performance. For many practical purposes, the design of this part of the electron beam device must take into account the fact that each beam electron is influenced not only by the electric field applied externally but also by the electric fields of all the other beam electrons. The greater the charge density in the beam, the greater will be the mutually repulsive electrostatic forces. The result of these forces is a spreading of the beam outward from its central axis, thus negating the desirable collimating effects of the accelerating region. There are, of course, some simple techniques that can be employed to reduce this space-charge effect. One technique is to reduce the beam density, reducing the amount of power handled by the device. Another is to reduce the length of the drift space. In situations where beam spreading is undesirable, some type of focusing mechanism is usually employed to counteract it. Focusing may be achieved by the application of electric or magnetic fields of sufficient strength and with the proper spatial characteristics. In analogy to optics, these field configurations are called electrostatic or magnetic lenses.

Cathode-Ray Tube

One common device utilizing an electron beam is the cathode-ray tube (CRT) used for the visual display of electronically processed information. A schematic representation of the essential features of a CRT is shown in Fig. 4-2. A beam is formed and then focused by an electrostatic or a magnetic

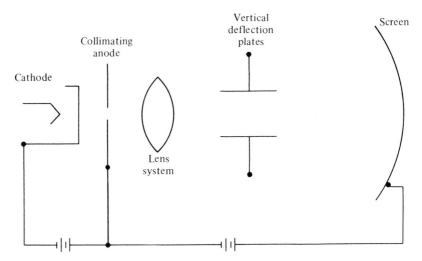

Fig. 4-2 Elements of a cathode-ray tube.

lens. The beam then passes through a deflection region where either electric or magnetic fields are applied that act on the beam to deflect it from its original path. The point of impact of the beam on the two-dimensional screen is controlled by the application of suitable signals to both the vertical and horizontal deflection systems. Figure 4-2 shows an electric deflection system that operates by attracting electrons to either of a pair of charged deflection plates. The horizontal deflection plates are not shown in the sketch. The beam energy is transferred on impact to the phosphor deposited on the inside of the glass screen, which then radiates this energy as light. Most phosphors are excited only by higher-energy electrons, 1–30 keV depending on the materials used. Since focusing and deflection are accomplished more efficiently on slower electrons, the overall acceleration of the beam in a CRT is usually performed in two stages. To facilitate the acceleration of the beam toward the screen, the inside of the phosphor is covered by a thin layer of aluminum. The aluminum is thin enough so that the high-energy electrons pass through it to reach the phosphor. The aluminum acts not only as a current drain for the electrons but also as a heat sink for the phosphor and as a mirror to reflect the emitted light back out toward the viewer.

The screen of a CRT used in the standard U.S. television system is divided into 195,000 phosphor spots arranged in 525 lines. In a color-television CRT, each spot consists of an array of three different phosphor dots, each one of which emits either red, green, or blue light. The color spot is small enough so that the individual dots cannot be distinguished by the unaided eye. Three separate electron beams are employed in the U.S. color-television system. Each beam is aligned to strike only one of the three dots, the same one at each spot. Thus each of the beams is responsible for the excitation of one color. By varying the intensity of each beam separately, the intensity of each of the emitted primary colors is varied. Theoretically, an appropriate combination of the three primaries will produce any color desired. Obviously, the art of electron beam optics has been highly refined in color picture tube technology.

High-Frequency Electron Beam Devices

Another class of device application of electron beams is amplifiers and oscillators that operate at high frequencies, that is, frequencies greater than 10^9 Hz or 1 GHz. The operating frequencies of conventional semiconductor transistors are limited to less than several gigahertz. Several reasons for this limitation were mentioned in the introduction to this chapter and will be considered in detail when we discuss these devices in later chapters. Another reason conventional electronics is not applicable to high-frequency problems is that the wavelength of the electromagnetic waves present may be comparable to the dimensions of the system. In such cases the electronics must be treated as a distributed system rather than as an ordinary circuit consisting of lumped-constant elements and devices. Both of these problems are commonplace, for instance, at the X-band of microwave frequencies extending from 8.2 to 12.4 GHz. Broadcast uses of X-band radiation include airborne radars, communications and space research satellite links, and amateur uses. At 10 GHz the oscillation period is 10^{-10} sec and the wavelength of signals is 3 cm.

In what follows, the physical principles of the common high-frequency, high-power electron beam devices will be treated descriptively rather than quantitatively. While this class of devices is of economic importance in modern electronics, the physical nature of electron-field interactions is beyond the scope of this book. These devices ordinarily are studied in detail in intermediate and advanced courses in electromagnetics. For our purposes simple descriptions of their physical configurations and of the fundamental interactions should suffice at least to familiarize the reader with them if he does not encounter them elsewhere.

Two general types of electron beam devices operate at high frequencies. In one type the beam electrons interact with the electric fields resonant in tuned cavities, while in the other the electrons interact with the electric field of a propagating electromagnetic wave. To the first type belong the various types of klystrons and the ordinary magnetron, while to the second belong the traveling wave tubes.

Klystrons and magnetrons are devices that depend for their operation on the interaction of the beam electrons with fields in resonant cavities. A schematic representation of a two-cavity klystron is shown in Fig. 4-3. A high-voltage electron beam passes first through holes in opposite walls of a narrow

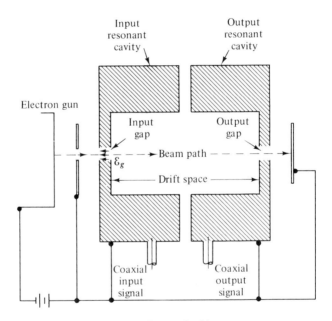

Fig. 4-3 Two-cavity klystron.

neck in a resonant cavity. An input signal drives this cavity and produces an electromagnetic field inside it. The field in the gap across which the electrons travel is denoted by $\mathcal{E}_g$ in the figure. Because the distance across the gap is very small and because the electron velocity is high, each electron crosses the gap in a small fraction of a period and experiences an electric field which is essentially constant as it does. This electric field alters the speed of each electron, either increasing it or decreasing it, depending on the phase of the field at the time the electron traverses the gap. The beam leaves the input cavity spatially uniform but with a modulated speed distribution. The electrons then enter a field-free drift space. As the beam travels across the drift space, that part of the beam that passed through the input cavity in such a phase as to be accelerated by the gap field will catch up with and overtake that part of the beam that was slowed down in the first gap. Thus alternate regions of higher and lower charge density will be created in the beam after it has traversed some of the drift space. This process is called *bunching*. The length of the drift space is selected to optimize the bunching process. A nonuniform electric field will be associated with the spatially nonuniform charge density in the bunched beam. As the bunched electrons pass through the second or output gap, some of the field energy of the beam will be transferred to high-frequency fields in the output cavity. Because of the nature of both the bunching process and the excitation of the output cavity, the output signal will be at a higher power level than the input signal if the device is properly tuned. The extra power has come from the electron beam supply and has been converted from d.c. to high-frequency a.c. power by the velocity modulation in the input gap and the attendant bunching. After passing through the output cavity, the beam is absorbed by an anode or collector. A typical high-power UHF klystron has the following operating characteristics: operating frequency, 470–680 MHz; output power of 13 kW at 30-db gain; beam voltage of 18.5 kV and current of 2 A; and collector voltage of 13.5 kV.

The reflex klystron, illustrated in Fig. 4-4, has only one cavity and is used as an oscillator. The electron beam actually passes through the cavity twice. On the first pass the beam velocity distribution is modulated by the field in the gap. The electrons begin to bunch after they enter the drift region. The beam then encounters a repulsive field at the reflector electrode and is reflected back toward the gap in the cavity.

The reflector potential is adjusted so that the bunched beam enters the cavity the second time in just the proper phase to reinforce the oscillation in the cavity. Reflex klystrons have been built to operate in many different frequency ranges, from 5 to 150 GHz, although power levels are generally on the order of 1 W or less.

Yet another type of device untilizing an interaction between electrons and cavity fields is the *magnetron*, shown schematically in Fig. 4-5. Electrons are emitted from a cylindrical cathode, the axis of which is parallel to a strong magnetic field. The electrons are accelerated toward the cylindrically symmetric anode block by a radial electric field, but the presence of the magnetic field causes the net motion of the electrons to be in the azimuthal direction. While traveling azimuthally around the device between the cathode

94 Electron Beams and Plasma Electronics Chap. 4

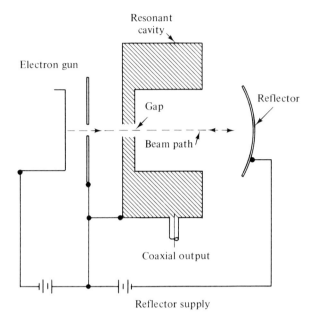

Fig. 4-4 Reflex klystron.

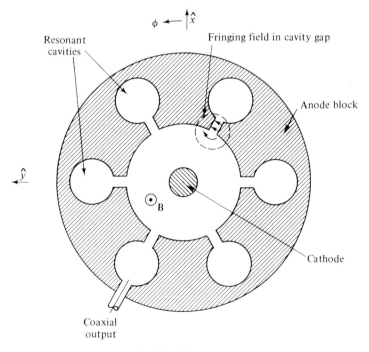

Fig. 4-5 Magnetron.

and anode the electrons encounter the fringing fields of resonant cavities recessed in the walls of the anode. These fields modulate the electron velocities to create a spatially nonuniform electron cloud analogous to a bunched beam. If the device is properly tuned, synchronism between the electron cloud and the a.c. field may be achieved, in which case the electrons eventually give energy to the fields and are collected at the anode.

Magnetrons operate at frequencies from 1 to 100 GHz with either pulsed or continuous wave output, depending on the design. Continuous wave output powers range to as much as 5 kW, and pulsed output powers to as much as megawatts.

The second class of high-frequency device that utilizes an electron beam is the *traveling wave tube* (TWT) in which the beam electrons interact with a propagating electromagnetic field. A strong interaction will occur if the electron speed is the same as the speed of propagation of the wave. Since electron speeds are always less than the speed of light, a mechanism must be found to slow the wave velocity. This is accomplished by introduction of a periodic structure made of a conducting material into the region where the beam and the waves are to interact. Frequently this structure is a helical coil of copper wire, and the electron beam travels down the axis of the helix parallel to the direction of wave propagation. If the electron speed is very close to the wave speed, the electrons will experience a slowly varying electric field and will be alternately accelerated and decelerated by it. If the device is properly tuned, however, the electrons will spend more time being decelerated than being accelerated and hence will give net energy to the wave field, thus amplifying it. Beam-wave interactions are usually the subjects of advanced courses in electromagnetics and microwave techniques.

There are many other types of TWTs in addition to the forward wave amplifier described above. Their main advantage over klystrons and magnetrons is their broad range of operating frequencies. A single TWT may operate over an octave (a factor of 2 in frequency) or more, whereas klystrons and magnetrons are rarely tunable in frequency by more than 10%. Traveling wave tubes are available with output powers as low as milliwatts or as high as megawatts depending on design.

4.2 SPACE-CHARGE LIMITED CURRENT

When large number of charged particles are present in a region of space, such as electrons in a beam, the electric and magnetic fields in that region consist not only of the externally applied fields but also of the fields due to the presence and the motion of the charges themselves. Furthermore, the net behavior of any particle is a result of its interaction with all the fields of the other particles as well as with the external fields. Self-defocusing of high-intensity electron beams is a simple example of the so-called space-charge problem. In more complicated cases solution for the dynamic behavior of all the particles using Newton's equation of motion is hopelessly complex. There are two reasons for the complexity of such a problem: First, the particle behavior depends on the fields that can be determined only after the particle behavior is known, and, second, there are too many particles. However, self-consistent solutions for the fields and the particle behavior can be obtained using the techniques of statistical mechanics in conjunction with Maxwell's equations.

The behavior of electrons between parallel plane electrodes in a vacuum tube is a good example of this type of space-charge problem. Two or more metallic electrodes inside a glass envelope that has been evacuated to a pressure of the order of 10^{-7} torr (about 10^{-10} atm) constitute what is known as a vacuum tube. At these pressures there are only about 10^9 residual gas molecules 1 cm^3 inside the tube. A *diode* is a vacuum tube with only two electrodes, a cathode and an anode. The cathode is caused to emit electrons by a variety of methods, two of which are heating it to a high temperature or illuminating it with an appropriate light source. A diode with a heated cathode is called a *thermionic diode*, while one whose cathode is sensitive to light is called a *photodiode*.

A schematic representation of a thermionic diode along with a typical current-voltage characteristic are shown in Fig. 4-6. The *I-V* characteristic

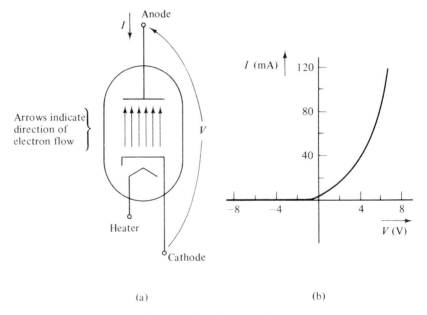

Fig. 4-6 Vacuum tube thermionic diode: (a) schematic representation, (b) typical *I-V* characteristic near the origin.

Sec. 4-2 Space-Charge Limited Current

reasonably resembles the idealized curve of a simple switch, as discussed in Chapter 1. Notice first, however, that with no voltage applied across the tube and even for small values of reverse bias a small forward current flows. This effect is due to the fact that some electrons are emitted from the cathode with sufficient energy to reach the anode even overcoming the retarding force of the electric field in the case of the reverse bias. Second, notice that under no condition does a reverse current flow through this type of diode. Third, the forward current-forward voltage region of the characteristic is a nonlinear curve. The normal operation of the device depends on the biases applied to it by the external circuitry. Analyses of the circuit applications of the diode may be performed using load line techniques and incremental equivalent circuits.

The actual shape of the I-V curve near the origin is characteristic of thermal emission of electrons from metals, to be studied in detail in Chapter 7. For higher values of forward currents, effects due to the presence of the charges themselves in the space between the cathode and anode determine the I-V relationship. The space charge limits the amount of current that can be drawn through the diode. This space-charge limited current is the result of the self-consistent field of the electrons and is the subject of this section. At very high forward voltages the current drawn may again be determined by cathode conditions, the emission being limited by the cathode temperature. This effect, saturated thermal emission, is not illustrated in this figure but will be studied in Chapter 7.

Consider first the diode with no charge between the cathode and anode, as shown in Fig. 4-7(a). If the plates are large compared with their separation, the electric field between the electrodes is uniform, and fringing fields can be neglected. Since the field is proportional to the gradient of the electrostatic potential, the potential varies linearly between the electrodes, as indicated in Fig. 4-7(c). Now let us consider what happens when charge is introduced into the region in front of the cathode. Figure 4-8 shows both the vacuum field and the field due to the presence of negative charge in front of the cathode. The net field on the anode side of the charge is greater than it was in the absence of charge, while that on the cathode side is less. Thus the electric field between the electrodes is no longer uniform, and the potential no longer varies linearly with distance. The effect of the space charge for practical devices is to reduce the field at the cathode. If the charge density is sufficiently high, the field due to the space charge may even be greater than the applied field actually reversing the field at the cathode. In other words, there may be so many electrons in the space-charge cloud that other electrons in the space between the cathode and the cloud are repelled by the cloud and return to the cathode. In this way the net number of electrons emitted from the cathode is limited by the presence of the space charge so that the net diode current is also limited. The maximum current that can flow is called the *space-charge limited current*.

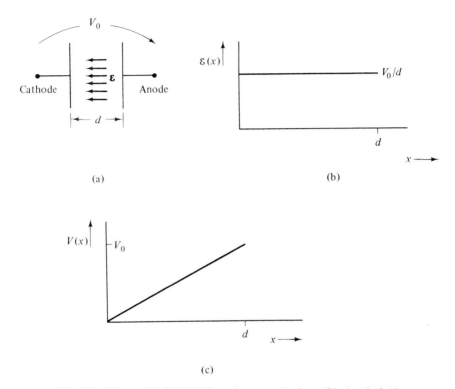

Fig. 4-7 Vacuum diode: (a) schematic representations, (b) electric field, (c) electric potential between the electrodes.

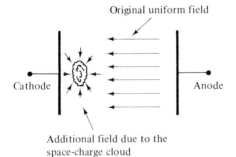

Fig. 4-8 Electric fields inside a diode with space charge.

Let us now obtain mathematical solutions for the potential distribution and the space-charge limited current for a diode in the steady state. To do this we need to know the charge density and electric potential at every point inside the diode. Since the charge density and the electric potential depend on each other, we must obtain a self-consistent solution. Such a solution can be obtained if we know two equations relating the two functions. One of

Sec. 4-2 Space-Charge Limited Current

these relations is Poisson's equation of electrostatics,

$$\nabla^2 V = -\frac{\rho}{\varepsilon_0} \tag{4-3}$$

where $\rho = -ne$ is the electronic charge density and ε_0 is the permittivity of free space. Poisson's equation is a direct result of Maxwell's equations of electrodynamics and the relationship between the electrostatic potential and the electric field. One of Maxwell's equations is $\nabla \cdot \mathbf{D} = \rho$, where $\mathbf{D} = \varepsilon_0 \mathcal{E}$ in the vacuum of the diode. Equation (4-3) results when the electric field is written as $\mathcal{E} = -\nabla V$.

The second relation can be obtained from the principles of charge and energy conservation. The conservation of charge is expressed by the continuity equation:

$$\frac{\partial \rho}{\partial t} + \nabla \cdot \mathbf{J} = 0 \tag{4-4}$$

For the static case, this relation implies that the current density, for one-dimensional flow, is constant. The current density $\mathbf{J}$ is related to the charge density and average electron velocity by

$$\mathbf{J} = -ne\langle \mathbf{v} \rangle \tag{4-5}$$

The average velocity of the electrons $\langle \mathbf{v} \rangle$ is a function of position inside the diode because of the speed gained by each electron as it is accelerated through a potential $V = V(x)$. If we assume that the electrons are all emitted from the cathode with no velocity, their speed as a function of position between the electrodes is given by Eq. (4-2):

$$v(x) = \left[\frac{2eV(x)}{m} \right]^{1/2} \tag{4-6}$$

This equation is a consequence of the conservation of energy. By combining Eqs. (4-5) and (4-6), we can obtain another relationship between n and V in terms of the constant current density J:

$$J = -ne \left[\frac{2eV}{m} \right]^{1/2} \tag{4-7}$$

Equations (4-3) and (4-7) are the two relations required for the self-consistent solution since J is constant between the electrodes.

The solution is effected in the following way. We shall eliminate the electron density n from Eq. (4-3) by substitution from Eq. (4-7) with the result, in one dimension,

$$\frac{d^2 V}{dx^2} = -\frac{J}{\varepsilon_0} \left[\frac{m}{2e} \right]^{1/2} V^{-1/2} \tag{4-8}$$

Equation (4-8) can be solved for $V(x)$ if the boundary conditions are known.

The boundary conditions, it will be recalled, are specified values for V and dV/dx at some point in space. For simplicity, we shall consider in detail only the case of a grounded cathode with just the right amount of space charge so that the electric field at the cathode is also zero. The boundary conditions we shall use are therefore $V = 0$ and $dV/dx = 0$ at $x = 0$. It is possible, of course, to do a more general case. However, solution of the general case tends to obscure the principles with more complicated mathematical technique. One technique for solving Eq. (4-8) is to multiply both sides by $2(dV/dx)$ and recognize that

$$\frac{d}{dx}\left(\frac{dV}{dx}\right)^2 = 2\frac{dV}{dx}\frac{d^2V}{dx^2}$$

Equation (4-8) becomes

$$\frac{d}{dx}\left(\frac{dV}{dx}\right)^2 = 2KV^{-1/2}\frac{dV}{dx} \qquad (4\text{-}9)$$

where a group of constants has been written as

$$K = -\frac{J}{\varepsilon_0}\left(\frac{m}{2e}\right)^{1/2} \qquad (4\text{-}10)$$

Equation (4-9) can be integrated with respect to position from the cathode, $x = 0$, to some general position in the diode, x, since

$$\int_0^x V^{-1/2}\frac{dV}{dx}dx = \int_0^{V(x)} V^{-1/2}\,dV$$

The result is

$$\left(\frac{dV}{dx}\right)^2 = 4KV^{1/2}$$

where the boundary conditions have been applied. The desired solution for V can be obtained by taking the square root of this relation and integrating again. The functional form that satisfies the resulting integral equation is

$$\tfrac{2}{3}V^{3/4} = K^{1/2}x \qquad (4\text{-}11)$$

This equation is plotted as curve B in Fig. 4-9 and is the desired solution for this special case. Equation (4-11) is valid everywhere between the cathode and anode and specifically at the anode is

$$\tfrac{2}{3}V_0^{3/4} = K^{1/2}d \qquad (4\text{-}12)$$

where V_0 is the total tube voltage and d is the electrode spacing. By combining Eqs. (4-10) and (4-12), we can obtain an expression relating the current density to the tube voltage and electrode spacing:

$$\boxed{\,J = -\frac{4}{9}\varepsilon_0\left(\frac{2e}{m}\right)^{1/2}\frac{V_0^{3/2}}{d^2}\,} \qquad (4\text{-}13)$$

This equation is known as the Child-Langmuir relation for the space-charge

Sec. 4-3   Plasma, Sheaths, and the Contact Potential   101

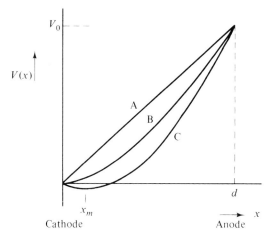

Fig. 4-9 Potential distribution in a parallel-plate diode. Curve A is for conditions of no space charge, curve B for space charge but no initial velocity, and curve C for space with initial velocity.

limited current with zero initial velocity. It is also called the space-charge equation and the three-halves power law. Since the tube current is proportional to the current density, the current-voltage characteristic of a thermionic diode under conditions of space-charge limited flow satisfies the three-halves power relationship. Practical diodes are normally operated under these conditions. The minus sign in Eq. (4-13) merely indicates that the current flows in the negative x-direction in the geometry of Fig. 4-7.

A more realistic problem than the one discussed above would include the effects of a real distribution of initial speeds and a nonzero electric field at the cathode whose value is dominated by the space charge. The result for this case will be a potential distribution like curve C in Fig. 4-9. The solution of the self-consistent equations for this case is facilitated by choosing as the *boundary* the position x_m of the potential minimum, where the electric field is zero. The location x_m must also be determined self-consistently. It turns out that the value of the space-charge limited current assuming a Maxwellian distribution of initial velocities is not very much different from that predicted by the simple Child-Langmuir model of zero initial speed, especially at higher voltages.

Practical diodes usually employ a cylindrical rather than a planar geometry. For cylindrical geometries the theory still predicts the current to be proportional to the three-halves power of the voltage. The constant of proportionality depends on the geometrical configuration.

4.3 PLASMA, SHEATHS, AND THE CONTACT POTENTIAL

Plasma is a state of matter composed at least partly of high densities of mobile charged particles. Gases, liquids, and solids can exhibit plasma behavior. Those that do exhibit such behavior have properties that are so different from the more conventional materials that plasma is frequently referred to as the *fourth* state of matter. The electric and magnetic properties are the most prominent ones that distinguish plasmas from the other states. One difference is that a plasma, while essentially neutral on the whole, consists of electrically charged particles that may move freely under the influence of electric and magnetic fields. The plasma is neutral in the sense that it is made up of equal numbers of positive and negative charges and thus has no net electric charge. Because highly mobile charges move under the influence of an electric field in such a way that a current results, plasmas are generally good electrical conductors. A second important feature that distinguishes a plasma is that the densities of the charge carriers, both positive and negative, are sufficiently high that the behavior of the plasma is influenced not only by the electric and magnetic fields imposed externally but also by the fields generated internally due to the presence and the motions of the plasma charges themselves. For example, the motion of a single charged particle may be determined as much by the fields of its neighboring particles as by any external fields. Such is the case in the regime of space-charge limited current of an electron beam as described in the previous section, although the beam, being charged, is not strictly speaking a plasma.

Plasmas occur widely in nature, and the characteristic behavior of gaseous and of solid-state plasmas has been applied to a variety of technological problems. Gaseous, liquid, and solid conductors are plasmas, as are, in some cases, semiconducting materials. The stars, the solar wind, and the ionosphere are also plasmas. Sun spots and solar flares are manifestations of plasma behavior that are not yet fully understood. The ionospheric plasma interacts with short electromagnetic radio waves in such a way as to be responsible for communication over the horizon. Plasmas will soon be utilized in power generation schemes, either in direct magnetohydrodynamic conversion or in controlled thermonuclear reactions. A class of electronic devices known as space-charge devices utilize the plasma properties of both the gaseous and the solid state. The Gunn-effect oscillator is a solid-state space-charge device.

The high-density, highly mobile charge in a plasma is responsible for a very important electrical property: A plasma cannot sustain internally a large low-frequency electric field. When an electric field is imposed on a plasma from an external source the internal charges redistribute themselves in such a way that the field of the charges opposes the applied field. If the charge densities are sufficiently high, the net field inside the plasma will essentially vanish. Thus, there is no potential difference between different parts of a metallic conductor.

In the remainder of this chapter we shall consider just three characteristics of plasmas: contact potentials, ambipolar diffusion, and plasma oscil-

Sec. 4-3 Plasma, Sheaths, and the Contact Potential 103

lations. Each is a result of the neutralizing effects of the mobile charge densities and has specific application to solid-state electronics. Other properties of plasmas have also found application in electronics and in other technologies. Important among these are magnetic effects and the interaction of plasmas with electromagnetic waves. The Bibliography contains a variety of references for further study.

Sheaths and the Contact Potential

When an external agent attempts to impose an electric field on a plasma, the mobile plasma charges are redistributed so as to cancel out the imposed field throughout the bulk of the plasma. An observer inside the plasma would be unaware of anything electrical going on outside. This isolation from external effects is achieved by the formation of a region at the surface of the plasma through which the transition from "external" to "internal" occurs. This transition region is called a *sheath*.

To illustrate the formation of a sheath, let us consider what happens when a foreign object comes in contact with a plasma. That foreign object might be a wall of the container that surrounds the plasma, an antenna inserted into it, or even another plasma. The contact may be accidental and troublesome or desirable and functional. A plasma surrounding an antenna may prohibit radio communications via that antenna. This occurs during the re-entry of spacecraft into the atmosphere. On the other hand, the behavior of the contact region between *p*-type and *n*-type semiconductors is fundamental to the electronics of solid-state junction devices.

A plasma and a foreign body are represented schematically in Fig. 4-10,

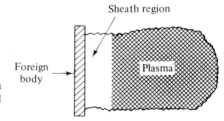

Fig. 4-10 Illustration of the sheath region between the bulk of a plasma and a foreign body.

which also shows the sheath region between the two. To simplify our considerations of the problem, we shall assume that we have a classical gaseous plasma, the electrons and ions of which are of equal densities in the bulk of the plasma and are in thermodynamic equilibrium. The ideas we shall consider and the results we shall develop are applicable to a much broader class of problems than this simple example. We use this classical model so that we can gain insight into the phenomenon without getting into the more complicated formulations of quantum mechanics.

Because the electrons are less massive than the ions, the mean squared speed of the electrons is greater than that of the ions since

$$\tfrac{1}{2}m_e\overline{v_e^2} = \tfrac{1}{2}m_i\overline{v_i^2} = \tfrac{3}{2}\kappa T$$

in thermodynamic equilibrium. If the foreign body is suddenly thrust into the vicinity of the plasma, electrons will strike the object initially and most often owing to their greater speed. Let us assume, again for simplicity, that the foreign body is isolated from an electrical circuit, in which case it quickly builds up a negative charge. This negative charge has associated with it an electric force that attracts ions to the body but repels additional electrons from it. Only the very fast or energetic electrons can overcome the repulsive force and hit the object. In the steady state, no net current can flow between the plasma and the body because the body is insulated from an electrical ground. Therefore, the electric field of the charge that has accumulated on the object must repel a sufficient number of electrons so that the electron current that overcomes the repulsive force just cancels the ion current attracted by it. Thus, in the steady state equal currents of electrons and ions strike the foreign body, and the net current to it is zero. The controlling factor is the electric field built up in the region between the bulk of the plasma and the foreign object. The region in which the electric field exists is called the sheath.

Let us now develop a theory for the sheath region. Rather than deal with the electric force field directly, we shall consider the electric potential

$$V = V(x)$$

which is a function of distance from the foreign body. The electric field may be obtained from the negative of the gradient of the potential function, $\mathcal{E} = -\nabla V$. The density of classical particles in a force field in which the potential energy of the particles is $U = U(x)$ is given by Eq. (2-58):

$$n(x) = n_0 \epsilon^{-U(x)/\kappa T} \tag{4-14}$$

where n_0 is the density at a place where $U = 0$. The potential energy of a charged particle in an electric field is related to the electric potential function by $U = qV$, where q is the charge of a particle. Therefore, the electron density in the sheath field is

$$n_e = n_{e_0} \epsilon^{eV/\kappa T} \tag{4-15}$$

since $q = -e$ for an electron. The ion density is

$$n_i = n_{i_0} \epsilon^{-ZeV/\kappa T} \tag{4-16}$$

where $Z = 1$ for singly charged ions, $Z = 2$ for doubly charged ions, and so forth. Equations (4-15) and (4-16) relate the densities of the charge species to the electric potential.

The electric potential, on the other hand, is related to the charge densities by Poisson's equation of electrostatics, which has the one-dimensional

Sec. 4-3 Plasma, Sheaths, and the Contact Potential

form in free space,

$$\frac{d^2V}{dx^2} = -\frac{\rho}{\varepsilon_0} \qquad (4\text{-}17)$$

where ρ is the net charge density,

$$\rho = Zen_i - en_e \qquad (4\text{-}18)$$

Equations (4-15) and (4-16) may be inserted into Eq. (4-18) to obtain a self-consistent equation for the potential function:

$$\frac{d^2V}{dx^2} = -\frac{e}{\varepsilon_0}[Zn_{i_0}\epsilon^{-ZeV/\kappa T} - n_{e_0}\epsilon^{eV/\kappa T}] \qquad (4\text{-}19)$$

In principle this equation can be solved to obtain $V = V(x)$ from which the particle densities and the electric field can be obtained. In practice, however, Eq. (4-19) is very difficult to solve without employing a simplifying assumption: The exponentials on the right-hand side can be expanded, provided $(ZeV/\kappa T) \ll 1$, using the series

$$\epsilon^{\pm s} = 1 \pm s, \qquad s \ll 1$$

If this assumption is valid, the equation becomes

$$\frac{d^2V}{dx^2} = -\frac{e}{\varepsilon_0}\left[Zn_{i_0}\left(1 - \frac{ZeV}{\kappa T}\right) - n_{e_0}\left(1 + \frac{eV}{\kappa T}\right)\right] \qquad (4\text{-}20)$$

If we utilize the fundamental charge neutrality of the plasma interior $Zn_{i_0} = n_{e_0}$, Eq. (4-20) reduces further to

$$\frac{d^2V}{dx^2} = \frac{1}{\lambda^2}V \qquad (4\text{-}21)$$

where

$$\boxed{\frac{1}{\lambda^2} \equiv \left[\frac{Z^2e^2n_{i_0}}{\varepsilon_0\kappa T} + \frac{e^2n_{e_0}}{\varepsilon_0\kappa T}\right]} \qquad (4\text{-}22)$$

Equation (4-21) has the solution

$$V(x) = V_0\epsilon^{-x/\lambda} \qquad (4\text{-}23)$$

which satisfies the boundary condition that the potential and the electric field vanish far from the foreign body. Equation (4-23) represents a potential function that decays exponentially into the plasma with scale length λ which is called the *Debye screening length*. The foreign body, located at $x = 0$, is at a potential V_0 relative to the interior of the plasma.

The screening length is an important parameter of any plasma because it is a measure of the distance over which external fields actually do penetrate into the plasma. The sheath region can be considered to be one or two Debye lengths thick, since the external fields are much reduced beyond these distances. A fundamental restriction on the definition of a plasma is that it be large in extent compared with its own Debye length. Otherwise, it would be all sheath and no plasma, external fields being able to penetrate throughout its volume.

The Debye length is the fundamental minimum distance over which collective plasma effects occur. In the definition of the screening distance, Eq. (4-22), it is frequently customary to treat the two terms on the right-hand side separately, defining one as a screening length for electrons and the other for the positive charges. The electron screening length λ_e is

$$\lambda_e = \left[\frac{\varepsilon_0 \kappa T}{e^2 n_{e_0}}\right]^{1/2} \qquad (4\text{-}24)$$

If the charged particles exist in a medium other than free space such as in a metal or semiconductor, the appropriate dielectric constant should be used in Eq. (4-22) or (4-24).

Two more comments about the sheath formulation are in order. First, the densities of the charged species in the sheath do not lead to charge neutrality in the sheath as exists in the interior of the plasma. This is evident from Eqs. (4-15), (4-16), and (4-23), which when combined yield the charge densities as functions of position in the sheath. Second, because for the problem we have discussed the electrons reached the object first and charged it negatively, the potential of the body with respect to that of the bulk of the plasma, V_0, is negative. The potential difference across the sheath is also called the *contact potential* because it arises naturally when a plasma and a foreign body come in contact. When the sheath potential is negative, measured with respect to the potential of the plasma interior, there is an excess of ions in the sheath and a deficiency of electrons. Physically this is a reasonable result since the field repels electrons from the object and attracts ions to it.

The principal problem with the foregoing mathematical development is the approximation made to facilitate the solution of Eq. (4-19), $(ZeV/\kappa T) \ll 1$. First, the physical mechanism of the sheath formation depends on the thermal properties of the plasma particles. The electrons reach the foreign body by virtue of their thermal motion. The steady-state field in the sheath region has a self-adjusted value that regulates the electron current from the plasma, overcoming the kinetic energy of some electrons and returning them to the plasma. Thus, it stands to reason that the potential energy associated with the net sheath potential eV_0 ought not to be much smaller than κT but rather ought to be on the order of κT, thus violating the approximation. In fact, the approximation, and hence the solution, is valid only on the plasma side of the sheath where the potential is small. While the mathematical solution obtained in Eq. (4-23) is not strictly valid everywhere in the sheath, the physical interpretations of the sheath and its formation are, and these interpretations form the basis for understanding a wide variety of plasma effects that occur when a foreign object comes in contact with a plasma.

4.4 BIASED SHEATHS

A second, practical aspect of the nature of the sheath between a plasma and its surroundings arises when an object that is connected to an external electrical circuit comes in contact with a plasma.

Historically, this problem arose during the search for means to determine the actual internal characteristics of plasma. Plasma diagnostics is obviously complicated by the fact that a plasma screens itself from all external influence by the formation of its sheath. For example, the plasma temperature cannot be measured by inserting one lead of a thermocouple into it, since the thermocouple will be surrounded by, and hence will sample the characteristics of, the sheath and not the plasma. As a result of this difficulty a wide variety of indirect techniques has been developed to measure plasma properties. Of principal interest to electronics, however, are the techniques using electrical circuitry, and chief among these is the so-called electrostatic probe. Our interest lies not so much in its function as a probe as in its characteristic behavior as part of an electrical circuit that can apply a bias to the plasma sheath. Viewed as a problem of driven sheaths, probe theory has important practical application in solid-state electronics, as, for example, in the contact between a conductor and a semiconductor or between two semiconductors.

For the purposes of discussion, let us consider a small metal flag of exposed area A inserted into a plasma and connected to an external biasing circuit via an insulated lead as indicated in Fig. 4-11. The dashed line is meant to indicate the edge of the sheath. The variable voltage supply is used to adjust the potential of the probe with respect to the electrical ground. When the probe voltage is varied, the potential of the probe with respect to

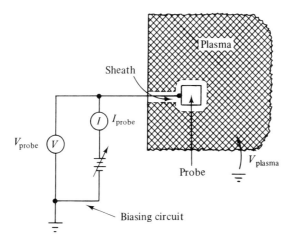

Fig. 4-11 Electrostatic probe with its biasing circuit.

the bulk of the plasma, that is, the potential difference across the sheath, is also varied. The sheath potential and the probe potential need not be the same since the plasma might be at some arbitrary potential with respect to ground. It is assumed that the potential of the plasma with respect to ground does not change when the probe is biased.

At some applied probe potential, the voltage drop across the sheath will be the same as that across a sheath between the plasma and an ungrounded, isolated object, i.e., the contact potential V_0 of Section 4-3. When this condition applies, no net current flows because of the detailed balance between the electron current and the ion current to the probe from the plasma. The applied probe voltage at which this balance occurs is called the *floating potential:* When the probe is at the floating potential, no current flows in the external circuit.

When the probe is changed slightly from the floating potential the sheath voltage will also change, assuming that the presence of the biased probe does not affect the plasma. This change in the sheath potential upsets the balance between the electron and ion currents and allows a net current to flow. If the probe is still negative with respect to the plasma, the ion current will not be affected by a small change in the sheath potential since all the ions will continue to be collected anyway. However, even a small change in the repulsive potential that the electrons experience in the sheath can affect considerably the number of electrons that ultimately reach the probe.

The essential features on which an understanding of the behavior of biased sheaths depends are the thermal distribution of velocities of the plasma particles and the existence of the potential barrier created by the electric field in the sheath. The velocity distribution of the particles can be translated into an energy distribution. For classical particles in equilibrium any energy is allowed, but the average energy of all the particles is described by the temperature. If the probe is negative with respect to the plasma, the electric field in the sheath acts as an energy barrier to electrons moving from the plasma toward the probe. Thus only the more energetic of the electrons will be able to overcome the repulsive force. By varying the sheath potential we can vary the "height" of the barrier and thus vary the number of electrons that can overcome it. The electrons that do overcome the repulsive force are collected by the probe. The probe electron current is a sensitive function of the sheath potential because, according to Eq. (4-15), the electron density at the probe is an exponential function of the sheath potential.

If the probe is biased slightly positive with respect to the floating voltage, the electrons will experience a smaller repulsive force that more of the electrons will be able to overcome. The additional electron current to the probe will not be matched by an ion current so that net current flows. If the probe is biased more negatively, the electrons will experience a greater repulsive force, and fewer will have sufficient energy to overcome it to reach the probe.

Sec. 4-4 Biased Sheaths

Thus, the electron current would not be large enough to balance the ion current and again a net current would flow, carried, in this case, by the ions. An idealization of the entire current-voltage characteristic of an electric probe in contact with a plasma is sketched in Fig. 4-12. This characteristic

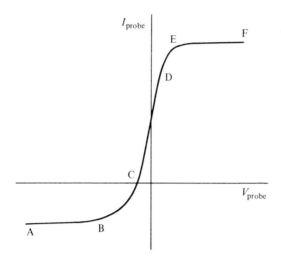

Fig. 4-12 *I-V* characteristic of an electrostatic probe in a plasma (not drawn to scale).

is obtained by varying the power supply of Fig. 4-11 and plotting the probe current as a function of probe voltage. At point C on the plot, no net current flows, so V_c is the floating potential.

When the probe potential is more negative than the floating potential, the current that flows is due to the uncompensated ion current. The more negative the bias, the smaller the electron current and the greater the uncompensated portion of the ion current. When the negative bias is sufficiently large, essentially all of the electrons in the sheath region will be returned to the plasma and none will reach the probe. At this bias, which is that of point B in Figure 4-12, the current is due only to the positive ions. Since the probe collects all the available ions that enter the sheath because of their random motion, this current is called the *ion saturation current*. In the ideal case, further negative increases in the probe voltage do not result in changes in the probe current, as indicated in curve section A-B in Fig. 4-12, since the probe is collecting as many ions as possible.

Our idealization neglects some effects, however, that do result in slight increases in ion current with negative probe voltage. These effects, which can become important under certain conditions in both gaseous and solid-state plasmas, include acceleration of the ions in the increased electric field in the sheath and an increase in the effective sheath volume as the higher fields penetrate more deeply into the plasma. We shall reserve discussion of both of these phenomena for consideration with the application of the sheath characteristics in solid-state semiconductor junction devices.

When the probe voltage is biased positively with respect to the floating voltage V_c, more electrons are collected due to the reduced sheath potential resulting in net current flow. At a sufficiently high probe bias, the sheath potential may be reduced to zero, which means there would not be any electric force on either the electrons or ions, and the probe would collect all the electrons and all the ions that wander to the probe as a result of their random motion. When this occurs, the probe is at the same potential with respect to ground as is the plasma, there being no drop across the sheath. Point D in Fig. 4-12 indicates the plasma potential. Collecting all electrons and ions results in a net electron current because, for gaseous plasmas of equal charge densities and in thermal equilibrium, the electrons have a larger average speed. Further increase in the probe voltage causes the probe to become positive with respect to the plasma, provided, of course, the characteristics of the plasma are not affected by the presence of the probe. A positive probe collects all electrons but repels some ions, depending on the probe voltage and the distribution of ion energies. At point E, and between points E and F, all the ions are being repelled and the probe collects the *electron saturation current*.

The *I-V* characteristic of an electrostatic probe in a gaseous plasma is used principally as a diagnostic tool. The saturation currents depend on the charge densities and average speeds, and the detailed shape of the characteristic between points B and D depends on the electron distribution function. The characteristic, then, can be used to measure these plasma parameters indirectly.

In semiconductor plasmas, an analogous characteristic is obtained for the sheath between two semiconductors at what is called a *p-n* junction. We shall study these effects again, with more practical application, in conjunction with the *p-n* junction in Chapter 9.

4.5 AMBIPOLAR DIFFUSION

Diffusion, as described in Chapter 3, is the process by which a net migration of neutral particles from regions of high concentration to regions of low concentration occurs. Diffusion is a result of the random motions of the individual particles. Ambipolar diffusion is the simultaneous diffusion of two types of particles, one carrying positive charge and the other negative. The two species diffuse from a region where the densities of both are high to one where both are low. As in the ordinary case, ambipolar diffusion is a result of the random motions of the individual particles. However, it differs from

Sec. 4-5 Ambipolar Diffusion

the diffusion of neutral particles in that the electric effects due to the fields of the charges alter the diffusive flow.

Let us consider an elementary model that will serve to accent the salient features of ambipolar diffusion. We shall assume that we have a plasma comprised of two charged species the densities of which are everywhere essentially equal but not uniform. In other words, the densities may vary as a function of position, but they are almost equal so that the plasma maintains overall charge neutrality. The negative charges might be electrons in a solid-state plasma or electrons or negative ions in a gaseous plasma. The positive charges would be positive ions in a gaseous plasma or the so-called holes in a semiconductor. The nonuniform density arises from the manner in which the plasma is formed locally. The plasma, like everything else, is always nonuniform at its surface, there being plasma inside and none outside, but the interior may also be nonuniform. For the purposes of discussion we shall concentrate on diffusion at the surface and assume that the system is in local thermodynamic equilibrium.

Before attempting to establish a mathematical model for ambipolar diffusion that would permit us to calculate an ambipolar diffusion coefficient, let us consider what happens physically in the region of space where the plasma density decreases outward. We shall assume that one species, the negative charges, say, is less massive than the other so that its thermal speed will be greater. Then the lighter species would diffuse faster than the more massive one according to the simple theory of diffusion developed in Chapter 3. If the negative charges diffuse outward faster than the positive ones, a net electric field will be set up where the diffusion is taking place because the charge densities there will no longer be equal. This internal electric field will be in such a direction as to slow down the negative charges and speed up the positive ones so that in the steady state the two species will diffuse together. The rate of ambipolar diffusion should then be less than the ordinary diffusion of the lighter particles but greater than that of the heavier ones.

The mathematical model for ambipolar diffusion must include not only the particle current flows due to the density gradients but also those due to the internal electric field that arises from differences in the diffusion coefficients of the two species. We shall use n and p to denote the densities of the negative and positive charges, respectively. The particle current density of the negative charges Γ_n is the result of diffusion in the density gradient and mobility in the electric field and is, in one dimension,

$$\Gamma_n = -D_n \nabla n - n\mu_n \mathcal{E} \qquad (4\text{-}25)$$

where D_n and μ_n are the diffusion coefficient and mobility of the negative charge carriers, respectively. The positive particle current density is

$$\Gamma_p = -D_p \nabla p + p\mu_p \mathcal{E} \qquad (4\text{-}26)$$

The electric field $\mathcal{E}$ is the self-consistent, internal steady-state field. In the steady state the two species diffuse together, and hence the particle current densities are the same. The condition $\Gamma_n = \Gamma_p$ leads to the result

$$\mathcal{E} = \frac{-D_n \nabla n + D_p \nabla p}{p\mu_p + n\mu_n} \quad (4\text{-}27)$$

Having obtained the steady-state electric field in terms of the gradients, we can eliminate it from either of the particle current densities to obtain something that has the form of a Fick's law, relating a particle current density to a density gradient via an effective diffusion coefficient. Let us substitute Eq. (4-27) into, for example, Eq. (4-25) to obtain

$$\Gamma_n = \frac{-n\mu_n D_p \nabla p - p\mu_p D_n \nabla n}{p\mu_p + n\mu_n} \quad (4\text{-}28)$$

While Eq. (4-28) evaluates the particle currents given the density gradients, it is not yet in its simplest form. The density gradients ∇p and ∇n are related to one another so that ∇p can be expressed in terms of ∇n, thus simplifying Eq. (4-28). To accomplish this objective, we note that from differential calculus

$$\nabla\left(\frac{p}{n}\right) = \frac{1}{n^2}(n\nabla p - p\nabla n)$$

which can be rearranged to yield

$$\nabla p = n\nabla\left(\frac{p}{n}\right) + \frac{p}{n}\nabla n$$

If we assume that the ratio p/n is uniform even though p and n individually are not uniform, then the relationship between the gradients becomes

$$\nabla p = \frac{p}{n}\nabla n \quad (4\text{-}29)$$

Inserting Eq. (4-29) into Eq. (4-28) to eliminate ∇p and rearranging yields

$$\boxed{\Gamma_n = -\left[\frac{\mu_n D_p + \mu_p D_n}{\mu_p + (n/p)\mu_n}\right]\nabla n} \quad (4\text{-}30)$$

which has the form of Fick's law. The term in brackets is the effective diffusion coefficient for ambipolar diffusion.

In gaseous plasmas and intrinsic semiconductors, the mobile charge densities, n and p, are essentially equal, so that for these cases the ambipolar diffusion coefficient is

$$D_a = \frac{\mu_n D_p + \mu_p D_n}{\mu_n + \mu_p} \quad (4\text{-}31)$$

Furthermore, for particles that can be described by Maxwellian statistics, the diffusion coefficient is related to the mobility by Einstein's relation, Eq.

(3-80),

$$D = \frac{\kappa T}{e}\mu$$

For this type of particle the ambipolar diffusion coefficient reduces to

$$D_a = \frac{\kappa(T_n + T_p)}{e}\left[\frac{\mu_n \mu_p}{\mu_n + \mu_p}\right] \quad (4\text{-}32)$$

An interpretation of Eq. (4-32) is that ambipolar diffusion occurs at a rate characteristic of the sum of the temperatures of the species but with a mobility dominated by the less mobile of the two, which supports our arguments based on physical intuition.

4.6 PLASMA OSCILLATIONS

Another class of plasma phenomena of interest to applied electronics is that of natural oscillation. Plasma particles can exhibit large-scale collective oscillatory motion in addition to the usual random thermal motion. These oscillations occur because of the interaction of the charged particles through their internal electric and magnetic fields. Electric and magnetic fields appear in conjunction with the particle motion, and these fields may, under certain circumstances, grow and propagate as an unstable electromagnetic wave. Such a plasma wave may contribute to undesirable noise in an electronic device. On the other hand, purposeful excitation of an unstable plasma mode might create an oscillating device that could be useful as a source of high-frequency electromagnetic energy.

Plasma oscillations arise from a number of effects. They may be induced by application of an appropriate disturbance such as an electric or magnetic field or an electron beam or a laser. Or they may arise accidentally by the chance motion of the particles.

Such a wide variety of interesting oscillations can exist in a plasma that just to categorize all of them would, for the purposes of this text, require so much space as to distract us from our primary objectives. However, one, the electrostatic oscillation of electrons, is of such fundamental importance to the behavior of plasmas that at least a brief description of it is required. This effect occurs because of the disposition of the plasma to maintain local charge neutrality and thus to exclude electric fields.

Suppose for a moment that by some mechanism the internal charge balance is upset so that two adjacent locations have net charges, one positive and the other negative. This type of imbalance can and does arise as a fluctuation in the macroscopic state of the plasma without violating the laws of thermodynamics. It is also possible to induce such a spatial imbalance by applying appropriate electric and magnetic fields. In any event, an internal electric field arises between the two regions that acts on the particles to draw them together. In response to this field the charges overshoot the neutral position and create an imbalance in the opposite direction in much the same way that a displaced pendulum overshoots its position of minimum energy and swings to the other side. Unless some damping mechanism interferes, the charges will oscillate back and forth without end. This oscillation is called an electrostatic one because the motion of the particles is colinear with the electric field and no magnetic field is involved.

A simple mathematical model will serve to derive the principal results. Let us assume that a plasma consisting of equal uniform charge densities n and p exists in free space in the form of a slab of thickness L. Now we allow the uniform density of negative charges (electrons, say) to be displaced with respect to the positive charges (ions, say) by a small distance x in the direction perpendicular to the slab, leaving thin sheets of positive and negative charges on either side of a neutral plasma. The configuration is shown in Fig. 4-13. The model is something like two uniform slabs of charge, positive

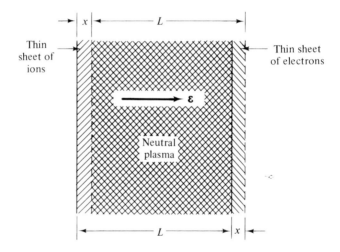

Fig. 4-13 Configuration of the slab model for electrostatic plasma oscillations.

and negative, that can move through one another easily. The force on the electron slab that has been displaced to x is given by the Lorentz force,

$$F = q\mathcal{E} = m\frac{dv}{dt} \qquad (4\text{-}33)$$

where v is the velocity of the electron slab moving as a rigid body and q and m are its total charge and mass. The motion of the electrons can be calculated from this expression of Newton's law of motion using $v = dx/dt$ provided we can determine the electric field.

Sec. 4-6 Plasma Oscillations

If the displacement x is very small compared with L, the electric field acting on most of the electron slab is the same as that of a parallel-plate capacitor charged with surface charge densities $\pm nex$. This field (see any introductory text in electromagnetics) has a value

$$\mathcal{E} = \frac{nex}{\varepsilon_0} \tag{4-34}$$

in the direction indicated in Fig. 4-13. The field of Eq. (4-34) is spatially uniform and its magnitude is directly proportional to the displacement of the electron slab.

The electric field of Eq. (4-34) acts on the electrons in a slab of thickness $L - x$. The total charge q of the electron slab of thickness $L - x$ and cross-sectional area A is approximately $q = -neLA$ since $x \ll L$. The mass of the slab is $m = nm_eLA$, where $-e$ and m_e are the charge and mass of a single electron, respectively. Equation (4-33) then becomes

$$(-neLA)\left(\frac{nex}{\varepsilon_0}\right) = (nm_eLA)\frac{d^2x}{dt^2}$$

which reduces to

$$\frac{d^2x}{dt^2} = -\left(\frac{ne^2}{\varepsilon_0 m_e}\right)x \tag{4-35}$$

Equation (4-35) is an equation of simple harmonic motion, its solution being an oscillatory motion of the electron slab about the ion slab at a frequency called the *electron plasma frequency* ω_{pe}:

$$\boxed{\omega_{pe}^2 = \frac{ne^2}{\varepsilon_0 m_e}} \tag{4-36}$$

If the plasma electrons exist in a medium other than free space such as in a metal or semiconductor, the appropriate dielectric constant should be substituted in Eqs. (4-34) – (4-36).

The model we have used to determine the frequency of electrostatic oscillation of a plasma assumes that the positive charges are stationary and has neglected the random motions of the particles due to their nonzero temperature. Corrections to the model can be made for these effects but the results are not changed significantly, at least in most applications.

The principal consequence of the oscillations at the plasma frequency concerns the motion of the electrons that tend to eliminate internal electric fields. The plasma frequency is in some sense a measure of the inertia of the electrons. When a d.c. electric field is applied to a plasma, the electrons set up a sheath to exclude the field from the interior of the plasma. If a time-varying electric field is applied, the electrons can move to cancel this field also, provided its frequency is less than the plasma frequency. If the applied frequency is higher than the plasma frequency, then the electrons do not have sufficient time to move to cancel the field, and an electromagnetic wave will propagate through the plasma. A plasma, then, is a high-pass filter for electromagnetic waves, passing radiation at frequencies above the plasma frequency and rejecting radiation at frequencies below it. A number of well-known phenomena can be explained on the basis of this characteristic of plasmas including the reflection of short radio waves from the earth's ionospheric plasma and the loss of communications with spacecraft during re-entry into the atmosphere, at which time a high-density plasma is created around the vehicle.

Because electrostatic plasma oscillations can arise by virtue of the random thermal motion of the particles, the scale length of the oscillation is determined by the distance an average particle can travel during a portion of a plasma oscillation. Roughly speaking, this distance should be $\bar{v}/\omega_{pe}$, where for $\bar{v}$ we shall take $(\kappa T/m)^{1/2}$, and for ω_{pe}, Eq. (4-36). This scale length turns out to be

$$\frac{\bar{v}}{\omega_{pe}} = \left[\frac{\varepsilon_0 \kappa T}{ne^2}\right]^{1/2} \quad (4\text{-}37)$$

which is the electron screening length λ_e, as comparison with Eq. (4-24) indicates. Thus, sheaths are formed on the time scale of a plasma oscillation.

BIBLIOGRAPHY

For details on electronic motion in beams, guns, and lenses, see

HARMON, W. W., *Fundamentals of Electronic Motion*. New York: McGraw-Hill Book Company, 1953.

PIERCE, J. R., *Theory and Design of Electron Beams*. New York: Van Nostrand Rinehold Company, 1954.

EL-KAREH, A. B , and J. C. J. EL-KAREH, *Electron Beams, Lenses, and Optics*, Vols. 1 and 2. New York: Academic Press Inc., 1970.

Application of electron beams to microwave devices is discussed in

COLLIN, R. E., *Foundations of Microwave Engineering*. New York: McGraw-Hill Book Company, 1966.

HUTTER, R. G. E., *Beam and Wave Electronics in Microwave Tubes*. New York: Van Nostrand Rinehold Company, 1960.

REICH, H. J., J. G. SKALNIK, P. F. ORDUNG, and H. L. KRAUSS, *Microwave Principles*. New York: Van Nostrand Rinehold Company, 1957.

An excellent, nontechnical introduction to plasma behavior can be found in paperback:

ARZIMOVICH, L. A., *Elementary Plasma Physics*. Waltham, Mass.: Ginn/Blaisdell, trans. from Russian by Scripta Technica, Inc., 1965.

Two introductory textbooks on plasma physics are

UMAN, M. A., *Introduction to Plasma Physics*. New York: McGraw-Hill Book Company, 1964.

GLASSTONE, S., and R. H. Lovberg, eds., *Controlled Thermonuclear Reactions*. New York: Van Nostrand Rinehold Company, 1960.

More advanced treatments may be found in

KUNKEL, W., ed., *Plasma Physics in Theory and Application*. New York: McGraw-Hill Book Company, 1966.

STIX, T. H., *Theory of Plasma Waves*. New York: McGraw-Hill Book Company, 1962.

HUDDLESTONE, R. H., and S. L. LEONARD, eds., *Plasma Diagnostic Techniques*. New York: Academic Press, Inc., 1965.

Plasma effects in solids and their application to devices are the subjects of

STEELE, M. C., and B. VORAL, *Wave Interactions in Solid State Plasmas*. New York: McGraw-Hill Book Company, 1969.

HARTNAGEL, H., *Semiconductor Plasma Instabilities*. New York: American Elsevier Publishing Company, Inc., 1969.

PROBLEMS

4-1. Calculate the speed of an electron that has been accelerated through a potential of 100 and 10,000 V. Compare these velocities with the thermal velocities of electrons at room temperature and at a thermal emission temperature of 2400°K.

4-2. An electron beam is injected halfway between two parallel deflection plates. The beam has been accelerated by a 300 V electron gun and is directed parallel to the plates. The plates are 8.0 cm long in the direction of the beam and spaced 2.0 cm apart. Find the maximum potential that can be applied to the plates to deflect the beam without causing it to strike the positive plate.

4-3. Find the forms of the solutions to the equation of motion for an electron in the geometry of a magnetron, that is, in cylindrical coordinates with an axial magnetic field $\mathbf{B} = B_0 \hat{z}$ and a radial electric field $\mathcal{E} = -\mathcal{E}_0 \hat{r}$. Assume electron emission from the cathode with zero initial velocity. (*Note:* The azimuthal component $\mathcal{E}_\phi$ of the fringing fields of the cavities of the magnetron in Fig. 4-5 causes the net radial drift of the electrons from cathode to anode.)

4-4. Find the electric potential in a space-charge-limited thermionic diode as a function of position between the electrodes. Express your result in terms of the tube voltage V_0 and electrode spacing d.

4-5. A thermionic diode with electrode spacing of 5 mm operates in a space-charge limited regime with a tube voltage of 8 V. Assuming that the electrons are emitted at the cathode with zero velocity, find the electron density at the anode. If the electrode areas are each 4 cm², what is the current through the device?

4-6. Average values for the electron density and temperature in the ionosphere are 10^6 cm^{-3} and $10^{3\circ}$K, respectively. Calculate the electron screening length and the number of electrons in a sphere of radius λ_e.

4-7. Find the electron screening length in copper at room temperature. Take the electron density to be 8.5×10^{22} cm^{-3} and the dielectric constant to be the same as that of free space.

4-8. Find the electron screening length in the semiconducting compound indium antimonide (InSb) at room temperature and for an electron density of 1.35×10^{16} cm^{-3}. The relative dielectric constant of InSb may be taken to be 17.5.

4-9. Redraw the data of Fig. 4-12 on new axes to represent a plot of the electron current drawn to an electrostatic probe as a function of the sheath potential. Label all the features of the sketch.

4-10. Calculate the electron plasma frequency ω_{pe} for the ionospheric conditions of Problem 4-6. For nonzero temperature plasmas, the electrostatic plasma oscillations occur at the frequency

$$\omega^2 = \omega_{pe}^2 + \left(\frac{3\kappa T}{m}\right)\left(\frac{2\pi}{\lambda}\right)^2$$

where λ is the wavelength of the oscillation. What would the oscillation frequency be if the wavelength were that of a free-space electromagnetic wave of frequency ω_{pe}?

4-11. Calculate the electron plasma frequency in hertz for copper assuming that the mass of an electron in copper is the same as that of a free electron, $m_e = 9.107 \times 10^{-31}$ kg. See Problem 4-7.

4-12. Calculate the electron plasma frequency in hertz for the sample of InSb of Problem 4-8. The effective mass of an electron in InSb is about 1.4% of that of a free electron.

4-13. Radio communications with spacecraft is accomplished at frequencies near 3 GHz. On re-entry a spaceship creates a plasma in the atmosphere around it by frictional heating. Blackout of radio communications during this phase of re-entry is a well-known phenomenon. Estimate the electron density required for blackout. When are communications restored?

5

Elements of Quantum Mechanics

A physical theory is the mathematical formulation of a physical picture used to explain natural phenomena. The physical picture is based on fundamental concepts that may be intuitively obvious, such as the principle of the conservation of energy, or that may originally be hypothetical. The value of a useful theory lies in its ability not only to explain the phenomena for which it was developed but also to predict the results of future experiments. A theory is a guide to understanding physical phenomena. It is also a tool that can be used to design new materials and to apply natural phenomena toward the development of technology. Quantum mechanics is just such a theory. Knowledge of its basic elements is the keystone to understanding the electronic properties of solid-state materials and the devices that can be created using these materials. It is, therefore, fundamental to the objectives of this book.

The quantum mechanical picture was developed because of the failure of classical Newtonian mechanics to explain certain natural phenomena. Historically the important observations that could not be explained classically and that lead to the development of quantum theory were the frequency spectrum of black-body radiation, photoemission of electrons from metals, the characteristic optical line spectra of atoms, and the wave-like behavior of particles evidenced under certain circumstances. We shall review the observations of the photoelectric effect and the so-called wave-particle duality in Section 5.1. The optical spectra can be understood from the material that will be studied in Section 5.4. Black-body radiation is beyond the interest of our study because it does not pertain directly to our objectives.

In this chapter we shall develop the concepts of quantum mechanics that are necessary for an understanding of the models of solid-state electronic materials. Our approach will be that of wave mechanics based on the Schrödinger equation. Our objectives will be to develop physical pictures based on the quantum mechanical hypotheses that will explain, for instance, quantization, tunneling, and the behavior of particles in the wave description. Finally, we shall consider the effects of the quantum idea on statistical mechanical models.

A word of caution before we proceed. Quantum mechanics is only a theory, as is classical mechanics. It is useful in describing certain phenomena but not others. It has its range of applicability as well as its limitations. Naturally, the same is true of Newtonian mechanics. The beauty of the quantum description is that the classical picture can be obtained from it under specific circumstances. Strictly speaking the behavior of electrons should always rightly be described by quantum mechanics. But for practical purposes in special cases the classical description provides an accurate explanation of observed phenomena. Such was the case with the material of Chapter 4 and such will also be the case with some of the material to follow in later chapters. To be able to decide when or when not to use the classical approach, we must have a competent knowledge of the quantum mechanical one.

5.1 WAVE-PARTICLE DUALITY

In Newtonian mechanics the behavior of a particle is described by the position **r** and momentum **p** of its center of mass as well as its angular momentum about the center of mass. When the particle is small it is sometimes convenient to consider it to be a point-mass object. Electrons in the classical picture are considered as point masses. The velocity of a classical particle is proportional to its momentum, and its kinetic energy is proportional to the square of its momentum.

On the other hand, the classical picture of radiation such as electromagnetic radiation is that of a wave. A pure sinusoidal wave is described by its amplitude A_0, frequency ω, and wave number k according to the convention

$$A(z, t) = \text{Re}\{A_0 \epsilon^{\pm j(kz \pm \omega t)}\} \tag{5-1}$$

where Re{ } is an instruction to take the real part of the complex number inside the braces. The wavelength of the wave is the minimum distance between locations of equal phase and is given by $\lambda = 2\pi/k$. The frequency may by expressed as ω (radians/sec) or as ν (Hz), where $\omega = 2\pi\nu$. Equation (5-1)

Sec. 5-1 Wave-Particle Duality 121

represents a uniform infinite plane wave propagating in the z-direction. A more realistic wave would be nonuniform and finite in extent. Such a finite wave can be thought of as being composed of a number of uniform plane waves in the form of a wave packet. Classically the energy of a wave is distributed throughout its extent, the energy density being proportional to the square of the amplitude A_0. The energy of a wave packet propagates at the so-called *group velocity*,

$$v_{\text{group}} = \frac{\partial \omega}{\partial k} \tag{5-2}$$

where the functional relationship between ω and k for a given medium is called the dispersion relation. Equation (5-2) for quantum mechanical waves is derived in Section 5.4. Because the energy in a wave is associated with its amplitude only, waves of appropriate phases can interfere with one another either constructively or destructively, causing interference patterns.

In the classical physics that had been developed by the end of the nineteenth century electromagnetic radiation was described as a wave phenomenon and electrons were described as point-mass particles. But the classical picture was unable to explain the results of certain experiments, including the spectral frequency of black-body radiation, the optical line spectra of atoms, and photoemission. Some success had been obtained in describing black-body radiation and the atomic spectra by making the assumption that particles could change their energy only in discrete amounts called quanta. It was Albert Einstein in 1905 who postulated that radiation could also deliver its energy only in discrete quanta. This assumption was sufficient to explain the phenomenon of photoemission of electrons from metals.

Einstein's Theory of Photoemission

When monochromatic light of frequency ω and intensity I shines upon a metallic electrode in an evacuated chamber, electrons may be emitted that can then be collected by another electrode. The effect is known as *photoemission*, and the emitting electrode is called the *photocathode*. A schematic representation of an experiment that demonstrates the photoemission effect is sketched in Fig. 5-1(a). The emitted electrons are called *photoelectrons*. The photoelectrons leave the cathode with a certain initial kinetic energy $mv_i^2/2$ that can be measured by applying a bias to the anode. It has been found that current flow to the anode ceases at a particular repulsive potential called the cutoff voltage V_0. The electrons do not have sufficient kinetic energy to overcome the repulsive force. Therefore their initial kinetic energy must be $mv_i^2/2 = eV_0$.

A number of additional observations have been made about this experiment. First, the emission of a photoelectron occurs almost instantaneously

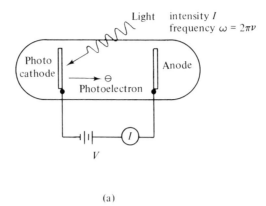

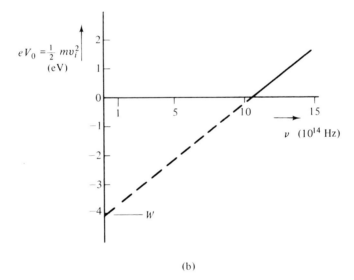

Fig. 5-1 Photoemission experiment: (a) schematic of the apparatus, (b) experimental data.

even in very weak light. Second, the initial kinetic energy of the electrons as measured by the cutoff voltage is independent of the intensity of the radiation but does depend on its frequency $\omega = 2\pi\nu$. A plot of this dependence is shown in Fig. 5-1(b). Third, the magnitude of the current collected at the anode does depend on the light intensity. Fourth, the low-frequency intercept of the experimental data with the energy axis of the plot of Fig. 5-1(b) depends on the photocathode material.

Einstein's hypothesis that radiant energy is delivered in discrete amounts explains all of the details of this phenomenon. The minimum amount of energy is called a *quantum* or a *photon*. Einstein postulated that the energy of a photon is $E = h\nu$. The constant of proportionality h is called Planck's constant. Its value has subsequently been determined as $h = 6.625 \times 10^{-34}$ J-sec. If one quantum of energy is delivered to an electron and it subsequently loses W electron volts of energy while leaving the metal, the maximum initial kinetic energy that the electron can have will be

$$\left(\frac{1}{2}mv_i^2\right)_{\max} = h\nu - W \quad (5\text{-}3)$$

The photon is assumed to give up either all or none of its energy when it interacts with an electron. Equation (5-3) is just the straight line plotted in Fig. 5-1(b). The number of photons but not their energy is proportional to the intensity. Therefore the number of photoelectrons is proportional to the intensity, but their energy is not. Emission of one photoelectron can occur if and when one photon falls on the cathode. How the cathode material determines W, the work function, will be described in Chapter 7.

The consequence of the Einstein theory, for which he won the Nobel Prize in 1921, is that electromagnetic waves have a particle-like behavior when interacting with matter, their energy being delivered in discrete quanta. The wave can be thought of as being composed of "particles" called photons even though all of the laws determining its propagation, interference phenomena, and so forth are characteristic of wave motion.

Wave-Particle Duality

In 1924 Louis de Broglie postulated that if photons are associated with wave motion, then wave motion ought to be similarly associated with the behavior of what we ordinarily think of as particles. In other words he assumed a duality to exist, that phenomena exhibited by light, for instance, might be explained in one instance on the basis of a wave theory and in another on the basis of a corpuscular theory. The same should also be true for the behavior of matter.

We shall use the following reasoning to argue for the mathematical relationships required between waves and particles. A particle is described by its energy E and momentum p, which are constants of its motion in the absence of forces. A wave is described by its frequency ω and wavelength λ or wave number k. If a wave packet is to be descriptive of particle motion, the rate at which energy is transmitted by the wave should be the same as the equivalent particle velocity.

The group velocity of a wave was stated in Eq. (5-2):

$$v_{\text{group}} = \frac{\partial \omega}{\partial k} \quad (5\text{-}2)$$

From the explanation of the photoemission effect we can assume with Einstein (being in good company) that the photon energy is

$$E = h\nu = \hbar\omega \qquad (5\text{-}4)$$

where $\hbar$ is shorthand notation for $h/2\pi$. Thus

$$v_{\text{group}} = \frac{\partial E}{\hbar \, \partial k} \qquad (5\text{-}5)$$

On the other hand, a classical nonrelativistic particle moves with velocity $v = p/m$ and has energy $E = p^2/2m$. It is easy to show that the particle velocity is

$$v_{\text{particle}} = \frac{\partial E}{\partial p} \qquad (5\text{-}6)$$

Comparing Eq. (5-6) with Eq. (5-5), we can relate the particle momentum to the wave number or wavelength via

$$p = \hbar k = \frac{h}{\lambda} \qquad (5\text{-}7)$$

Equations (5-4) and (5-7) constitute de Broglie's hypothesis of wave-particle duality.

The wavelengths of laboratory-sized particles, as given by Eq. (5-7), are not very great. However, for electrons accelerated through a potential of V volts the wavelengths are

$$\lambda_{\text{electron}} = \frac{h}{mv} = \frac{h}{(2meV)^{1/2}} = \frac{12.13 \times 10^{-8}}{V^{1/2}} \text{ cm} \qquad (5\text{-}8)$$

For voltages on the order of 100 V, the electron wavelength is comparable with interatomic distances in solids.

Bragg Reflection

When electromagnetic waves propagate through a periodic structure, reinforced reflections of the waves occur in certain directions. A defraction grating is such a structure and so is a solid crystal. A crystal can be considered as a series of effective atomic planes each of which partially reflects incident electromagnetic radiation. If the spacing of the planes is just right, the reflected rays will interfere constructively, creating a condition of strong reflection. The condition for reflection can be illustrated with the help of Fig. 5-2. Constructive interference occurs when rays 1 and 2 are in phase after reflection. An equivalent condition is that the difference in the path lengths for the rays between wave fronts prior to and after reflection be an integral number of wavelengths. From Fig. 5-2 it is evident that the path lengths differ

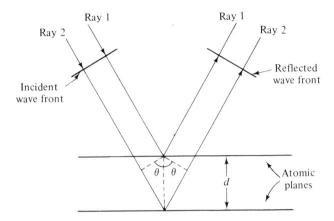

Fig. 5-2 Illustration of Bragg reflection of waves from parallel planes.

by $2d \sin \theta$, where d is the separation of the reflecting planes and θ is the angle of incidence. The condition for constructive interference upon reflection is therefore

$$2d \sin \theta = n\lambda \tag{5-9}$$

The phenomenon is called Bragg reflection, and Eq. (5-9) is known as Bragg's law.

Bragg reflection occurs in crystals but does not depend on the crystal composition. It depends only on the symmetry of the effective reflecting planes and the wavelength and angle of incidence of the radiation. According to Eq. (5-9), for Bragg reflection to occur the wavelength must be less than twice the separation of the planes.

The Davisson-Germer Experiment

If, as the de Broglie hypothesis indicates, matter waves accompany the motion of particles, then it should be possible to observe wave-like phenomena such as constructive and destructive interference of these matter waves. In fact for matter waves of appropriate wavelengths it should be possible to observe a reflection phenomenon in crystals. Such an experiment was performed by C. Davisson and L. H. Germer in 1927. In the experiment an electron beam was directed at a nickel crystal. The reflected beam formed a diffraction pattern typical of wave motion. From the observed diffraction pattern it was possible to calculate the wavelength of the matter waves associated with the electron beam. This wavelength was found to be in agreement with the de Broglie hypothesis, thus confirming the notion of wave-particle duality.

5.2 SCHRÖDINGER'S EQUATION

The quantum mechanical description of physical phenomena is based on matter waves. To develop this picture we shall need a mathematical formulation of the behavior of these waves analogous to Newton's laws, which describe the behavior of classical particles. The appropriate relation that governs matter wave phenomena is called *Schrödinger's equation* or the wave equation. In this section we shall derive the equation, present a physical interpretation of the wave functions that satisfy it, and consider the general forms of possible solutions.

The Wave Equation

Waves can be represented mathematically either by sinusoidal or exponential functions of the forms indicated in Eq. (5-1). The form appropriate for matter waves is determined by the equation that governs their behavior. To derive Schrödinger's equation we shall employ de Broglie's hypotheses, Eqs. (5-4) and (5-7), and the assumption that the wave function has either an exponential or a sinusoidal form.

A uniform plane wave propagating in the positive z-direction has either of the forms

$$\sin(kz - \omega t), \quad \cos(kz - \omega t)$$
$$\epsilon^{+j(kz-\omega t)}, \quad \epsilon^{-j(kz-\omega t)} \qquad (5\text{-}10)$$

where the exponentials are complex numbers in general. The wave is characterized by ω and k. The clue to our derivation of the wave equation is the recognition that for all of the forms of Eq. (5-10) the frequency ω can be obtained by performing a differentiation of the function with respect to time and dividing the result by the wave function. The propagation constant k can be obtained in a similar manner but in this case with a spatial derivative. The frequency is also determined by the de Broglie relation $E = \hbar\omega$, where E is the energy of a free particle, that is, a particle not in a force field. The relation $p = \hbar k$ also determines k in terms of the particle momentum. The energy of a free particle is

$$\frac{p^2}{2m} = E \qquad (5\text{-}11)$$

Let us substitute Eqs. (5-4) and (5-7) into Eq. (5-11) to obtain a relationship between ω and k for a matter wave. The result is

$$\frac{\hbar k^2}{2m} = \omega \qquad (5\text{-}12)$$

By the argument at the beginning of this paragraph, the right-hand side of

Sec. 5-2 Schrödinger's Equation

Eq. (5-12) is proportional to the time derivative of the matter wave function, while the left-hand side is proportional to its second spatial derivative. Therefore the equation for the wave function must be of the form

$$\nabla^2 \Psi(\mathbf{r}, t) = \gamma \frac{\partial \Psi(\mathbf{r}, t)}{\partial t} \quad (5\text{-}13)$$

in three dimensions, where $\Psi(\mathbf{r}, t)$ is the wave function, $\nabla^2 \Psi$ is its second spatial derivative in three dimensions, and γ is a constant of proportionality.

The sinusoidal functions of Eq. (5-10) do not satisfy Eq. (5-13), but either of the exponential functions will. We shall adopt as a convention the solution of the form

$$\Psi(\mathbf{r}, t) = \Psi_0 \epsilon^{+j(kz-\omega t)} \quad (5\text{-}14)$$

for a wave function of amplitude Ψ_0 propagating in the positive z-direction if k is positive. To obtain the proportionality constant γ we need only perform the operations indicated in Eq. (5-13) on the function of Eq. (5-14). The result is

$$\gamma = \frac{k^2}{j\omega} \quad (5\text{-}15)$$

By relating both k^2 and ω to the energy E using the de Broglie relations and Eq. (5-11) we can obtain an alternative expression for γ from Eq. (5-15),

$$\gamma = \frac{2m}{j\hbar} \quad (5\text{-}16)$$

depending only on the mass of the particle and the constant $\hbar = h/2\pi$. Equation (5-13) can now be written in its common form,

$$\frac{\hbar^2}{2m} \nabla^2 \Psi = -j\hbar \frac{\partial \Psi}{\partial t} \quad (5\text{-}17)$$

which is Schrödinger's equation in the absence of forces.

It is an easy matter to include the effects of forces that can be derived from potential functions in the wave equation. By substitution of the form of the wave function, Eq. (5-14), into the left-hand side of Eq. (5-17) the latter becomes

$$\frac{\hbar^2}{2m} \nabla^2 \Psi = -\frac{\hbar^2 k^2}{2m} \Psi = \frac{-p^2}{2m} \Psi \quad (5\text{-}18)$$

By the same substitution, the right-hand side becomes

$$-j\hbar \frac{\partial \Psi}{\partial t} = -\hbar\omega \Psi = -E\Psi \quad (5\text{-}19)$$

so that Schrödinger's equation in the absence of forces is a statement of Eq. (5-11):

$$\frac{p^2}{2m} \Psi = E\Psi \quad (5\text{-}11)$$

Of course, this is where we started in our derivation. For a particle with potential energy $U(\mathbf{r})$ and total energy E, the *kinetic energy* is proportional to p^2 so that

$$\frac{p^2}{2m} = E - U(\mathbf{r}) \qquad (5\text{-}20)$$

By multiplying each term in Eq. (5-20) by $-\Psi$ and using Eqs. (5-18) and (5-19), we find

$$\frac{\hbar^2}{2m} \nabla^2 \Psi = -j\hbar \frac{\partial \Psi}{\partial t} + U\Psi$$

or the more usual form

$$\boxed{\frac{\hbar^2}{2m} \nabla^2 \Psi - U\Psi = -j\hbar \frac{\partial \Psi}{\partial t}} \qquad (5\text{-}21)$$

Equation (5-21) is Schrödinger's equation governing the wave of a particle in a potential field.

Separation of Variables in the Wave Equation

Before proceeding to an interpretation of the wave function $\Psi(\mathbf{r}, t)$ we shall begin the process of obtaining solutions to the wave equation, Eq. (5-21). Schrödinger's equation is a partial differential equation that can be separated into two differential equations using the technique of separation of variables. To do so, we assume that $\Psi(\mathbf{r}, t)$ can be written as a product of two functions

$$\Psi(\mathbf{r}, t) = \psi(\mathbf{r})\phi(t) \qquad (5\text{-}22)$$

The function $\psi(\mathbf{r})$ depends only on $\mathbf{r}$, while $\phi(t)$ contains only the time dependence of $\Psi(\mathbf{r}, t)$. The procedure is to substitute Eq. (5-22) into the wave equation and then divide each term by $\Psi(\mathbf{r}, t)$. For example, the term on the right-hand side of Eq. (5-21) would become

$$\frac{-j\hbar}{\Psi} \frac{\partial \Psi}{\partial t} = \frac{-j\hbar \psi(\partial \phi/\partial t)}{\psi\phi} = \frac{-j\hbar}{\phi(t)} \frac{d\phi(t)}{dt}$$

since $\psi(\mathbf{r})$ depends only on $\mathbf{r}$. All of Eq. (5-21) becomes

$$\frac{\hbar^2}{2m} \frac{1}{\psi(\mathbf{r})} \nabla^2 \psi(\mathbf{r}) - U(\mathbf{r}) = -j\hbar \frac{1}{\phi(t)} \frac{d\phi(t)}{dt} \qquad (5\text{-}23)$$

The term on the right-hand side of Eq. (5-23) depends only on time and therefore at a given instant has the same value at every point in space. The terms on the left depend only on the spatial coordinates and therefore at a given position have the same value for all time. The only way an equation like this one can have a solution is if each side is equal to a constant, independent of both space and time. That constant we shall call the separation constant, denoted

Sec. 5-2 Schrödinger's Equation

by $-E$. That is,

$$\frac{d\phi(t)}{dt} = -j\frac{E}{\hbar}\phi(t) \qquad (5\text{-}24)$$

and

$$\frac{\hbar^2}{2m}\nabla^2\psi(\mathbf{r}) - U\psi(\mathbf{r}) = -E\psi(\mathbf{r}) \qquad (5\text{-}25)$$

Equations (5-24) and (5-25) are the two separate differential equations whose solutions $\phi(t)$ and $\psi(\mathbf{r})$ can be combined to form $\Psi(\mathbf{r}, t)$.

Equation (5-24) is a familiar form that has the solution

$$\phi(t) = \phi_0 \epsilon^{-jEt/\hbar} \qquad (5\text{-}26)$$

This result is central to the physical interpretation of the magnitude of the separation constant E. According to Eqs. (5-26) and (5-22), $\Psi(\mathbf{r}, t)$ has temporal dependence of the form

$$\epsilon^{-jEt/\hbar}$$

whereas Eq. (5-14) indicates that its time dependence is

$$\epsilon^{-j\omega t}$$

Obviously $E = \hbar\omega$ is also the total energy of the particle, as in the de Broglie hypothesis.

Equation (5-25) is known as Schrödinger's equation for the stationary state, where $\psi(\mathbf{r})$ is the *stationary state wave function*. The solution for ψ depends on the form of the potential function $U(\mathbf{r})$. In Section 5.3 we shall consider several examples of specific potentials. Before doing so, however, we shall consider more general problems such as the physical interpretation of $\Psi(\mathbf{r}, t)$ and the general conditions of solution for $\psi(\mathbf{r})$.

Physical Interpretation of $\Psi(\mathbf{r}, t)$

The function $\Psi(\mathbf{r}, t)$ is the matter wave of a real particle. Because the particle exists, in the sense that its behavior can be measured, we need a real mathematical function to predict that behavior. The wave function Ψ is generally a complex number, but a real number can be obtained from the product of Ψ with its complex conjugate Ψ^*. Not only is $\Psi^*\Psi$ real, but it is also independent of time.

A sketch of possible values of $\Psi^*\Psi$ as a function of position is shown in Fig. 5-3. From the sketch one would suppose that the particle with which the wave Ψ is associated would be located somewhere between z_1 and z_3, but most likely near z_2. In fact, $\Psi^*\Psi$ is generally interpreted as a *probability density* where the probability of observing the particle in a range dz in the

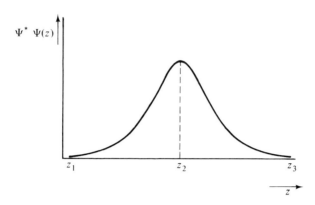

Fig. 5-3 Probability density $\Psi^*\Psi(z)$.

vicinity of z_0 is $\Psi^*(z_0)\Psi(z_0)\,dz$. For three-dimensional problems the probability of finding the particle in the volume $d\mathbf{r} = dx\,dy\,dz$ in the vicinity of $\mathbf{r}_0$ is $\Psi^*(\mathbf{r}_0)\Psi(\mathbf{r}_0)\,d\mathbf{r}$. If Ψ represents the wave function of a single particle, the integral of $\Psi^*\Psi$ over all space must be

$$\int_{\text{all space}} \Psi^*\Psi\,d\mathbf{r} = 1 \qquad (5\text{-}27)$$

since the particle, being real, must be found someplace. Equation (5-27) is the *normalization condition*. It is a restriction on the magnitude of $\Psi(\mathbf{r}, t)$ given its functional form.

The need for a normalizing condition can be seen mathematically from the wave equation itself, Eq. (5-21). Schrödinger's equation is homogeneous in Ψ so that in principle there are an infinite number of functions that satisfy it, differing only by constants. Such a situation is not physical so that some type of restriction on Ψ is necessary. The normalization condition is such a restriction that lends physical meaning to $\Psi^*\Psi$ as a probability density.

Several additional restrictions apply to the solution Ψ of the wave equation. First, so that $\Psi^*\Psi$ will be a probability density that can be used to predict the position of a particle, Ψ must be finite and single-valued. Second, it can also be proved mathematically that, as long as U remains finite, both Ψ and its spatial derivative are continuous functions. The latter restrictions are often termed boundary conditions because they are usually applied at a position where there is a discontinuity in the potential U.

5.3 FREE STATES AND POTENTIAL BARRIERS

The distinguishing feature that characterizes the solution for the wave function $\psi(\mathbf{r})$ of a stationary state is the difference between the total energy of the

Sec. 5-3 Free States and Potential Barriers

particle and its potential energy, $E - U(\mathbf{r})$. Equation (5-25) governs $\psi(\mathbf{r})$ and can be rewritten for a one-dimensional problem as

$$\frac{1}{\psi}\frac{d^2\psi}{dz^2} = \frac{-2m}{\hbar^2}(E - U) \qquad (5\text{-}28)$$

where U may be a function of the coordinate z. In classical physics, the total energy E must always be greater than the potential energy, the difference being the kinetic energy of the particle. Classical kinetic energy is always positive. However, Eq. (5-28) has solutions for E less than U as well as for E greater than U. In this section we shall explore the consequences of this mathematical fact for the physical interpretation of waves in terms of particle behavior. Our procedure will be first to study the general forms of the solutions to Eq. (5-28) and then to apply these to some problems with simple potentials.

Characterization of ψ According to the Sign of $E - U$

According to Eq. (5-28) the sign of $E - U$ determines the sign of the ratio of the spatial second derivative of ψ to ψ itself. The sign of this ratio is a very useful tool for characterizing the mathematical behavior of the wave function. Frequently it is possible to sketch solutions for ψ graphically using this tool along with the continuity of ψ and its derivative. Then, knowing the form of the solution, one need only apply the normalization condition to obtain the complete wave function.

The key to understanding how the sign of $E - U$ characterizes ψ is the second derivative $d^2\psi/dz^2$. The second derivative is called the curvature of ψ. When the curvature is positive, the plot of ψ vs. z is always concave upward. A bowl positioned so that it holds water is concave upward. For negative curvature, ψ is concave downward. A bowl that is concave downward will not hold anything.

When $E - U > 0$, the right-hand side of Eq. (5-28) is negative. Then if ψ is positive, its curvature is negative and ψ is concave downward. If ψ is negative, its curvature must be positive so that ψ is concave upward. These two possibilities, sketched in Fig. 5-4, indicate that the plot of ψ is always concave toward the abscissa for $E - U > 0$. A curve having this characteristic oscillates between positive and negative values. One example of such a curve is the sine function. Where $E > U$, the stationary-state wave function always exhibits an oscillatory behavior.

When $E - U < 0$, the right-hand side of Eq. (5-28) is positive. This condition is not permitted to exist in the classical view, but the quantum mechanical wave function may exist. If the right-hand side of Eq. (5-28) is positive and ψ is positive, then the curvature is also positive, in which case the plot of ψ is concave upward. When ψ is negative the curvature must also be negative so that ψ is concave downward. Figure 5-5 indicates these forms

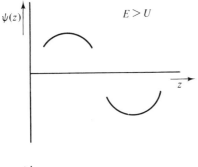

Fig. 5-4 Demonstration of the curvature of $\psi(z)$ when $E - U > 0$.

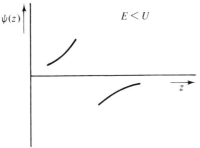

Fig. 5-5 Demonstration of the curvature of $\psi(z)$ when $E - U < 0$.

and shows that ψ is always concave away from the abscissa for $E < U$. Therefore ψ can never change sign under these conditions. If it is positive somewhere it is positive everywhere provided U is greater than E. The curve of exponential decay has this characteristic behavior.

Having classified the solutions for ψ as being oscillatory or nonoscillatory according to whether $E - U$ is positive or negative, we shall turn now to several examples of potentials that will point out salient features of the process of the mathematical solution for ψ. These examples will also permit us to expand upon our physical interpretation of the wave function.

Wave Functions for a Particle in a Uniform Potential

Suppose that a particle is moving in the field of a uniform potential $U = U_0$, as sketched in Fig. 5-6. The equation for the stationary state, Eq. (5-25),

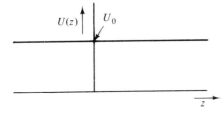

Fig. 5-6 Uniform potential.

Sec. 5-3 Free States and Potential Barriers

is then

$$\frac{d^2\psi}{dz^2} = \frac{-2m}{\hbar^2}(E - U_0)\psi$$

which can be rewritten as

$$\frac{d^2\psi}{dz^2} + k^2\psi = 0 \qquad (5\text{-}29)$$

where

$$k^2 = \frac{2m}{\hbar^2}(E - U_0) \qquad (5\text{-}30)$$

Equation (5-29) is a second-order homogeneous equation that has two linearly independent solutions that depend on the parameter k. This parameter, in turn, depends on $E - U_0$.

First we shall consider the case $E - U_0 > 0$ so that k is a real number. The solution to Eq. (5-29) is

$$\psi(z) = Ae^{jkz} + Be^{-jkz} \qquad (5\text{-}31)$$

where A and B are equivalent to the two constants of integration. The complete solution for $\Psi(z, t)$ is obtained by multiplying $\psi(z)$ by $\phi(t)$ obtained in the previous section:

$$\Psi(z, t) = Ae^{j(kz-\omega t)} + Be^{-j(kz+\omega t)} \qquad (5\text{-}32)$$

The constants of integration have all been absorbed into the amplitude factors A and B. Equation (5-32) represents the superposition of two uniform infinite traveling waves, one moving toward increasing z and the other toward decreasing z. The result, of course, is an oscillatory wave that extends throughout all space. The wave function (5-32) cannot be normalized, and therefore it is not a realistic wave function. However, the potential was not a very realistic one either, being constant throughout all space. Therefore we should not be overly concerned about our inability to interpret $\Psi^*\Psi$ as a probability density for this particular case. Notice that mathematical solutions exist, however, for any value of the particle energy E provided it is greater than U_0.

For the special condition when $U_0 = 0$, the propagation constant k is related to the particle energy by

$$E = \frac{\hbar^2 k^2}{2m}, \qquad U_0 = 0 \qquad (5\text{-}33)$$

and Eq. (5-32) still applies. A particle moving in a uniform zero potential field is said to be a *free particle*. Equation (5-33) is the free-particle relationship between energy and wave number. A free particle may have any value of energy provided $E > 0$, as might have been expected.

Next, let us consider the case $E - U_0 < 0$. According to Eq. (5-30), k must be an imaginary number if $k^2 < 0$. If we write $k = j\alpha$, the solutions to

Eq. (5-29) are

$$\psi(z) = C\epsilon^{-\alpha z} + D\epsilon^{\alpha z} \qquad (5\text{-}34)$$

Both terms in Eq. (5-34) are concave away from the z-axis. For this particular case we can evaluate the coefficients C and D immediately by recalling that ψ must be finite everywhere. This requirement dictates that both C and D must be zero; therefore $\psi = 0$ everywhere. There is no solution for a particle with energy less than U_0 in a uniform potential field extending throughout all space.

Wave Functions for a Particle Near a Finite Potential Barrier

The next problem we shall consider is that of a finite potential barrier such as that represented in Fig. 5-7. For simplicity we let $U = 0$ for $z < 0$

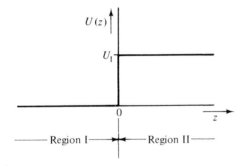

Fig. 5-7 Finite potential barrier.

and $U = U_1$ for $z > 0$. It is not necessary that the potential be strictly discontinuous at $z = 0$. However, the mathematical picture is simplified if it is. Formally a particle near such a barrier might have its total energy E in one of three ranges. These are $E < 0$, $0 < E < U_1$, and $E > U_1$. There is, however, no nonzero solution for ψ for $E < 0$, just as in the example of the uniform potential. The other energy ranges do have solutions, which we shall now examine.

In either case, we can write Schrödinger's equations in each of the regions of space where the potential is uniform. That is,

$$\frac{d^2\psi_\mathrm{I}}{dz^2} + k_\mathrm{I}^2 \psi_\mathrm{I} = 0 \qquad (5\text{-}35)$$

and

$$\frac{d^2\psi_\mathrm{II}}{dz^2} + k_\mathrm{II}^2 \psi_\mathrm{II} = 0 \qquad (5\text{-}36)$$

where

$$k_\mathrm{I}^2 = \frac{2m}{\hbar^2} E \qquad (5\text{-}37)$$

Sec. 5-3 Free States and Potential Barriers   135

and

$$k_{II}^2 = \frac{2m}{\hbar^2}(E - U_1) \tag{5-38}$$

When $E - U_1 > 0$, both k_1 and k_2 are real so that the solutions to Eqs. (5-35) and (5-36) are oscillatory in nature and of the form of Eq. (5-31). Solutions for the total wave function are obtained in this case by ensuring that ψ and its spatial derivative are everywhere continuous. At $z = 0$ we must have $\psi_I = \psi_{II}$ and similarly for the derivatives. The results indicate that solutions exist for any energy provided it is greater than U_1, although again the solutions do not satisfy the normalization condition. A wave function for a particle with $E > U_1$ in the potential of Fig. 5-7 is shown in Fig. 5-8.

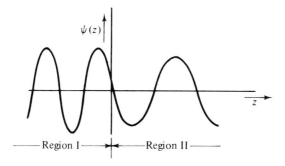

Fig. 5-8 Wave function for a particle in the potential of Fig. 5-7 with $E > U_1$.

Notice that since $k_{II} < k_I$, the wavelength in region II is greater than that in region I. Furthermore, the momentum and kinetic energy of the particle are greater in the region $z < 0$ than for $z > 0$.

When E_1 is less than U_1 and greater than zero, k_I is real but k_{II} is imaginary. Therefore the solution in region I is oscillatory, like Eq. (5-31), and that in region II is exponential, like Eq. (5-34). Again ψ and its derivative must be continuous at $z = 0$ and ψ must remain finite as $z \to \infty$. It turns out that a solution can be obtained for any energy in the range, although these solutions cannot be normalized either. A wave function for a particle with energy greater than zero and less than U_1 is shown in Fig. 5-9. Notice that a nonzero wave function exists in region II. The wave function decays exponentially in this region with a coefficient of

$$\alpha = \left[\frac{2m}{\hbar^2}(U_1 - E)\right]^{1/2}, \quad U_1 > E \tag{5-39}$$

which depends on the particle energy. According to the interpretation of $\Psi^*\Psi$, we should expect that there is a nonzero probability of finding such a particle in region II, "under" the potential barrier. This result is definitely a

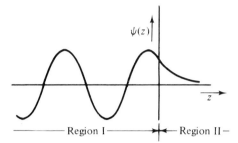

Fig. 5-9 Wave function for a particle in the potential of Fig. 5-7 with $E < U_1$.

nonclassical one. The phenomenon is called *tunneling*. Tunneling means that a quantum mechanical particle might be found where a classical one could not. A number of physical phenomena can be understood using a tunneling picture, thus confirming quantum theory. Of particular importance is the tunneling effect at a thin potential barrier.

Wave Functions for a Particle Near a Thin Potential Barrier

Figure 5-10 illustrates a thin potential barrier of height U_2 and width b. The space can be divided into three regions in which Schrödinger's equation

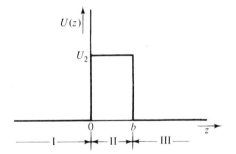

Fig. 5-10 Thin potential barrier.

is of the form of Eq. (5-29). For the configuration of Fig. 5-10, $k_I = k_{III}$ and the solutions in regions I and III are oscillatory. However, if $E < U_2$, k_{II} is imaginary, and the solution in region II is exponential. For this case matching the boundary conditions at $z = 0$ and $z = b$ is not easy to do, but it is not impossible either, especially by using numerical techniques.

One possible wave function is sketched in Fig. 5-11. This particular wave function was obtained assuming that the particle most likely was in region I, but notice that there is a real probability that the particle will be found in region III. In classical physics, a particle with this energy that was originally in region I could never overcome the barrier to get to region III. In quantum physics a particle can be found in region III, although the theory says nothing about how it got there. We could suppose, perhaps, that the particle tunneled

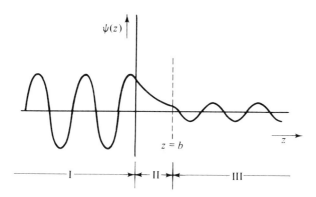

Fig. 5-11 Wave function for a particle in the potential of Fig. 5-10 with $0 < E < U_2$.

under the barrier, but we have no real way of knowing. We do know, however, that this model explains the operation of the tunnel diode and in part the phenomenon of high-field emission of electrons from metals.

One more comment about tunneling. Because the probability of finding a particle is given by $\Psi^*\Psi$, the probability of finding the particle represented by the wave function of Fig. 5-11 is much less in region III than in region I. The relative probability can be thought of as a probability for tunneling. The relative amplitude of ψ in region III is determined by the attenuation coefficient α and the barrier width b. The decay coefficient, according to Eq. (5-39), is determined by the height of the barrier above the energy of the particle. The tunneling probability is therefore determined by the "volume" of the barrier, meaning a function of its height and width.

The potential functions studied in this section all yield wave functions that represent free particles in one form or another. Free particles are distinguished by the fact that they may have any energy provided it is greater than their minimum potential energy. Furthermore, the wave function of a free particle for a given E cannot be normalized because it extends throughout all space. This feature does not, however, represent a difficulty because an actual particle is more realistically represented by a linear combination of wave functions called a wave packet. Wave packets, the subjects of Section 5.4, are well localized in space and can therefore be normalized.

5.4 WAVE PACKETS

Because the wave functions of the free particles discussed in the previous section cannot be normalized, they do not by themselves represent real particles. One reason for this is that the wave functions vary periodically and extend to infinity, implying a finite probability of finding the particle almost anywhere. Experience dictates that this is unrealistic.

More realistic wave functions can, however, be constructed from these infinite waves by the mathematical process of superposition. Schrödinger's equation is linear in Ψ so that any linear combination or superposition of different wave functions will satisfy the equation provided each of the constituent wave functions do so. Such a combination may form a *wave packet*. The mathematical analysis of wave packets is based on the principles of Fourier analysis of nonperiodic functions. The wave-packet theory also predicts the motion of the particle, which cannot be determined from the analysis of Section 5.3. Another important consequence of the representation of a particle by a wave packet is a statement of fundamental limitations in making precise measurements, the uncertainty principle.

Fourier Superposition of Infinite Plane Waves

To obtain a matter wave that represents a free particle localized to a relatively small region of space we must add together a group of infinite free-particle waves. Each wave of the group must have an amplitude and phase so that they will add constructively only over the region where the particle is likely to be found. Outside this region the destructive interference of the plane waves yields a net wave-packet amplitude that approaches zero as one moves away from the central location of the packet. Plane waves of every value of k may be required to construct a wave packet representative of a single particle.

Suppose that a particle is represented by the net wave function $\Psi_{\text{packet}}(z, t)$. Figure 5-12 shows the magnitude of such a wave function as well as its real part plotted against position at an instant of time. The wave function

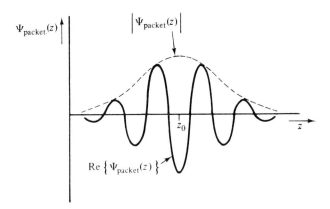

Fig. 5-12 Wave function of a localized particle.

Sec. 5-4 Wave Packets

itself can be represented *mathematically* by an integral over infinite free-particle waves of different wave numbers and amplitudes. Adopting the Fourier integral representation, we have

$$\Psi_{\text{packet}}(z, t) = \frac{1}{\sqrt{2\pi}} \int_{-\infty}^{\infty} A(k)\epsilon^{j(kz-\omega t)} \, dk \tag{5-40}$$

where $A(k)$ is the amplitude spectrum of the constituent waves of the packet. The amplitude spectrum can be evaluated using the inverse Fourier relationship

$$A(k) = \frac{1}{\sqrt{2\pi}} \int_{-\infty}^{\infty} \Psi_{\text{packet}}(z, t = 0)\epsilon^{-jkz} \, dz \tag{5-41}$$

We have arbitrarily chosen to use the value of Ψ at $t = 0$ to evaluate $A(k)$ at that time, but its value at any other time would also suffice as long as A does not change in time. Equations (5-40) and (5-41) are called a *Fourier transform pair*. The square roots of 2π are normalizing factors.

Equation (5-41) implies that if we know the form of a wave packet at some instant of time we can determine the spectrum of waves that comprise it. The spectrum for the wave packet of Fig. 5-12 is sketched in Fig. 5-13.

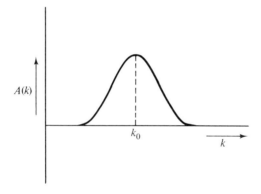

Fig. 5-13 Amplitude spectrum for the wave function of Fig. 5-12.

Equation (5-40) is the prescription for composing the wave packet at any other time having once determined $A(k)$. Although the division of a finite pulse into infinite free-particle waves is purely a mathematical model, it has been successful in explaining so many diverse phenomena that it has become accepted as a physical model as well.

There are two very important consequences of the wave-packet analysis. One results from the physical interpretation of $\Psi^*\Psi$ and the other from the mathematical formalism of the transform pair. We shall next consider the physical interpretation that is stated in the uncertainty principle. Then we shall examine the formulation of the wave packet again to investigate the motion of the wave packet.

The Uncertainty Principle

The probability density $\Psi^*\Psi$ is interpreted as the probability per unit volume of finding the particle as the result of an observation. Clearly, the most probable or expected location of the particle represented by the wave function of Fig. 5-12 will be near z_0 because this is the location for the maximum value of $|\Psi|^2$. On the other hand, there is a finite probability that the particle will be located elsewhere. In other words, because the wave packet extends over a small but nonvanishing region of space, the location of the particle is not known exactly. The uncertainty in the location of the particle is measured by the width Δz of the wave packet.

By the same token, the wave number k of the wave that represents the particle is also not known precisely. The plane wave amplitude spectrum $A(k)$ shows that there are many values of k present in the packet. Since the particle momentum is $p = \hbar k$, a measurement of the particle momentum will give a value somewhere in the range of momenta proportional to the range of k. Hence the momentum of a quantum mechanical particle is also uncertain. The uncertainty in momentum is measured by $\hbar \Delta k$, where Δk is the width of the amplitude spectrum.

It can be shown that if the wave packet is made more narrow so as to locate the particle with greater accuracy, the amplitude spectrum will become broader, making the determination of the momentum less accurate. The *uncertainty principle* is the statement of the fact that the uncertainties in position and momentum are inversely related. Its mathematical statement is

$$\Delta z \, \Delta p_z \geq \hbar \qquad (5\text{-}42)$$

The product of the uncertainties in position and in the momentum in the corresponding direction is greater than or equal to $\hbar$. As a consequence of the uncertainty in position and momentum, a similar relationship holds between the uncertainty in particle energy as measured by an observer and the uncertainty in the time required for the measurement. The energy uncertainty is related to the momentum uncertainty. The uncertainty in time is the interval required for the wave packet to pass the observer. The product of these uncertainties is also greater than $\hbar$:

$$\Delta E \, \Delta t \geq \hbar \qquad (5\text{-}43)$$

The constant $\hbar$ in Eqs. (5-42) and (5-43) is a very small number by ordinary standards. For laboratory-sized objects the uncertainties may therefore be

Sec. 5-4 Wave Packets

negligibly small. In fact, classical mechanics is regarded as the limit in quantum mechanics when h can be considered to be negligibly small. In this limit the de Broglie hypothesis, Eq. (5-7), indicates that $h/p = \lambda = 0$ for a particle with any value of momentum. Such a particle shows no wave-like phenomena such as diffraction and always behaves classically.

The Motion of a Wave Packet

We now turn to the mathematical problem of the motion of a wave packet that represents a moving particle. The wave packet and its spectrum are given in Eqs. (5-40) and (5-41), respectively. The motion of such a packet can be derived from the formalism by recognizing that ω is a function of the wave number k. For a particle in a zero potential field the relationship is obtained from Eq. (5-33), $\omega = \hbar k^2/2m$. Whatever the form of this relationship, it can be generalized to $\omega = \omega(k)$. For simplicity we shall assume that ω does not vary greatly with k over the range of values of k in the wave packet. We then can expand ω as a Taylor series about its value ω_0 for the wave with wave number k_0,

$$\omega(k) = \omega_0 + \left.\frac{\partial \omega}{\partial k}\right|_{k=k_0} (k - k_0) + \cdots \tag{5-44}$$

keeping only the first-order term in the expansion. If we substitute Eq. (5-44) into Eq. (5-40), we obtain

$$\Psi_{\text{packet}}(z, t) = \frac{1}{\sqrt{2\pi}} \int_{-\infty}^{\infty} A(k) e^{j[kz - [\omega_0 + (\partial \omega/\partial k)|_{k_0}(k-k_0)]t]} \, dk$$

Extracting the constant terms from inside the integral we get

$$\Psi_{\text{packet}}(z, t) = \epsilon^{-j[\omega_0 + (\partial \omega/\partial k)|_{k_0}k_0]t} \frac{1}{\sqrt{2\pi}} \int_{-\infty}^{\infty} A(k) e^{jk[z - (\partial \omega/\partial k)|_{k_0}t]} \, dk \tag{5-45}$$

The integral on the right-hand side of Eq. (5-45) can be interpreted by using Eq. (5-40) to write an expression for $\Psi_{\text{packet}}(z_0, t = 0)$, where z_0 is now the expected position of the wave packet at $t = 0$:

$$\Psi_{\text{packet}}(z_0, t = 0) = \frac{1}{\sqrt{2\pi}} \int_{-\infty}^{\infty} A(k) e^{jkz_0} \, dk \tag{5-46}$$

Comparison of Eq. (5-46) with Eq. (5-45) permits us to rewrite Eq. (5-45) as

$$\Psi_{\text{packet}}(z, t) = \epsilon^{-j[\omega_0 + (\partial \omega/\partial k)|_{k_0}k_0]t} \Psi_{\text{packet}}(z_0, t = 0) \tag{5-47}$$

Equation (5-47) says that the wave function at time t has the same amplitude, though different phase, as it did at $t = 0$ *and* that it has moved in that time from position z_0 to position z. Comparing Eqs. (5-46) and (5-45), we see that

$$z_0 = z - \left.\frac{\partial \omega}{\partial k}\right|_{k_0} t$$

or

$$z - z_0 = \left.\frac{\partial \omega}{\partial k}\right|_{k_0} t \qquad (5\text{-}48)$$

From Eq. (5-48) it is evident that the wave packet moves with a velocity

$$v_{\text{group}} = \left.\frac{\partial \omega}{\partial k}\right|_{k_0} \qquad (5\text{-}49)$$

called the group velocity. For a free particle, ω is related to k by $\hbar k^2/2m$ so that

$$\frac{\partial \omega}{\partial k} = \frac{\hbar k}{m} = \frac{p}{m} = v, \qquad \text{free particle}$$

That is, the group velocity for a wave packet of free-particle waves is just the classical particle velocity.

All of the concepts of this section extend to situations where the solutions to Schrödinger's equation are more complicated than the free-particle solutions. Examples of such potential configurations are the finite potential barriers of Section 5.3. Wave packets can always be formed by the linear combination of solutions to the wave equation, whatever they may be.

5.5 BOUND PARTICLES

The free particles of Section 5.3 exhibit some interesting quantum mechanical effects, especially tunneling. Their behavior can be explained using wave packets. Yet many particles are not free. Instead they are localized to a region of space by virtue of a force that acts upon them. For instance, the electrons in an atom are bound to the nucleus by the electrostatic force. Some of the electrons in a metal are held within the metal by forces exerted on them at the surfaces of the material. The forces that hold the particles locally usually can be expressed in terms of a potential function that has the form of a local potential minimum or well. In this section we shall study some examples of particles bound in potential wells. The foremost result of our study will be the development of the concept of the quantization of the energy that such a particle may have. Mathematically, only certain discrete values of the separation constant E will permit the solution of Schrödinger's equation.

As an example of the problem of bound or trapped particles we shall first consider qualitatively the potential of Fig. 5-14. The energy ranges for which nontrivial solutions to Eq. (5-25) exist are $0 < E < U_1$, $U_1 < E < U_2$, and $E > U_2$. The second and third ranges present problems of the kind considered in Section 5.3. However, solutions for $\psi(z)$ with energy between zero and U_1 are of a different type and represent a bound particle. The potential

Sec. 5-5 Bound Particles 143

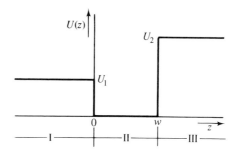

Fig. 5-14 Potential illustrating the possibility of a bound particle.

between $z = 0$ and $z = w$ in Fig. 5-14 is in the form of a potential well. A classical particle with energy less than U_1 is confined in this region. As we shall see, a quantum mechanical particle is also "confined," but in addition it can only exist in the well with specified values of energy.

Qualitatively the wave functions for this condition can be sketched using the results of Section 5.3. We know, for example, that the wave functions will be oscillatory in region II with, according to Eq. (5-30),

$$k_{II}^2 = \frac{2m}{\hbar^2} E$$

In regions I and III the wave functions will be nonoscillatory, decaying exponentially both to the right and to the left. Since region I has negative values of z, the decay coefficients will be, from Eq. (5-39),

$$\alpha_I = -\left[\frac{2m}{\hbar^2}(U_1 - E)\right]^{1/2}$$

and

$$\alpha_{II} = \left[\frac{2m}{\hbar^2}(U_2 - E)\right]^{1/2}$$

The main difficulty in effecting a solution to the wave equation lies in maintaining the continuity of ψ and $d\psi/dz$ at the boundaries $z = 0$ and $z = w$. Not every choice of energy E will permit $d\psi/dz$ to be continuous at both boundaries and still give a wave function that decays exponentially in both directions from the well. Figure 5-15 shows two attempts at solution for the potential of Fig. 5-14. In Fig. 5-15(a) the energy was chosen so that the boundary conditions could be satisfied. Remember that E determines the wavelength of the oscillation in region II. Figure 5-15(b) shows the results of a choice that fails to satisfy the boundary condition at $z = w$ given the necessary form of the wave function in region III. Failure means there is no solution for this energy; hence the particle cannot have this energy: It is not allowed. In both attempts at solution illustrated in Fig. 5-15, we have started to construct the wave function in region I and worked our way toward region III. There is no surefire way of guessing an allowed energy. Finding the allowed

144　　Elements of Quantum Mechanics　Chap. 5

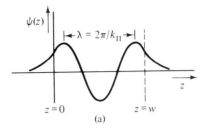

(a)

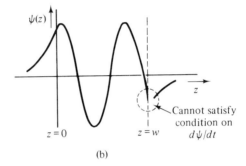

(b)

Fig. 5-15 Two attempts to solve Schrödinger's equation for the potential of Fig. 5-14: (a) success, (b) failure.

values for a problem as complicated as this one is a matter of trial-and-error hunting.

The One-Dimensional Infinite Square Well

There are, however, some potential well problems that can be solved analytically. The simplest of these is the so-called infinite square well. The potential for this problem is infinite outside a region of width L, as indicated in Fig. 5-16. The infinite potential simplifies the solution for the allowed ener-

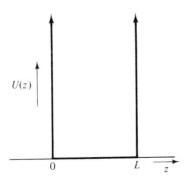

Fig. 5-16 Potential of a one-dimensional infinite square well.

Sec. 5-5 Bound Particles

gies and the corresponding wave functions in two ways. First, the wave function outside the well must be zero because the attenuation factor α, as determined by Eq. (5-39), is infinite. Second, although ψ is always continuous, $d\psi/dz$ need not be continuous at an infinite discontinuity in the potential. Continuity of $d\psi/dx$ is required only at finite discontinuities in $U(z)$. Therefore the boundary conditions at $z = 0$ and $z = L$ are simplified, both being $\psi = 0$ with no restriction on $d\psi/dz$.

Inside the well, Schrödinger's equation for the stationary state has the form

$$\frac{d^2\psi}{dz^2} + k^2\psi = 0 \qquad (5\text{-}50)$$

where

$$k^2 = \frac{2m}{\hbar^2} E \qquad (5\text{-}51)$$

Equation (5-50) has solutions of either of the following forms:

$$\psi(z) = A_1 \epsilon^{jkz} + B_1 \epsilon^{-jkz} \qquad (5\text{-}52)$$

$$\psi(z) = A_2 \sin kz + B_2 \cos kz \qquad (5\text{-}53)$$

or

$$\psi(z) = A_3 \sin(kz + \theta) \qquad (5\text{-}54)$$

Each of these solutions has two constants of the integration that must be evaluated using the boundary conditions. Because the boundary conditions for the infinite square well are periodic ones, $\psi = 0$ at $z = 0$ and $z = L$, it is convenient to use either Eq. (5-53) or Eq. (5-54) for $\psi(z)$. For either choice, application of the boundary condition at $z = 0$ leaves

$$\psi(z) = A \sin kz \qquad (5\text{-}55)$$

since it is necessary that $B_2 = 0$ and $\theta = 0$. The amplitude factor A will be determined by the normalization condition.

Application of the boundary condition $\psi = 0$ at $z = L$ requires that

$$kL = n\pi \qquad (5\text{-}56)$$

where n is a positive or negative integer. The value $n = 0$ yields a wave function that is zero everywhere; it is therefore the trivial solution, which we shall neglect. Equation (5-56) is an extremely important result: Only the values of k that satisfy this equation give wave functions that are proper solutions to the wave equation. The values of k are said to be quantized. Furthermore, because of Eq. (5-51), the values of energy E for which realistic wave functions exist are also quantized. Combining Eqs. (5-51) and (5-56), we obtain

$$E_n = \frac{\hbar^2 \pi^2}{2mL^2} n^2, \qquad n = \pm 1, \pm 2, \cdots \qquad (5\text{-}57)$$

The discrete values of allowed energy E_n are called *eigenvalues* or *eigener-*

gies and the integer n is called the *quantum number*. The wave functions that correspond to the eigenenergies are called *eigenfunctions*:

$$\psi_n(z) = A_n \sin\left(\frac{n\pi z}{L}\right), \quad 0 < z < L \quad (5\text{-}58)$$

where we have indicated that the amplitude A_n may depend on the quantum number. Figure 5-17 shows a diagram of the allowed energies and several of the eigenfunctions.

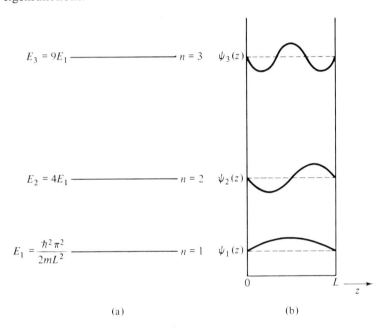

Fig. 5-17 Solutions to the one-dimensional infinite square well: (a) the energy diagram, (b) eigenfunctions.

The complete set of functions that satisfy the time-dependent Schrödinger equation is

$$\Psi_n(z, t) = A_n \sin\left(\frac{n\pi z}{L}\right) \epsilon^{-jE_n t/\hbar}, \quad 0 < z < L \quad (5\text{-}59)$$

The probability per unit length of finding a particle as the result of a measurement is $\Psi^*\Psi$. Notice that, since $\sin(-\theta) = -\sin\theta$, the probability density is the same for the quantum numbers $\pm n$. Notice, too, that the eigenenergies, Eq. (5-57), do not depend on the sign of n. Therefore the negative integer quantum numbers do not contribute new wave functions and so are ordinarily neglected. The normalization condition, Eq. (5-27), requires that

$$\int_{z=0}^{L} \Psi_n^* \Psi_n \, dz = 1$$

Sec. 5-5 Bound Particles

from which it is not difficult to show that

$$A_n = \sqrt{\frac{2}{L}}, \quad n = 1, 2, 3, \cdots \tag{5-60}$$

Having obtained the solutions to the one-dimensional infinite square well, let us now interpret these results in light of the uncertainty principle, Eq. (5-42). For instance, the eigenfunction corresponding to the lowest eigenenergy E_1 is

$$\Psi_1(z) = \sqrt{\frac{2}{L}} \sin\left(\frac{\pi z}{L}\right) \epsilon^{-jE_1 t/\hbar}, \quad 0 < z < L$$

so that

$$\Psi^*\Psi = \frac{2}{L} \sin^2\left(\frac{\pi z}{L}\right), \quad 0 < z < L$$

This probability density assures us only that the particle is somewhere between $z = 0$ and $z = L$. Our uncertainty in the location of the particle represented by this state is of the order of L, the width of the well. On the other hand, the particle momentum is $p_1 = (2mE_1)^{1/2}$. However, we do not know in which direction the particle is moving so that our uncertainty in momentum is $\Delta p = 2p_1$. By substitution from Eq. (5-57) we find that $\Delta p = h/L$. The product of uncertainties is then

$$\Delta z \, \Delta p_z \simeq L\left(\frac{h}{L}\right) = h$$

which is greater than $\hbar$ by a factor of 2π.

The Three-Dimensional Infinite Square Well

The square well problem can be extended to three dimensions with the only complications being mathematical ones and those not difficult at all. The three-dimensional square well potential is, however, a very important one because it is used as a model to describe the confinement of certain electrons in solids. Furthermore, the electrons for which this model applies are just the ones that participate in the flow of electric current, the so-called conduction electrons.

The potential of the three-dimensional square well is infinite outside the region defined by $0 < x < L_x$, $0 < y < L_y$, $0 < z < L_z$. That is, the well has dimensions L_x, L_y, L_z and volume $L_x L_y L_z$. The three-dimensional form of the wave equation for the stationary state inside the well is obtained from Eq. (5-25),

$$\nabla^2 \psi(x, y, z) + \left[\frac{2mE}{\hbar^2}\right]\psi(x, y, z) = 0 \tag{5-61}$$

Equation (5-61) is a partial differential equation that can be solved using the technique of separation of variables. First we assume that

$$\psi(x, y, z) = \psi_x(x)\psi_y(y)\psi_z(z) \qquad (5\text{-}62)$$

where $\psi_x(x)$ is a function of x alone, and so forth. Then, substituting Eq. (5-62) into Eq. (5-61) and dividing by $\psi(x, y, z)$, we obtain

$$\frac{1}{\psi_x}\frac{d^2\psi_x}{dx^2} + \frac{1}{\psi_y}\frac{d^2\psi_y}{dy^2} + \frac{1}{\psi_z}\frac{d^2\psi_z}{dz^2} = \frac{-2mE}{\hbar^2} \qquad (5\text{-}63)$$

The first term on the left-hand side of Eq. (5-63) depends only on x, the second only on y, and the third only on z. For the equation always to be true each of these terms must be a constant. For each of the three terms we can write

$$\nabla^2\psi_i = -k_i^2\psi_i, \qquad i = x, y, z \qquad (5\text{-}64)$$

where the k_i are the separation constants and

$$E = \frac{\hbar^2}{2m}(k_x^2 + k_y^2 + k_z^2) \qquad (5\text{-}65)$$

The appropriate solutions to Eq. (5-64) are of the form of Eq. (5-53) or Eq. (5-54). When the boundary conditions are applied, quantization of the k_i appears:

$$k_i = \frac{n_i\pi}{L_i}, \qquad i = x, y, z \qquad (5\text{-}66)$$

Notice that in the three-dimensional case there are three quantum numbers ($n_i = 1, 2, 3, \ldots$) corresponding to quantization of the three components of momentum $p_i = \hbar k_i$. The complete normalized eigenfunctions for the stationary state are

$$\psi_{n_x n_y n_z}(x, y, z) = \left(\frac{8}{L_xL_yL_z}\right)^{1/2} \sin(k_x x)\sin(k_y y)\sin(k_z z) \qquad (5\text{-}67)$$

The eigenvalues corresponding to these wave functions are

$$E_{n_x n_y n_z} = \frac{\pi^2\hbar^2}{2m}\left(\frac{n_x^2}{L_x^2} + \frac{n_y^2}{L_y^2} + \frac{n_z^2}{L_z^2}\right) \qquad (5\text{-}68)$$

The Hydrogen Atom

Another important class of bound-particle problems is that of an electron in the electrostatic field of an atomic nucleus. The quantum mechanical treatment of such problems yields, as for the square well potential, discrete values of allowed energy for the electron. An electron may be in any of the eigenstates available to it. By eigenstates we mean the description of an electron by an eigenfunction with a discrete eigenenergy. The eigenstates and the transition of an electron from one eigenstate to another are fundamental

Sec. 5-5 Bound Particles

to understanding the characteristic optical spectra of the elements and the operation of the laser.

The simplest atomic system is represented by the hydrogen atom, the field being that of a positively charged nucleus with spherical symmetry. The potential energy of an electron in such a Coulomb field is

$$U(r) = \frac{-e^2}{4\pi\varepsilon_0 r} \qquad (5\text{-}69)$$

which is sketched in Fig. 5-18. That U is taken as negative should cause no

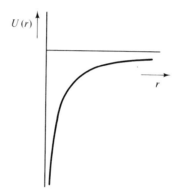

Fig. 5-18 Potential of an electron in the field of a hydrogen nucleus.

trouble because potential energy is known only with respect to some reference value. According to Eq. (5-69), the potential approaches zero as the electron moves away from the nucleus toward infinity. The potential at infinity is our reference value.

Schrödinger's equation for the stationary state, Eq. (5-25), should be written in spherical coordinates because of the symmetry of the potential:

$$\nabla^2 \psi(r, \theta, \phi) + \frac{2m}{\hbar^2}[E - U(r)]\psi(r, \theta, \phi) = 0 \qquad (5\text{-}70)$$

In spherical coordinates the ∇^2 operator is

$$\nabla^2 \psi(r, \theta, \phi) = \frac{1}{r^2} \frac{\partial}{\partial r}\left(r^2 \frac{\partial \psi}{\partial r}\right)$$
$$+ \frac{1}{r^2 \sin \theta} \frac{\partial}{\partial \theta}\left(\sin \theta \frac{\partial \psi}{\partial \theta}\right)$$
$$+ \frac{1}{r^2 \sin^2 \theta} \frac{\partial^2 \psi}{\partial \phi^2} \qquad (5\text{-}71)$$

Equation (5-70) with the Coulomb potential of Eq. (5-69) can actually be solved exactly using, again, the technique of separation of variables. Assuming that

$$\psi(r, \theta, \phi) = R(r)\Theta(\theta)\Phi(\phi) \qquad (5\text{-}72)$$

Eq. (5-70) separates into three equations, one each for $R(r)$, $\Theta(\theta)$, and $\Phi(\phi)$. The nature of the separation is such that two separation constants appear, denoted by l and m. Both l and m must have integer values with $l \geq 0$ and $|m| \leq l$ in order that the equations have solutions. It also turns out that for $\psi(r, \theta, \phi)$ to be a well-behaved function both at $r = 0$ and as $r \rightarrow \infty$ the solution of the equation for $R(r)$ exists only when the energy E takes on the discrete values

$$E_n = \frac{-me^4}{32\pi^2\varepsilon_0^2 h^2} \frac{1}{n^2} \tag{5-73}$$

where n is an integer that satisfies

$$n \geq l + 1 \tag{5-74}$$

The integers n, l, and m are quantum numbers, with the *principal quantum number* n determining the energy of the state. The state corresponding to $n = 1$ is called the *ground state*. The integer l is interpreted as quantizing the total vector angular momentum of the orbit of the electron. The integer m quantizes an arbitrary component of the orbital angular momentum. Each eigenfunction $\Psi_{nlm}(r, \theta, \phi)$ is characterized by the three integers n, l, and m, although the eigenenergy for all states with the same value of n are not significantly different. For each value of n there are, by Eq. (5-74), n possibilities for l and for each value of l there are $2l + 1$ possibilities for m. An allowed energy for which there is more than one eigenfunction is said to be *degenerate*. Figure 5-19 shows a diagram of the allowed energies, given by Eq. (5-73), superimposed on the Coulomb potential of Fig. 5-18. The lines representing the allowed energy levels in Fig. 5-19 are drawn in such a way as to represent the general trend of increasing spatial extent of the wave functions with higher values of n. The radial factors of some representative wave functions are sketched in Fig. 5-20.

The functions $\psi_{nlm}(r, \theta, \phi)$ are a mathematically complete set of solutions to the wave equation. Yet it has been found that the behavior of electrons is not described completely in the physical sense by these functions. The difficulty lies in the physical model of particles on which the development of the wave description was based. Briefly, our model was that of a point-mass particle in some external field. We neglected the possibility of structure within the particle itself. In particular, it has been found that a more accurate description of quantum particles must attribute to them an inherent magnetic moment. Associated with the magnetic moment is an angular momentum or *spin*. Particle spin is quantized. The electron spin quantum number s may only have values $\pm\frac{1}{2}$. These two spin values are interpreted as being the two possible orientations of the spin angular momentum of the electron with respect to an imposed coordinate system, called spin-up and spin-down.

The general theory of spin allows some types of particles to have integer spins, but others, such as the electron, must have half-integer spins. We shall

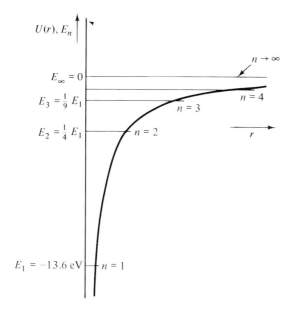

Fig. 5-19 Energy diagram for the hydrogen atom.

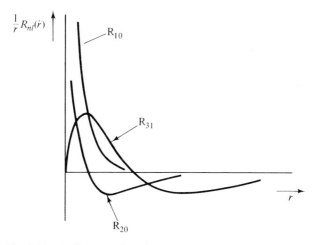

Fig. 5-20 Radial wave functions $R(r)/r$ for hydrogenic electrons.

have more to say about this later. An important consequence of the theory is that no two particles of half-integer spin may occupy the same three-dimensional eigenstate at the same time when the eigenstate is described by three spatial quantum numbers and one spin quantum number. For the case of an electron in the hydrogen atom the four quantum numbers are n, l, m, and s. An equivalent statement is that no more than two electrons can occupy the same state described by only the three spatial quantum numbers. One of these electrons must have $s = \frac{1}{2}$ and the other $s = -\frac{1}{2}$. Either of these equivalent statements is known as the *exclusion principle*. The exclusion principle implies that no two of these particles can occupy the same place with the same spin orientation. This does not apply to particles with whole-integer spins such as photons.

As a result of the exclusion principle, only two electrons may occupy the same eigenstate of an atomic system. Thermodynamically, the available electrons fill up the lowest energy states to minimize the energy of the system. However, fluctuations in the occupancy can occur because of the finite temperature of the system of particles. An electron can change to a higher-energy state by absorbing thermal energy from the whole system provided the higher state, including its spin description, is not already occupied. For an atomic system with many electrons, thermal fluctuation in the occupancy of the eigenstates is much more likely to occur in the higher-energy states, because there are many available unoccupied states at higher energies and the amount of energy that must be absorbed in order for the transition to occur is smaller, the energy states being more closely packed in energy space. With an electron in an elevated state the atomic system contains more than its minimum energy. The excited electron will return to the lower-energy state and radiate the excess energy as light. Occupation probabilities are described using the techniques of statistical mechanics, which will be modified for quantum systems in Section 5.6.

Optical Spectra and Lasers

We are now in a position to understand the phenomena of optical spectra. Each element of the Periodic Table is known to radiate light, the spectrum of each being composed of a distinctive set of discrete frequencies. These spectra are due to transitions of electrons between the characteristic eigenstates of the respective atoms. To conserve energy, when an electron makes a transition to a lower state electromagnetic radiation is emitted. The frequency of the emitted radiation is given by

$$E_i - E_f = h\nu \tag{5-75}$$

where E_i is the energy of the initial state before transition and E_f is that of the final state after transition. Obviously there may be very many transitions and hence very many frequencies in a radiated spectrum. And there are. However, it has been found experimentally that some transitions do not occur. To describe this empirical fact, selection rules have been formulated to describe the allowed transitions.

The probability of transition from one state to another depends on the wave functions of the two states and the availability of radiation quanta (photons) of the proper energy. The probability of excitation between two states by absorption of a photon is always equal to the probability of de-excitation between the same two states by emission of a photon provided the appropriate state is initially occupied and the radiation field is available. However, transitions between certain pairs of states occur more readily than between others. The relative ease of transition is measured by a lifetime, the average time an electron spends in an excited state. Some states have very long lifetimes; these are called *metastable* states.

The operating principles of the laser are based on these concepts of electronic transitions. Figure 5-21 shows three electronic states of the atoms of a

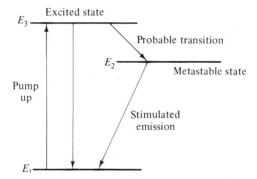

Fig. 5-21 Electronic states of a three-level laser.

hypothetical laser—the ground state, an excited state, and a metastable state. If light shines on the laser material and has photons of energy $h\nu_{31} = E_3 - E_1$, some of the electrons in the ground states of the various atoms will absorb this energy and be excited to state E_3. Once at E_3 they either reradiate back to E_1 or make another radiative transition to E_2, the metastable level. If the transition to E_2 is more probable than that back to E_1, the effect of the presence of the radiation at frequency ν_{31} will be the net transference of electrons to the metastable state. The frequency ν_{31} is called the pump frequency. As electrons begin to accumulate in level E_2 the internal energy of the material rises. Thermodynamically every system tends toward a state of minimum energy, and our laser is no exception. Although the metastable state has a long lifetime, the probability of transition between it and the ground state is enhanced if photons of frequency ν_{21} are present. In other words, one photon of energy $h\nu_{21}$ can stimulate the emission of another. If one photon is present, the process can avalanche, producing a tremendous amplification in the number of photons. A feature of the stimulated radiation is that it is in phase with the radiation that triggers it. Therefore the output of the laser is optically coherent. The word *laser* is an acronym for light amplification by stimulated emission of radiation.

5.6 QUANTUM STATISTICS

To apply the fundamentals of the quantum description to electronic systems, we must consider behavior of large numbers of quantum mechanical electrons. The most straightforward way of doing this is to develop a statistical description for the system. In Chapter 2 we reviewed the basic features of the statistical mechanics of classical particles. In this section we shall modify that statistical picture to account for quantum effects. Our prime objective will be to obtain the equilibrium probability function for the occupancy of electronic eigenstates, the Fermi-Dirac distribution function.

The features of quantum mechanics that must be included in a statistical description are the uncertainty principle, the discrete nature of the allowed states, the exclusion principle, and the fundamental property of indistinguishability of quantum particles.

Indistinguishability

Quantum particles are not distinguishable in the following sense. One picture of the classical limit of quantum mechanics of a many-bodied system is that the wave functions of classical particles do not overlap; that is, the wave functions are well separated in space. It is then possible to follow from instant to instant the behavior of each of the particles individually. While the positions and momenta are only approximately known, the expected values satisfy Newton's laws. We can follow each of the particles in its classical motion. On the other hand, the wave functions of quantum particles do overlap. Actually there is only one wave function for the whole system made up of the overlapping wave functions. At any instant the probability density of the global wave function indicates the probability of finding a particle here or there. At the next instant, it indicates new probabilities only. There is nothing in the formalism to specify which particle went where. Figure 5-22 shows a time-dependent global probability density for two particles at two successive instants of time. The wave function happens to be exactly the same at both t_1 and t_2. Suppose at time t_1 we measured a particle near z_1 and another at z_2. At time t_2 we would also measure a particle near z_1 and another near z_2. But whether the particle that was at z_1 at t_1 moved to z_2 in the interval or whether it stayed at z_1 (or whatever it did) is unknown. We cannot distinguish between the particles.

Another view of the property of indistinguishability is illuminated when we attempt to construct the global wave function from the individual over-

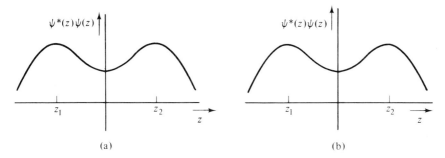

Fig. 5-22 Global probability density for two particles at two instants of time.

lapping wave functions. If the particles were distinguishable, we could interchange the four quantum members of two particles and thereby obtain a new wave function. It turns out that mathematically the interchange of two particles between two states leaves the global wave function either unchanged or changes only its sign, depending on the spin characterization of the type of particle. For half-integer spin quantum numbers the sign of Ψ is changed and the wave function is said to be antisymmetric to the interchange. For integer spins the sign of Ψ is unchanged, the wave function denoted as being symmetric. In either case the probability density $\Psi^*\Psi$ is unchanged so that the new state cannot be distinguished from the old. We cannot tell which particle is which; they are not distinguishable.

Quantization and the Density of States

In Chapter 2 we learned that the condition of thermodynamic equilibrium is described by the most probable macroscopic state of a system. A macrostate is in turn described by a distribution function, the density of representative points in the parameter space. Classically a particle may have any energy, and therefore its representative point may occupy any position in energy space or momentum space. Quantum mechanically a bound particle may have only certain discrete values of energy or momentum. Therefore its representative point may not occupy every position in energy or momentum space. The density of points representative of quantum particles is determined therefore not only by the statistical thermodynamic probability that a particle has, say, energy in a range dE about some value but also by the number of allowed energy states in that range. The latter factor is usually expressed in terms of the density of allowed states in parameter space. The density of allowed states depends on the confining potential. The probability of occupancy is determined using the approaches of Chapter 2, taking into consideration the indistinguishability of particles and the uncertainty principle.

For particles of half-integer spin this probability is the Fermi-Dirac probability function.

Before turning to the derivation of Fermi-Dirac statistics we shall consider the density of states for electrons in the potential of an infinite three-dimensional square well. We remarked earlier that this problem has widespread applicability to electrons in solids. According to the results of the previous section, the allowed components of momentum for such an electron are

$$p_i = \hbar k_i = \frac{n_i \pi \hbar}{L_i}, \quad i = x, y, z \quad (5\text{-}76)$$

where L_i is the dimension of the well in the ith direction and the n_i are positive, nonzero integers.

Suppose that we wish to determine the density of allowed states in momentum space. Each eigenstate can be represented in the space by a point at the momentum coordinate (p_x, p_y, p_z) corresponding to the three components of allowed momenta. The vector momentum of each state is a vector from the origin of momentum space to the coordinate. Figure 5-23 shows some of

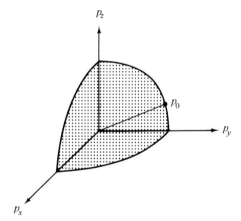

Fig. 5-23 Points in p-space representing eigenstates of electrons in a three-dimensional infinite square well.

these representative points. The points form a uniformly spaced array in the octant of positive values of all components. The volume of momentum space surrounding each point can be associated with that state. This volume is the same for each state, being

$$\frac{\pi^3 \hbar^3}{L_x L_y L_z}$$

Then the number dN_s of represented states in a small volume $d\mathbf{p} = dp_x \, dp_y \, dp_z$ of momentum space is

$$dN_s = \frac{d\mathbf{p}}{(\pi^3 \hbar^3 / L_x L_y L_z)}$$

The density of states in momentum space is the number of states per unit volume or

$$\frac{dN_s}{d\mathbf{p}} = \frac{V}{\pi^3 \hbar^3}, \quad \text{infinite square well} \quad (5\text{-}77)$$

where $V = L_x L_y L_z$ is the spatial volume of the well.

Two alternative forms of the density of states for electrons in an infinite square well are frequently encountered. Each depends on the parameter space employed in the description, and each can be derived from Eq. (5-77). In the wave picture it is common practice to use a wave number or **k**-space rather than **p**-space. Since $p_i = \hbar k_i$, the two parameter spaces are proportionally related. The density of states in **k**-space is

$$\frac{dN_s}{d\mathbf{k}} = \frac{V}{\pi^3}, \quad \text{infinite square well} \quad (5\text{-}78)$$

The second alternative representation in frequent use employs energy as the parameter of interest. According to Eq. (5-65), $E = p^2/2m$, where p^2 is the square of the magnitude of the vector **p**. Recall that the eigenstates are confined to that part of p-space where all components of momentum are positive and that there are two electronic states, remembering the spin, for each eigenstate. The number of electronic states in one octant of a spherical shell of radius p and thickness dp is

$$dN_s = 2\left[\frac{1}{8}(4\pi p^2 \, dp)\right]\frac{V}{\pi^3 \hbar^3} \quad (5\text{-}79)$$

where $V/\pi^3 \hbar^3$ is the density of states in momentum space and the factor of 2 takes spin into account. Equation (5-79) can be expressed in terms of energy using $E = p^2/2m$. The resulting density in compact form is

$$\boxed{\frac{dN_s}{dE} = \frac{V}{2\pi^2}\left(\frac{2m}{\hbar^2}\right)^{3/2} E^{1/2}} \quad \text{infinite square well} \quad (5\text{-}80)$$

The Fermi-Dirac Probability Function

To find the statistical probability that an energy level is occupied we can follow the same procedure that was used in Section 2.3. First we divide the phase space into volumes $d\mathbf{r}\, d\mathbf{p}$ so that the specification of the number of particles in each volume is a macroscopic description. We shall construct the volume $d\mathbf{r}\, d\mathbf{p}$ so that all states within it have the same energy. For notational purposes we shall assume that there are m such volumes and that there are G_i states in the ith volume occupied by N_i particles. For particles with half-integer spin, only one particle may occupy each state, so $N_i \leq G_i$. Our objective will be to find the equilibrium value of N_i from which we can determine the probability that a particle has energy E_i.

The procedure for finding the equilibrium values of N_i is the same as in Chapter 2 with modifications in details appropriate to quantum theory. We count the number of microstates W, which is the number of ways that indistinguishable particles can be arranged among the volumes; define the entropy S in terms of this number; and then maximize the entropy given the constraints of conservation of energy and of particles. The set of N_i that maximizes the entropy constitutes the equilibrium distribution.

To calculate the number of microstates that give the same macrostate let us concentrate on the ith volume and the number of particles N_i with energy E_i. These particles are arranged among G_i states. Suppose for the moment that the particles are distinguishable. The number of ways N_i electrons can be distributed among G_i states is

$$G_i(G_i - 1)(G_i - 2) \cdots (G_i - N_i + 1) = \frac{(G_i)!}{(G_i - N_i)!}$$

since there are G_i states available to the first one, $G_i - 1$ states available to the second, and so forth. However, the particles are in fact indistinguishable so that many of these permutations do not lead to new wave functions. We must divide the number above by the number of permutations of the particles among the states. The number of ways of arranging N_i indistinguishable particles among the G_i states is, therefore,

$$W_i = \frac{(G_i)!}{(N_i)!(G_i - N_i)!} \tag{5-81}$$

An alternative way of looking at W_i is that it is the way G_i states can be divided into two groups, a group of N_i occupied states and a group of $G_i - N_i$ unoccupied states. The total number of ways that the N indistinguishable electrons can be arranged among all of the m volumes is the repeated product

$$W = \prod_{i=1}^{m} W_i \tag{5-82}$$

According to Eq. (2-22), the entropy is $S = \kappa \ln W$ or

$$S = \kappa \sum \{\ln(G_i!) - \ln(N_i!) - \ln[(G_i - N_i)!]\} \tag{5-83}$$

We can now maximize S by making virtual changes δN_i in the N_i requiring

$$\frac{1}{\kappa} \delta S = 0 \tag{5-84}$$

under the constraints of conservation of the number of electrons

$$\delta N = \sum_{i=1}^{m} \delta N_i = 0 \tag{5-85}$$

and of energy

$$\delta E_{\text{total}} = \sum_{i=1}^{m} E_i \, \delta N_i = 0 \tag{5-86}$$

The method for solving the condition of Eq. (5-84) is that of the undeter-

Sec. 5-6 Quantum Statistics 159

mined Lagrangian multipliers. We multiply Eq. (5-85) by α and Eq. (5-86) by β and add them to Eq. (5-84) using Eq. (5-83), Stirling's approximation, and the fact that the number of states G_i is constant. As in Chapter 2 we have

$$\frac{1}{\kappa}\delta S + \alpha\,\delta N + \beta\,\delta E_{\text{total}} = 0$$

After making the appropriate substitutions, we get

$$\sum_{i=1}^{m}\left[\ln\!\left(\frac{G_i - N_i}{N_i}\right) + \alpha + \beta E_i\right]\delta N_i = 0 \qquad (5\text{-}87)$$

By the arguments of Section 2.3 the only solution to Eq. (5-87) is

$$\ln\!\left(\frac{G_i - N_i}{N_i}\right) = -\alpha - \beta E_i \qquad (5\text{-}88)$$

Equation (5-88) can be solved for N_i, yielding

$$N_i = \frac{G_i}{\epsilon^{-\alpha - \beta E_i} + 1} \qquad (5\text{-}89)$$

The probability of finding an energy state E_i occupied is given by the number of particles with that energy N_i divided by the number of available states:

$$P_i = \frac{N_i}{G_i} = \frac{1}{\epsilon^{-\alpha - \beta E_i} + 1} \qquad (5\text{-}90)$$

The undetermined multipliers can be found in the same manner as with the classical statistics. The multiplier β is the inverse of the thermodynamic temperature expressed in energy:

$$\beta = \frac{-1}{\kappa T} \qquad (5\text{-}91)$$

The multiplier α can be determined by a normalization procedure and depends on the partition function. However, the common practice is to define

$$\boxed{\alpha \equiv \frac{E_F}{\kappa T}} \qquad (5\text{-}92)$$

where E_F is known as the *Fermi energy*. As we shall shortly discover, the Fermi energy has an important physical interpretation from thermodynamics.

Generalizing Eqs. (5-90) – (5-92), the probability for finding any energy state occupied is

$$\boxed{P_{\text{FD}}(E) = \frac{1}{\epsilon^{(E - E_F)/\kappa T} + 1}} \qquad (5\text{-}93)$$

where P_{FD} is the *Fermi-Dirac statistical probability function*. The number of particles in a system with energy in the range dE about a value E is given by

$$dN = \left(\frac{dN_s}{dE}\right) P_{FD} \, dE \qquad (5\text{-}94)$$

where dN_s/dE is the density of states appropriate to the potential. Comparison of Eq. (5-94) with Eq. (2-4) indicates that $(dN_s/dE)P_{FD}$ is the energy distribution function. Particles that are statistically distributed according to P_{FD} are called *fermions*. For instance, protons as well as electrons are fermions.

Figure 5-24 shows the function P_{FD} plotted against energy for three

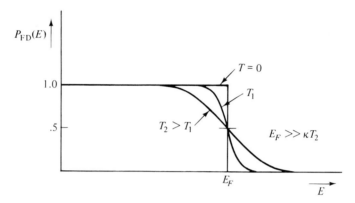

Fig. 5-24 P_{FD} vs. E for $T = 0$ and two nonzero values of temperature.

values of temperature. Notice that at the low-temperature limit P_{FD} has a value of unity for $E < E_F$ and a value of zero for $E > E_F$. Therefore at $T = 0$, all of the lowest-energy states are occupied by electrons, and the system is at its energy minimum. If a particle is added to the system at $T = 0$, E_F will change by the energy of the state of the new particle. At nonzero temperatures the occupation probabilities for states near the Fermi energy change. As the temperature is increased, more and more states with $E > E_F$ have greater and greater probabilities of being occupied. The electrons that occupy these higher-energy states come from states with $E < E_F$ since the occupation probabilities of these lower-energy states falls with increasing temperature. In fact the probability that the state $E_2 = E_F + \Delta E$ is occupied is equal to the probability that the state $E_1 = E_F - \Delta E$ is not occupied since

$$P_{FD}(E_F + \Delta E) = 1 - P_{FD}(E_F - \Delta E) \qquad (5\text{-}95)$$

Sec. 5-6 Quantum Statistics

The probability that a state at the energy level E_F is occupied is always

$$P_{FD}(E_F) = \tfrac{1}{2} \qquad (5\text{-}96)$$

However, we must remember that E_F or any other E need not be an allowed energy for an electron. The Fermi function $P_{FD}(E)$ only gives the probability that an electron would have energy E. The density of states dN_s/dE indicates whether or not E is an allowed energy.

The Fermi Energy

The Fermi energy is that for which the Fermi probability always has the value $\tfrac{1}{2}$. It is also the energy about which P_{FD} exhibits the odd symmetry expressed in Eq. (5-95). In addition, E_F is defined by the Lagrangian multiplier α, which is otherwise determined by a normalization procedure. The normalization property of distribution functions was stated in Eq. (2-7):

$$N = \int_0^\infty f(E)\,dE \qquad (2\text{-}7)$$

where N is the total number of particles in the system. For fermions, Eq. (5-94) can be integrated to obtain

$$N = \int_{\text{all energies}} \left(\frac{dN_s}{dE}\right) P_{FD}\,dE$$

or, using Eq. (5-93),

$$N = \int_{\text{all energies}} \left(\frac{dN_s}{dE}\right) \frac{1}{\epsilon^{(E-E_F)/\kappa T} + 1}\,dE \qquad (5\text{-}97)$$

The integral on the right-hand side of Eq. (5-97) depends only on E_F and T. After the integration, which generally is not easy, this relationship can be rewritten to express E_F as a function of N and T:

$$E_F = E_F(N, T) \qquad (5\text{-}98)$$

The Fermi energy depends not only on the number of particles but also on the thermodynamic temperature.

We now turn to the thermodynamic interpretation of E_F. The probability function was derived earlier in this section by maximizing the entropy while conserving the total number of particles and their total energy. The procedure was to add Eqs. (5-84) – (5-86) using the multipliers α and β. The sum of these equations was taken to be

$$\frac{1}{\kappa}\delta S + \alpha\,\delta N + \beta\,\delta E_{\text{total}} = 0 \qquad (5\text{-}99)$$

To interpret E_F we can compare Eq. (5-99) with the combined statement of the first and second laws of thermodynamics of systems with a variable number of particles

$$dE_{\text{total}} = T\,dS - P\,dV + \mu\,dN \qquad (5\text{-}100)$$

where P is pressure, V is the volume of the system, and μ is called the *chemical potential*. The chemical potential is the amount of energy by which the net energy of the system changes as particles are added to or subtracted from it. From Eq. (5-100), μ is also given by

$$\left(\frac{\partial S}{\partial N}\right)_{E,V} = \frac{-\mu}{T} \qquad (5\text{-}101)$$

Equation (5-101) is the usual thermodynamic definition of chemical potential as being proportional to the rate of change of entropy with the number of particles at constant energy and constant volume. From Eq. (5-99) we find for virtual changes in the fermion system

$$\frac{\delta S}{\delta N} = -\kappa\alpha \qquad (5\text{-}102)$$

Comparing Eq. (5-102) with Eq. (5-101), we see that

$$\alpha = \frac{\mu}{\kappa T}$$

or, comparing this with Eq. (5-92),

$$\boxed{\mu = E_F} \qquad (5\text{-}103)$$

in thermodynamic equilibrium.

The Bose-Einstein Probability Function

Bosons are particles with zero or whole-integer spins. They are indistinguishable, but the exclusion principle does not apply to their occupation of eigenstates. There is no limit to the number of particles in each state. It turns out that the number of ways N_i such particles can be arranged among the G_i states each with energy E_i so that each arrangement is a different microstate is

$$W_i = \frac{G_i(G_i + N_i - 1)!}{(G_i)!(N_i)!}$$

By the standard method the equilibrium solution for N_i is

$$\frac{N_i}{G_i} = \frac{1}{\epsilon^{(E_i - E_B)/\kappa T} - 1} \qquad (5\text{-}104)$$

where E_B is normalizing factor related to the partition function. Equation

(5-104) can be generalized to

$$P_{BE}(E) = \frac{1}{\epsilon^{(E-E_B)/\kappa T} - 1} \quad (5\text{-}105)$$

called the *Bose-Einstein probability function*. One interpretation of P_{BE} is as the average number of particles in a state of energy E. So that P_{BE} will be positive, E_B must be less than or equal to the energy of the lowest allowed eigenstate. Figure 5-25 shows plots of P_{BE} vs. energy for two temperatures, illustrating the so-called condensation of bosons into the lowest-energy state at low temperatures.

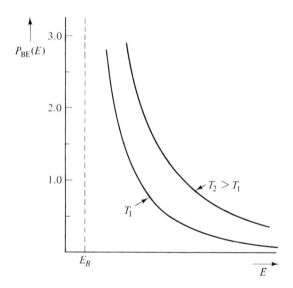

Fig. 5-25 P_{BE} vs. E for two temperatures.

Comparison of Statistics

We have studied three equilibrium statistics that describe systems of large numbers of particles, the classical Maxwell-Boltzmann distribution function and the quantum mechanical Fermi-Dirac and Bose-Einstein probability functions. To facilitate their comparison we shall define a parameter E_M analogous to E_F and E_B so that

$$P_{MB} = \epsilon^{-(E-E_M)/\kappa T} \quad (5\text{-}106)$$

This parameter E_M is also related to the appropriate partition function. Figure 5-26 compares the three statistics by plotting Eqs. (5-93), (5-105), and

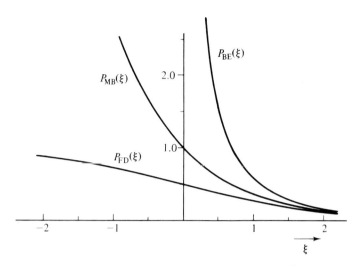

Fig. 5-26 Comparison of the three equilibrium statistics.

(5-106) as functions of ξ, where ξ alternatively is

$$\xi = \frac{E - E_F}{\kappa T} \quad \text{Fermi-Dirac}$$

$$\xi = \frac{E - E_B}{\kappa T} \quad \text{Bose-Einstein}$$

$$\xi = \frac{E - E_M}{\kappa T} \quad \text{Maxwell-Boltzmann.}$$

Notice that for large ξ the three functions are essentially equal. This means, for instance, that classical statistics can be used to approximate quantum statistics when $E - E_F$ or $E - E_B$ is large compared with κT. According to the sketch, the energy states with large positive ξ are sparsely populated. Particles in these states are said to be nondegenerate. For $\xi \lesssim 1$, the energy levels are heavily populated, and the particles are said to be degenerate. In the latter case the appropriate quantum statistic must be used to describe quantum mechanical effects.

BIBLIOGRAPHY

A good introductory text on quantum mechanics is

WICHMANN, E. H., *Quantum Physics.* New York: McGraw-Hill Book Company, 1967.

More advanced and thorough general texts are

DICKE, R. H., and J. P. WITTKE, *Introduction to Quantum Mechanics.* Reading, Mass.: Addison-Wesley Publishing Company, Inc., 1960.

MERZBACHER, E., *Quantum Mechanics.* New York: John Wiley & Sons, Inc., 1961.

BOHM, D., *Quantum Theory.* Englewood Cliffs, N.J.: Prentice-Hall, Inc., 1951.

Quantum statistical mechanics is discussed in

MORSE, P. M., *Thermal Physics.* Reading, Mass.: W. A. Benjamin, Inc., 1962.

Also see

LONGINI, R. L., *Introductory Quantum Mechanics for the Solid State.* New York: John Wiley & Sons, Inc. (Interscience, Division), 1970.

PROBLEMS

5-1. Tungsten has a work function W of 4.5 eV. What is the minimum frequency of the incident light that will cause photoemission of electrons from it? To what wavelength does this correspond?

5-2. A good fastball pitcher can throw a baseball about 90 miles per hour. Assuming a baseball weighs 9 ounces, what is its wavelength?

5-3. Calculate the wavelength associated with helium atoms (mass = 4 atomic mass units) at 300°K.

5-4. In the Davisson-Germer experiment an electron beam was observed to be reflected by a single layer of atoms spaced d Å (angstroms) apart. Show that for the configuration of Fig. 5P-4 the condition for the first constructive reflection is

$$d \sin \theta = \lambda$$

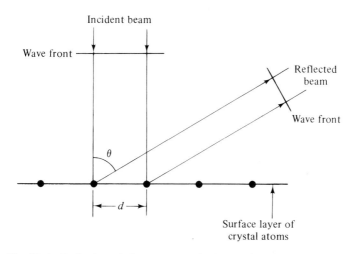

Fig. 5P-4 Reflection of electron waves in the Davisson-Germer experiment.

For a nickel crystal with $d = 2.15$ Å, a diffraction peak of the 54 eV electrons was observed at $\theta = 50°$. Use these data to verify the wave nature of matter.

5-5. Assume that the plane wave amplitude spectrum $A(k)$ of a wave packet has values

$$A(k) = \begin{cases} \sqrt{2\pi}, & k_0 - \delta \leq k \leq k_0 + \delta \\ 0, & \text{all other } k \end{cases}$$

Calculate the wave packet $\Psi(z, t = 0)$ and sketch $\Psi^*\Psi$ vs. z. What is the relationship between δ and the width of $\Psi^*\Psi$ at half its maximum value?

5-6. Show that Eq. (5-60) is true.

5-7. What are the degeneracies of the lowest four energy states of the hydrogen atom?

5-8. The Balmer lines in the spectrum of the hydrogen atom result from transitions of excited electrons from higher states to the first excited state, $n = 2$. Calculate the shortest wavelength of the Balmer series and indicate in what part of the radiation spectrum this line lies.

5-9. Show that Eq. (5-95) is true.

5-10. Find the dependence of E_F on N at $T = 0$ for electrons in a three-dimensional infinite square well.

5-11. Sketch $(dN_s/dE)P_{FD}$ vs. E for electrons in a three-dimensional square well for $T = 0$ and for $T \neq 0$ but $\kappa T \ll E_F$.

5-12. The Fermi-Dirac probability function changes from a value of 1 to 0 in a range of energies about E_F. How wide is the region in energy between values of P_{FD} equal to .9 and .1? Express your answer in terms of κT.

5-13. With what accuracy can we use the classical probability distribution for fermions when $E - E_F = 3\kappa T, 4\kappa T, 5\kappa T$?

6

Electrons in Solids

Since the development of the transistor in the late 1940s and early 1950s, engineering design of the electronic properties of solid-state materials has revolutionized the electronics industry. The creative design of new configurations of materials, their application to discrete and integrated devices, and the circuit modeling of the characteristics of these devices all require a fundamental understanding of the behavior of electrons in solids. In this chapter we shall develop the so-called energy band theory of solids to describe this behavior. In Section 6.1 we shall establish a physical picture of how energy bands arise based on the interaction of overlapping atomic wave functions. From the configuration of these energy bands we shall be able to categorize solid materials as being either metals, semimetals, semiconductors, or insulators. In Sections 6.2 and 6.3 we shall construct a mathematical model for a simplified physical picture based on the behavior of an electron in a periodic potential. Then in Section 6.4 we shall describe some of the complications that arise when dealing with real solids rather than with idealized mathematical models. The next two chapters deal with the particular properties of two classes of solids, metals and semiconductors.

Before proceeding we must understand what a solid is. Matter exists in three states. In order of increasing internal energy and usually decreasing density, these phases are solid, liquid, and gas. The internal energy of a state of matter is characterized by its thermodynamic temperature, which for point-mass objects is a measure of the average kinetic energy of the particles of the system. Now all particles, be they electrons, ions, atoms, or molecules, interact with each another through interaction forces. In the case of charged particles, interaction is through the long-range Coulomb force. For neutral particles or for charged-neutral interactions the forces are short-range. The potential function of a typical short-range interaction force as a function of the separation of the centers of the interacting particles is sketched in Fig. 6-1. If the particles are packed together with sufficient density and if they do

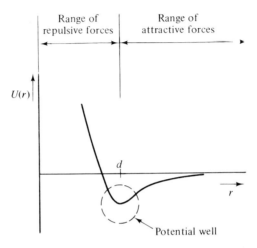

Fig. 6-1 Short-range particle interaction potential as a function of the separation of centers.

not have high kinetic energies, the particles may find themselves in a potential well similar to the one indicated in the figure, in which case all the particles of the system would be arranged in a regular array, each particle separated on the average by the distance d from its nearest neighbor. Particles thus formed at high density and low temperature constitute a crystalline solid.

Naturally, the density and temperature that the particles must have in order to form into a solid depends on the detailed form of their interaction potential. Weakly interacting molecules may be in the gaseous state at the same temperature that strongly interacting molecules form a solid. Because the particles that comprise solids are always densely packed and interacting strongly, any theoretical model of the behavior of solids must be based on quantum mechanics. Gases, on the other hand, are less dense, and the constituent particles do not interact strongly. Therefore the classical picture can be used to describe most gases. Liquids are materials in a state intermediate between solid and gas. In some cases, liquids can be thought of as dense gases, and in others as solids whose density is low and temperature high so that the aggregation of particles into a regular array exists only over relatively short distances. Amorphous materials are usually regarded as extremely viscous liquids, although there are indications that their electronic properties can be understood using an energy band theory analogous to that applicable to solids.

Bonds, Lattices and Crystals

The mechanisms that hold a closely packed aggregate of atoms or molecules into a regular array to form a solid are called *bonds*. Bonds are descriptive of the source of the interaction that creates the potential well of Fig. 6-1. In an ionic bond between two unlike atoms, electrons are transferred from one species to another, making positive ions of the former and negative ions of the latter. The attractive force in the ionic bond then is the Coulomb force between unlike charges. Crystals of sodium chloride are formed by an ionic bond. A covalent bond is usually one in which two atoms each share a pair of electrons, each atom contributing one electron to the pair. The electrons are found in the region between the two atoms. The binding force is that between the positive ions and the two electrons. Diamond is a covalently bound crystal of carbon. A metallic bond is a multivalent one, each atom in effect sharing one electron with very many other atoms. Other types of bonds are the van der Waals bond, due to the attractive force between the electric dipole moments of two polarizable molecules, and the hydrogenic bond, due to the attractive force between the nucleus of a hydrogen atom and two electronegative atoms. Some organic solids are bound by van der Waals' forces, while water is formed by a hydrogenic bond. The attractive forces in all of these binding mechanisms is fundamentally electrostatic. When the atoms, ions, or molecules get too close to one another, repulsive forces come into play, thus creating the potential well of Fig. 6-1. The repulsive force is also electrostatic, having its origin in the mutual repulsion of the like charges of the electrons orbiting their respective nuclei.

Because of the bonds formed by each atom to its neighbors, the atoms in a solid may form a regular, three-dimensional array called a *crystal*. The actual pattern of the array depends on the nature and form of the binding potential. We should note that the interaction potential does not have to be spherically symmetric as Fig. 6-1 implies. It can depend on direction in space so that, for instance, the equilibrium separation of centers may be different in three different directions. Furthermore, the preferred directions in which binding occurs need not be mutually perpendicular and therefore need not correspond to a cartesian geometry. The structures of actual crystals, the preferred directions for binding and the equilibrium spacing, can be determined experimentally using techniques such as Bragg x-ray diffraction.

Whatever the form of the crystalline array, it must exhibit some symmetry properties. The symmetry properties exist because of the symmetries of the electronic wave functions and the periodicity of the array of potentials. A *lattice* is an infinite three-dimensional array of mathematical points that has the same symmetry as the crystal. The lattice is a mathematical abstraction that has the geometrical properties of real crystals. A real crystal may have an atom or a group of atoms at each lattice point, but it may also have an atom or atoms located symmetrically between lattice sites. The lattice is a useful concept because it permits examination of properties of solids that do not depend on the actual arrangement of atoms at or between the lattice sites.

An important feature of a lattice is that every lattice point has identical surroundings. If an imaginary observer could move from site to site through a lattice, the view from each lattice point would be the same. The structure of the lattice can be described by the three vectors of translation, $\mathbf{d}_1$, $\mathbf{d}_2$, $\mathbf{d}_3$, that recreate the lattice. These vectors are called *basis vectors* and have the direction and magnitude of the minimum translation in space required from one lattice site to another. If the origin is a lattice site, then

$$\mathbf{r} = n_1 \mathbf{d}_1 + n_2 \mathbf{d}_2 + n_3 \mathbf{d}_3 \qquad (6\text{-}1)$$

is also a lattice site provided n_1, n_2, and n_3 are integers. The basis vectors need not be orthogonal. A parallelepiped formed from the basis vectors is called the *primitive cell*. A primitive cell contains lattice sites at each of its corners.

Equation (6-1) arises because of the inherent translational symmetry of all lattices. There are additional symmetries that lattices may possess, among which are rotation about a lattice point, reflection in a plane, and inversion. Each combination of symmetries defines a particular lattice. There are only 14 possible lattices, called *Bravais lattices*, that exhibit different symmetries. The smallest volume that has the symmetry of the lattice is known as the *unit cell*. For 6 of the Bravais lattices, the unit cells are the same as the primitive cells. For the other 8, the symmetries require that the unit cell be larger, containing more than one lattice site.

Whatever its form, every crystal ideally has the symmetry of 1 of the 14 Bravais lattices. In practice, however, solids do not always conform to this idealized picture. There are many types of *imperfections* that can disturb the regularity of the crystalline array as represented by the lattice. Some of these are unavoidable, such as a surface that the infinite Bravais lattice does not have. In principle, others can be controlled, such as the introduction of impurity atoms that distort the binding potentials. Furthermore, many solid materials come in polycrystalline form, not as one large crystal but as many very small crystals randomly oriented and closely packed together. For the present, however, we shall consider pure, regular crystals. We shall return to imperfections in Chapter 8 when we study doped semiconductors.

6.1 THE BAND THEORY OF SOLIDS

A crystalline solid is composed of a regular array of closely packed atoms. The wave functions for the electrons in the solid must then be related to the wave functions for the electrons in the individual atoms themselves. We should not, however, expect the eigenstates to be the same in both cases since the potential function for an electron in a solid may differ from that in an isolated atom. After all, the atoms in a solid are close enough to interact strongly with their nearest neighbors. Therefore it is likely that at least the wave functions of the higher-energy states, which extend far from the nucleus, will be altered by the presence of its neighbors. Consider the following argument.

State Splitting in Coupled Systems

Suppose that we have N identical isolated atoms. For the sake of simplicity we shall consider only three eigenstates for electrons in the atom. Figure 6-2 shows the electronic potential and the allowed energy levels for

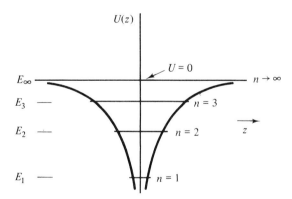

Fig. 6-2 Electronic potential and three eigenenergies of an atom.

this hypothetical case. The length of the line indicating an allowed energy E_n represents the spatial extent of the associated eigenfunction $\psi_n(z)$. When the atoms are isolated from one another, there will be $2N$ electrons with the ground-state energy E_1. If each atom ordinarily has six electrons, there will be $2N$ electrons in each of the three allowed energies. For the system as a whole each state will have a $2N$-fold degeneracy, there being $2N$ states available to electrons each with the same energy. Suppose now that the atoms are arranged in a one-dimensional array. If the spacing of the array is less than the spatial extent of the $n = 3$ wave function of the isolated atom, the wave functions of the isolated $n = 3$ electrons will no longer be valid solutions. When the isolated wave functions overlap, the potential of the electron is not just the sum of the potentials due to the individual atoms but includes a perturbation that is related to the binding forces. The coupling between atoms creates a potential function that alters the wave functions of the electrons. Figure 6-3 shows the electronic potential of a string of three atoms of the

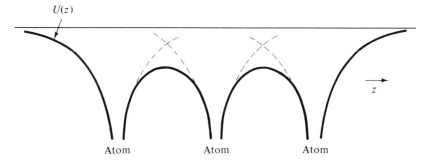

Fig. 6-3 Potential for a string of three atoms of the type shown in Fig. 6-2.

type of Fig. 6-2. The new wave functions have eigenenergies that are slightly different from the original energy E_3. In general if there are N interacting atomic systems, there will be N new wave functions, each with slightly different energy near E_3. Each new state will have only a twofold degeneracy due to spin. By coupling the atoms together, the original energy levels are split and the $2N$-fold degeneracy is reduced to N twofold degeneracies. Figure 6-4 shows the hypothetical energy levels of the three coupled atoms of Fig. 6-3 superimposed on the potential. Notice that the $n = 3$ state is split due to the coupling. The figure indicates that the new wave functions extend through a range encompassing all three atoms. Actually each new wave function will define a probability of finding an electron somewhere, but because of their

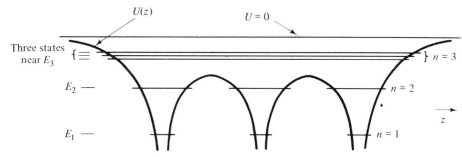

Fig. 6-4 Energy levels for the coupled atoms of Fig. 6-3.

indistinguishability, there is no way of telling which electron "came from" or "belongs to" which atom. Electrons in these states are shared by neighboring atoms, participating in the bond that holds the solid together.

Mechanical Analog

The most important aspect of the coupling of atomic systems for electronics is the splitting of the higher-energy states. Energy splitting due to coupling of systems occurs not only in quantum mechanics but also in classical physics. For example, consider two identical masses m moving on a frictionless plane and each attached separately to a rigid wall by identical springs k_0 as indicated in Fig. 6-5(a). It is well-known that each mass, once

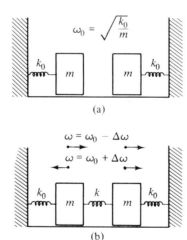

Fig. 6-5 Mechanical analog for state splitting of coupled systems: (a) uncoupled, (b) coupled spring-mass systems.

displaced from its neutral position, will oscillate independently with frequency $\omega_0 = (k_0/m)^{1/2}$. If the two mass-spring systems are coupled together via a third spring k as shown in Fig. 6-5(b), the two masses no longer oscillate independently. It can be shown that any motion of the coupled system is a linear combination of two eigenfunction solutions. Figure 6-5(b) shows the instantaneous velocity vectors for these two solutions, in one case compressing the coupling spring and in the other case not. The oscillation frequencies of these two states differ by $2\,\Delta\omega = (k/k_0)\omega_0$, which is proportional to the strength of the coupling. The twofold degeneracy in frequency that exists for the uncoupled system of Fig. 6-5(a) has been split into two nondegenerate frequencies by the coupling interaction. Furthermore, each of the new eigenstates describes a mode of behavior of the coupled system, not just of one mass or the other.

Energy Bands in Solids

We mentioned earlier that if N atoms were coupled together, the $2N$-fold degeneracy of the states of the isolated system of N atoms would be split into N twofold degenerate states each with slightly different energy depending on the strength of the coupling. In solids N is a very large number so that there are very many eigenstates for the solid with energies in a range near the eigenenergies of the isolated atoms. The range of energy in which these $2N$ discrete states fall is called an allowed *energy band*. Although the eigenenergies of these states are discrete, there are so many states in a band that it is sometimes convenient for practical purposes to consider the permitted energies to be essentially continuous in the range. The density of states in a band is ordinarily approximated to be that of electrons in a three-dimensional infinite square well, Eq. (5-80), even though the confining potentials at the surfaces of a solid are not infinite. The choice of an approximation to the actual density of states (dN_s/dE) is frequently dictated by mathematical convenience.

Figure 6-6 is an example of an energy band diagram. Coupling is stronger

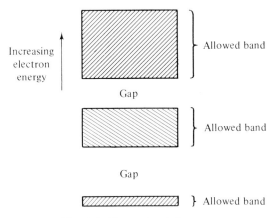

Fig. 6-6 Energy band diagram.

among the higher-energy states; therefore the higher-energy bands are wider, demonstrating the dependence of the splitting on the strength of the coupling. The regions of energy space between the allowed bands encompass forbidden energies and are called *gaps*. There need not, however, be a gap between adjacent bands. In some cases the state splitting is sufficiently strong that neighboring bands overlap, eliminating the gap between them.

In the description of the electrical properties of solids according to their band structure, the location of the Fermi energy E_F with respect to the bands is of critical importance. As we shall see later in this chapter, the aggregate of electrons that occupy *all* of the states in a distinct band cannot support an

Sec. 6-1 The Band Theory of Solids   175

electric current. A band must be partially filled or partially empty in order for current to flow. According to the Fermi-Dirac probability function, only those states with energy less than a few times κT greater than E_F have a high probability of being occupied. Therefore only these allowed states, if they exist, can support a current. It follows that for a solid with good conductive properties the Fermi level must be somewhere in the vicinity of permitted energy states.

The *valence band* is the highest-energy band that is completely filled with electrons at very low temperatures. Since at $T = 0$, $P_{FD} = 1$ for all $E < E_F$, the Fermi energy must lie above the valence band. It is the electrons in valence band states that are responsible, for instance, for covalent bonding. The next band of energy higher than the valence band is called the *conduction band*. Electrons that partially occupy a conduction band can support a conduction current in response to an applied electric field—hence the name of the band.

Figure 6-7 shows the location of the Fermi energy in typical band diagrams of materials classed as insulators, intrinsic semiconductors, metals,

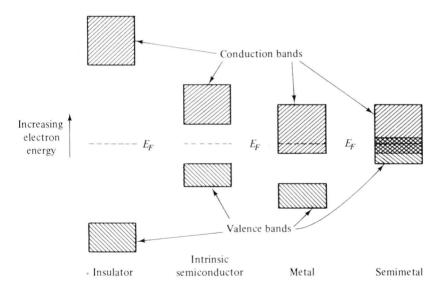

Fig. 6-7 Energy band diagrams with Fermi levels for materials classified as insulators, semiconductors, metals, and semimetals.

and semimetals. For insulators such as diamond, E_F lies in the middle of the gap between the valence and conduction bands. The same is true for intrinsic or pure semiconductors such as silicon or germanium. What distinguishes a semiconductor from an insulator is the width of the forbidden gap. In insulators the gap width is many times κT, so that there is essentially no probability that a state in the valence band is empty and one in the conduction band occupied. The band gap in diamond is about 6 eV wide. In semiconductors, the gap width is much smaller so that there is a nonzero probability depending on the temperature that some valence states will be empty and some conduction states occupied. When this is so, both the conduction electrons *and* the valence electrons can support a current. We shall study semiconductor physics in detail in Chapter 8.

The Fermi level in a metal such as copper lies within the conduction band. Many of the conduction band states are empty so that the conduction band electrons can carry current. A semimetal such as bismuth has overlapping bands, with the Fermi energy lying between the bottom of the upper band and the top of the lower band.

To proceed beyond this crude physical picture of energy bands to understand quantitatively the details of the conductive properties of solids we must develop a mathematical model for the behavior of electrons in solids.

6.2 ELECTRONS IN A PERIODIC POTENTIAL

A quantitative theory for the electronic behavior of solids would properly be based on the following ideal procedure. First assume a form for the potential experienced by the electrons in the regular array of atoms of a crystal including the potential interactions of the electrons themselves. Then solve the quantum mechanical wave equation for the electronic wavefunctions and the allowed energies of the system. From the eigenstates construct wave packets from which a description of the behavior of the electrons in the solid can be obtained. The foregoing is a difficult task, particularly because it attempts to solve a system of very many interacting electrons. A simpler procedure is to assume a periodic potential that accounts for the effects of the periodic array of the crystal atoms and all but one of the electrons. Then solve for the wave functions and allowed states of a single electron in such a system. From the single electron eigenstates, construct wave packets descriptive of the behavior of a particular electron. Next assume that the description of each of the other electrons is identical to that of this one electron, i.e., that all electrons are alike. Finally, distribute the electrons of the solid statistically through the single electron allowed states and calculate the behavior of the system. The latter approximate procedure, the single electron theory, utilizes the concepts of both quantum mechanics and statistical physics that we have already developed. It is the approach that we shall follow.

First we shall assume a periodic potential of amplitude U_0 to represent the periodicity of the lattice. For simplicity we shall restrict the mathematical development to a one-dimensional case. One possibility for the potential might be that sketched in Fig. 6-8(a). Whatever the solution for the station-

Sec. 6-2 Electrons in a Periodic Potential   177

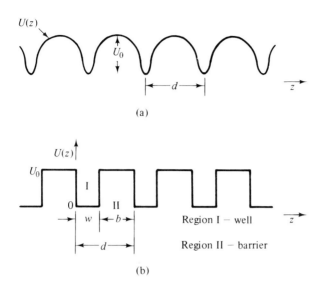

Fig. 6-8 Periodic potentials: (a) realistic potential, (b) the Kronig-Penney model.

ary-state wave function for the periodic potential is, $\psi(z)$ must reduce mathematically to the free-particle wave function, Eq. (5-31), when the amplitude U_0 goes to zero. Therefore we shall assume that $\psi(z)$ has the form

$$\psi(z) = \mu(z)\epsilon^{jkz} \qquad (6\text{-}2)$$

which is called the *Bloch wave function*, where $\mu(z)$ may be a complex number that depends on the amplitude of the potential and $\mu \to 1$ as $U_0 \to 0$. Because of the periodicity of the potential function, the probability density $\psi^*\psi$ must also be periodic. Hence

$$|\psi(z + nd)|^2 = |\psi(z)|^2$$

is true for all integers n, where d is the basis vector of the one-dimensional lattice. Because of the assumed form of $\psi(z)$, the periodicity of the potential also requires

$$|\mu(z + nd)|^2 = |\mu(z)|^2$$

So far the periodic lattices that we have studied are infinite in extent. Real crystals, of course, are finite in size. To simulate a finite crystal of length $L = Nd$, we shall arbitrarily impose periodic conditions on the wave functions of the infinite lattice. That is, we shall assume that

$$\psi(z + Nd) = \psi(z) \qquad (6\text{-}3)$$

This assumption permits us to consider only a segment of length L of the infinite crystal since other segments periodically displaced from it behave in exactly the same manner. Equation (6-3) imposes periodic boundary conditions on the segment of length L consisting of N atoms. Both the magnitude $\mu(z)$ and the phase ϵ^{jkz} of the complex wave function $\psi(z)$ must be periodic with periodicity $L = Nd$ in this model. That is, combining Eqs. (6-2) and (6-3),

$$\mu(z + Nd)\epsilon^{jk(z+Nd)} = \mu(z)\epsilon^{jkz}$$

Because the Bloch amplitude is periodic with periodicity nd for all integers n, it is also periodic for the specific integer N. The periodic boundary condition applied to the phase of $\psi(z)$ requires

$$kNd = 2\pi m, \qquad (6\text{-}4)$$

where m is another integer. The periodic boundary condition within the infinite lattice imposed to simulate a finite crystal therefore restricts k to discrete values. Each wave function $\psi_k(z)$ of the form of Eq. (6-2) has one discrete value of k determined by the boundary conditions of the finite segment of the lattice, but not by the form of $U(z)$ itself. The potential function $U(z)$ determines the Bloch amplitude $\mu(z)$.

The Kronig-Penney Model

Solution for the single electron wave functions $\psi(z)$ depends on the actual form of the periodic potential $U(z)$. Realistic cases are very difficult to solve in analytical form so we shall adopt the Kronig-Penney model of Fig. 6-8(b) as an approximation. The Kronig-Penney model consists of an infinite one-dimensional array of square well potentials of width w separated by potential barriers of height U_0 and width b. Assuming that a wave function of the form of Eq. (6-2) satisfies the wave equation for the stationary state,

$$\frac{d^2\psi}{dz^2} + \frac{2m}{\hbar^2}[E - U(z)]\psi = 0 \qquad (6\text{-}5)$$

we can obtain an equation for $\mu(z)$ by substituting Eq. (6-2) into Eq. (6-5). After some manipulation we find

$$\frac{d^2\mu}{dz^2} + 2jk\frac{d\mu}{dz} + \left[\frac{2m}{\hbar^2}(E - U) - k^2\right]\mu = 0 \qquad (6\text{-}6)$$

Solution for $\mu(z)$ from Eq. (6-6) can be facilitated by further assuming that $\mu(z)$ has the form

$$\mu = \epsilon^{\gamma z} \qquad (6\text{-}7)$$

Substitution of Eq. (6-7) into Eq. (6-6) yields an equation for γ:

$$\gamma^2 + 2jk\gamma + \left[\frac{2m}{\hbar^2}(E - U) - k^2\right] = 0 \qquad (6\text{-}8)$$

Sec. 6-2 Electrons in a Periodic Potential 179

Equation (6-8) has the standard quadratic form, the solution of which is

$$\gamma = -jk \pm \left[\frac{2m}{\hbar^2}(U - E)\right]^{1/2} \quad (6\text{-}9)$$

Substitution of Eq. (6-9) into Eq. (6-7) yields the general solution for the Bloch amplitude.

To solve the specific case of the Kronig-Penney model we must write the wave functions in each region of space and ensure that the necessary conditions on $\psi(z)$ are satisfied at the discontinuities in potential. Let us denote the well as region I and the barrier as region II, and define

$$\alpha \equiv \left[\frac{2mE}{\hbar^2}\right]^{1/2} \quad (6\text{-}10)$$

and

$$\beta \equiv \left[\frac{2m}{\hbar^2}(U_0 - E)\right]^{1/2} \quad (6\text{-}11)$$

Then, according to Eqs. (6-7) and (6-9), complete solutions for $\mu(z)$ in the two regions will be of the form

$$\mu_\mathrm{I}(z) = \epsilon^{-jkz}[A\epsilon^{j\alpha z} + B\epsilon^{-j\alpha z}] \quad (6\text{-}12)$$

$$\mu_\mathrm{II}(z) = \epsilon^{-jkz}[C\epsilon^{\beta z} + D\epsilon^{-\beta z}] \quad (6\text{-}13)$$

The four constants A, B, C, and D must be chosen so that $\psi(z)$ and $d\psi/dz$ are continuous everywhere, especially at the potential discontinuities $z = 0$, and $z = w$. Continuity of both functions at both positions yields the four equations necessary for the solution of the four unknown constants.

It turns out to be more productive to solve for A, B, C, and D by adapting the continuity requirements to $\mu(z)$ rather than to $\psi(z)$. Actually, $\mu(z)$ differs from $\psi(z)$ only by a phase factor that depends on position and not on the local potential. Thus the condition

$$\psi_\mathrm{I}(0) = \psi_\mathrm{II}(0) \quad (6\text{-}14)$$

implies

$$\mu_\mathrm{I}(0) = \mu_\mathrm{II}(0) \quad (6\text{-}15)$$

In addition to Eq. (6-15), continuity requires

$$\mu_\mathrm{I}(w) = \mu_\mathrm{II}(w) = \mu_\mathrm{II}(-b) \quad (6\text{-}16)$$

since $z = w$ and $z = -b$ are the same points by the periodicity. It is again more productive to equate the extreme left- and right-hand sides of Eq. (6-16) in order that the barrier width b enter into the solution directly. Continuity of the derivatives requires

$$\mu'_\mathrm{I}(0) = \mu'_\mathrm{II}(0) \quad (6\text{-}17)$$

and

$$\mu'_\mathrm{I}(w) = \mu'_\mathrm{II}(-b) \quad (6\text{-}18)$$

where μ' is an abbreviated notation for the spatial derivative of μ.

Substitution of μ_I and μ_II from Eqs. (6-12) and (6-13) into the conditions expressed in Eqs. (6-15) through (6-18) yields a set of four homogeneous equations in A, B, C, and D. A nontrivial solution for A, B, C, and D exists only if the determinant of the matrix of their respective coefficients is zero. After considerable algebra it can be shown that the determinant is zero provided

$$\frac{\beta^2 - \alpha^2}{2\alpha\beta} \sinh(\beta b) \sin(\alpha w) + \cosh(\beta b) \cos(\alpha w) = \cos(kd) \qquad (6\text{-}19)$$

where $\sinh(\beta b)$ and $\cosh(\beta b)$ are the hyperbolic functions and $d = b + w$. Equation (6-19) relates the wave number k to the energy and the potential height through α, β, and b. Not all combinations of k, β, and α will satisfy the equation. Those that do satisfy this condition permit the solution for A, B, C, and D and hence of $\mu(z)$ and finally $\psi(z)$. Acceptable wave functions exist only for the allowed values of energy. A sketch of $\Psi^*\Psi$ for a Kronig-Penney potential is displayed in Fig. 6-9 for $E = U_0/2$. The wave function has the form of trigonometric sine functions in the well regions and of hyperbolic sine functions in the barrier regions.

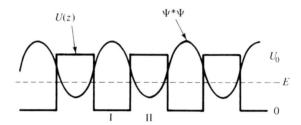

Fig. 6-9 Probability density in a Kronig-Penney potential for the eigenstate with $E = U_0/2$.

A Limiting Case of the Kronig-Penney Potential

Equation (6-19) is very difficult to solve. We can obtain a simpler problem by altering the Kronig-Penney model. Suppose that we let U_0 get large and at the same time let b become small in such a way that the product $U_0 b$ is a constant. The constraint that $U_0 b$ be constant ensures that in some sense the "volume" of the barrier remains constant. According to the discussion of Chapter 5, the probability for tunneling between adjacent wells depends on the barrier "volume," which is a function of the barrier height and width. Therefore, while we alter the height and width of the barriers, we shall not alter the coupling between the wells. In the limit we shall let $U_0 \to \infty$, $b \to 0$, such that $U_0 b$ is a constant.

Let us now perform these limiting operations on Eq. (6-19). Unfortunately

Sec. 6-2 Electrons in a Periodic Potential 181

U_0 does not appear explicitly in the equation. However, notice that, from Eq. (6-11),

$$\lim_{U_0 \to \infty} \frac{\hbar^2 \beta^2}{2m} = U_0 - E \to \infty$$

so that $\beta^2 \to \infty$ as $U_0 \to \infty$. In other words, we can also take the equivalent limit $\beta^2 \to \infty$, $b \to 0$, $\beta^2 b = $ constant. Let us define

$$\beta^2 b \equiv \frac{2P}{w} \qquad (6\text{-}20)$$

where P is a constant that is a measure of the well coupling. It will become clear shortly why we chose this value for $\beta^2 b$. Now Eq. (6-19) involves the factor βb, not $\beta^2 b$, so we must determine what happens to βb in the limit. According to Eq. (6-20), $\beta b = [2Pb/w]^{1/2}$, which goes to zero as b does. Then, for instance,

$$\lim_{\substack{\beta^2 \to \infty \\ b \to 0 \\ \beta^2 b = 2P/w}} \cosh(\beta b) = 1$$

The factor

$$\frac{\beta^2 - \alpha^2}{2\alpha\beta} \sinh(\beta b) \simeq \frac{\beta^2}{2\alpha\beta} \sinh(\beta b)$$

becomes in the limit

$$\lim_{\substack{\beta^2 \to \infty \\ b \to 0 \\ \beta^2 b = 2P/w}} \frac{P}{\alpha w} \frac{\sinh(\beta b)}{\beta b} = \frac{P}{\alpha w}$$

Equation (6-19) then has the limiting form

$$P \frac{\sin(\alpha d)}{\alpha d} + \cos(\alpha d) = \cos(kd) \qquad (6\text{-}21)$$

since $w = d$ in the limit $b \to 0$.

Only certain values of energy or α will satisfy Eq. (6-21). The allowed values depend on the coupling parameter P. For instance, if $P = 0$, Eq. (6-21) has the solution $k = \alpha$. From the definition of α, Eq. (6-10), we find that for $P = 0$ the allowed energies are $E = \hbar^2 k^2/2m$, the energies of a free electron. Thus $P = 0$ corresponds to the elimination of the potential barrier or perfect coupling between the wells. If $P \to \infty$, Eq. (6-21) can be satisfied only if sin $\alpha d = 0$ or $\alpha = \pm n\pi/d$. Again using Eq. (6-10), we find that the allowed energies are $E_n = \pi^2 \hbar^2 n^2 / 2md^2$, which are the energy levels of the infinite square well potential. Thus $P \to \infty$ corresponds to infinite barriers between wells or isolated wells.

Determination of the allowed states for intermediate values of P requires the formal solution of Eq. (6-21), which is a transcendental equation. Clearly, the equation has a solution only when the left-hand side equals the right-hand side. The right-hand side of Eq. (6-21) is limited to values between $+1$ and

−1; therefore solutions exist only when the left-hand side has values between +1 and −1. Figure 6-10 indicates the values of αd that permit the solution of Eq. (6-21) for the case $P = 5$. Since α^2 is proportional to E, the ranges of allowed αd represent permissible ranges of energy, the energy bands. The ranges of E for which αd does not allow a solution of Eq. (6-21) are the energy gaps. Notice that the higher-energy bands are wider than the lower bands, as our earlier notions implied.

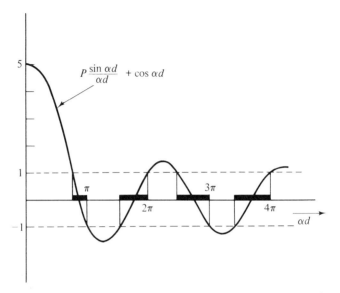

Fig. 6-10 Left-hand side of Eq. (6-21) vs. αd for $P = 5$. Values of αd that permit solution of the equation are indicated by the darkened segments of the αd axis.

The E vs. k Diagram

For each value of E for which a solution of Eq. (6-21) exists there are two values of k. A plot of E vs. k for the simplified Kronig-Penney potential is shown in Fig. 6-11. Notice that the diagram is symmetric in k and that the discontinuities in energy occur for values of $k = n\pi/d$. The ranges of k corresponding to the various energy bands are called *Brillouin zones*. The first Brillouin zone contains values of k satisfying

$$-\frac{\pi}{d} < k < \frac{\pi}{d}, \quad \text{first Brillouin zone.} \quad (6\text{-}22)$$

The second Brillouin zone has

$$\frac{\pi}{d} < |k| < \frac{2\pi}{d}, \quad \text{second Brillouin zone,} \quad (6\text{-}23)$$

and so forth.

Sec. 6-2 Electrons in a Periodic Potential

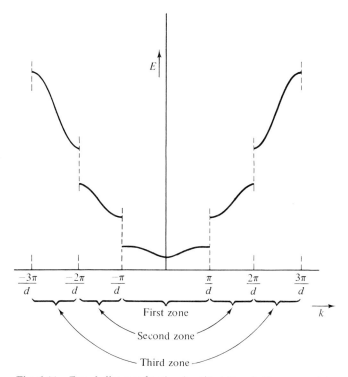

Fig. 6-11 E vs. k diagram for the simplified Kronig-Penney model.

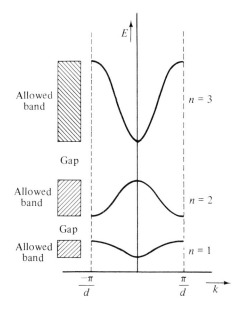

Fig. 6-12 E vs. k diagram for the simplified Kronig-Penney model reduced to the first Brillouin zone.

An alternative representation of the information in an E vs. k diagram such as Fig. 6-11 is possible. The so-called *reduced zone* E vs. k diagram is sketched in Fig. 6-12. Figure 6-12 is obtained from Fig. 6-11 by shifting the energy curves in k-space by factors of $n\pi/d$. The wave number k is unchanged by such a change because of the periodicity of $\psi(z)$. The result of these shifts is that the various bands are plotted against values of k in the first Brillouin zone. To designate a state in the reduced zone representation, it is necessary to specify both k and the zone number n.

Behavior of Electrons in a Periodic Potential

The E vs. k diagram for electrons in a periodic potential reveals their characteristic behavior. Let us suppose that an electron is in a state specified by k and a zone number n. Actually, of course, a particle is represented by a wave packet as described in Chapter 5. For particles in a periodic potential the packet is composed of the eigenfunctions $\psi_{k,n}$. If the amplitude spectrum $A(k)$ of the packet is sharply peaked about k_0, we can expect the particle to have the group velocity, which, from Eq. (5-49), is

$$v_g = \frac{1}{\hbar}\frac{\partial E}{\partial k}\bigg|_{k_0} \qquad (6\text{-}24)$$

Notice that the electron velocity is proportional to the slope of the E vs. k diagram evaluated at its expected state. This velocity is independent of position even though the potential is periodic in space since the state is descriptive of the whole system. If the wave-packet spectrum is sharply peaked about k_0, then the uncertainty relation indicates that the electron will not be well localized in space.

Furthermore, it should be remembered that not all values of k are permitted but that k is restricted by the periodicity conditions expressed in Eqs. (6-3) and (6-4). According to Eq. (6-4),

$$k = \frac{2\pi m}{Nd} \qquad (6\text{-}25)$$

where m is an integer, d is the basis vector of the lattice, and Nd is the distance over which ψ is required to be periodic. Another approach to Eq. (6-25) is to assume that a finite one-dimensional crystal that is N primitive cells long can be represented by any N cells of an infinite lattice provided ψ is periodic with period Nd. In the first Brillouin zone, k is further restricted by Eq. (6-22) so that $|m| \leq N/2$. Each state occupies a volume in the first Brillouin zone of $2\pi/Nd$. The density of states therefore is uniform and has the value

$$\frac{dN_s}{dk} = \frac{Nd}{2\pi} \qquad (6\text{-}26)$$

Sec. 6-3 The Pseudoclassical Description

The total number of states in each band is

$$\int_{-\pi/d}^{\pi/d} 2\left(\frac{dN_s}{dk}\right) dk = 2N \qquad (6\text{-}27)$$

having taken spin into account. In practice N is generally such a large number that energy bands are frequently considered to be ranges of continuous allowed energy when in fact they are ranges of energy each encompassing $2N$ discrete energy levels.

The electrical current carried by electrons in the states of a particular energy band is given by

$$\mathbf{J} = \sum_{\substack{\text{occupied} \\ \text{states} \\ \text{of band}}} (-e)\mathbf{v}_i \qquad (6\text{-}28)$$

where the sum is over the occupied states of the band and $\mathbf{v}_i$ is the expected velocity of an electron described as being in the ith state. The expected velocity is proportional to the slope of the E vs. k diagram. In its reduced representation, $E(k)$ always exhibits even symmetry, $E(-k) = E(k)$, and its derivative exhibits odd symmetry,

$$\left.\frac{\partial E}{\partial k}\right|_{-k} = -\left.\frac{\partial E}{\partial k}\right|_{k} \qquad (6\text{-}29)$$

From Eqs. (6-24) and (6-29) it is easy to see that a particle in the state $(-k, n)$ has an expected velocity equal to but directed oppositely from that of a particle in the state $(+k, n)$. It follows from this argument as applied to Eq. (6-28) that if all the states in a band are occupied, the electrons in these states do not support an electric current. Conversely, only the electrons in a partially empty or a partially filled band can conduct current.

6.3 THE PSEUDOCLASSICAL DESCRIPTION

Beginning with the quantum description of an electronic wave packet in a periodic potential it is possible to obtain a formal mathematical expression that looks like Newton's second law. In this section we shall do this, developing a very useful pseudoclassical description for these electrons. A word of caution to remember: Electrons in solids cannot properly be described as classical particles. That is why this method is called *pseudoclassical*. As we proceed we shall expand upon the difference between the pseudoclassical description and a classical one.

We shall start with the expected velocity of an electronic wavepacket, Eq. (6-24). The electron moves through the one-dimensional crystal with velocity v_g. Let us now apply a small electric field $\boldsymbol{\mathcal{E}} = -\nabla V$ for a short time

δt and assume that the energy states and wave functions are not affected by the field. The work done on the electron during the interval δt is

$$\delta E = -e\mathcal{E}\,\delta z = -e\mathcal{E} v_g\,\delta t \qquad (6\text{-}30)$$

On the other hand,

$$\delta E = \frac{\partial E}{\partial k}\,\delta k = v_g \hbar\,\delta k \qquad (6\text{-}31)$$

By eliminating δE from Eqs. (6-30) and (6-31), we find

$$\hbar\,\delta k = -e\mathcal{E}\,\delta t \qquad (6\text{-}32)$$

If we take the limit as $\delta t \to 0$, we get

$$\hbar\frac{dk}{dt} = F \qquad (6\text{-}33)$$

where we have denoted $-e\mathcal{E}$ by F, the external force on an electron in the field $\mathcal{E}$. Equation (6-33) looks like Newton's second law, where $\hbar k$ acts as an *effective momentum*. Actually $\hbar k$ is not the true momentum of an electron in a solid since the electron is subject to the internal forces of the crystal as well as the external force. Equation (6-33) is, however, a useful result.

The real acceleration of an electron would be manifested in the time rate of change of its expected velocity:

$$\frac{dv_g}{dt} = \frac{d}{dt}\left(\frac{1}{\hbar}\frac{\partial E}{\partial k}\right) = \frac{1}{\hbar}\frac{\partial^2 E}{\partial k\,\partial t} \qquad (6\text{-}34)$$

We can rewrite Eq. (6-34) using

$$\frac{\partial E}{\partial t} = \frac{\partial E}{\partial k}\frac{dk}{dt}$$

so that

$$\frac{dv_g}{dt} = \frac{1}{\hbar}\frac{\partial^2 E}{\partial k^2}\frac{dk}{dt}$$

or

$$\frac{dv_g}{dt} = \frac{1}{\hbar^2}\frac{\partial^2 E}{\partial k^2}\frac{d(\hbar k)}{dt} \qquad (6\text{-}35)$$

Combining Eqs. (6-35) and (6-33), we obtain the form

$$F = \left[\frac{\hbar^2}{\partial^2 E/\partial k^2}\right]\frac{dv_g}{dt} \qquad (6\text{-}36)$$

That is, when an external force is applied to an electron in a crystal, its expected velocity changes *as if* it had an *effective mass*:

$$\boxed{m^* = \frac{\hbar^2}{(\partial^2 E/\partial k^2)}} \qquad (6\text{-}37)$$

The effective mass takes into account the interaction of the particle with the crystal potential as well as the external field. The instantaneous value of the effective mass of an electron depends on the state that it occupies. The effective mass m^* need not be the same as the rest mass of a free electron $m_e = 9.10956 \times 10^{-31}$ kg. Because it depends on the curvature of the E vs. k diagram, m^* may be either positive or negative.

We shall use the concept of effective mass extensively in the rest of this book. Its usefulness is surprising. Once the effective masses are known we can essentially ignore the crystal and deal with particles of the appropriate m^* since the interactions with the periodic crystal are included in m^*. The effective mass appears in the transport coefficients of mobility and diffusion. And, importantly, m^* in real crystals is subject to direct measurement.

6.4 REAL CRYSTALS

The simplified one-dimensional Kronig-Penney model is useful because it is relatively easy to solve and it demonstrates many of the features that distinguish the behavior of electrons in solids. However, some additional features appear in real three-dimensional crystals.

The fundamental complication of three-dimensional crystals is that they are in general anisotropic. Properties of the medium such as mobility may depend on orientation with respect to the lattice structure. When an electric field is applied to an anisotropic crystal, the current that flows depends on the orientation of the field with respect to the crystal axes. In three dimensions the wave number is a vector **k** that points in the direction of propagation of the wave. The components of **k** are easier to work with than the vector itself so it is customary to define certain directions in the crystal and the components of **k** in these directions. The directions are chosen with consideration for the lattice symmetry and are not always in the direction of the basis vectors. Anisotropy is evidenced in different potentials and/or different lengths of periodicity in each direction. Therefore the $E(k_i)$ diagram depends on the lattice direction indicated by the subscript i.

Even though an E vs. k_i diagram may depend on lattice direction, it is still symmetrical in k_i. Therefore an $E(k_i)$ plot in only half of the first Brillouin zone contains the same information as a plot in the entire zone. Figure 6-13 shows the combined $E(k_i)$ diagram for the valence and conduction bands of germanium in two directions, labeled [100] and [111], respectively. The combined $E(k_i)$ diagram utilizes the symmetry property of $E(k_i)$ to pack more information into one illustration.

The $E(k_i)$ diagram of Fig. 6-13 demonstrates some additional features of the band structure of real crystals. First, in real crystals there are usually many overlapping bands in the same range of energies. Overlapping bands do not occur in the Kronig-Penny model because the uncoupled wave functions

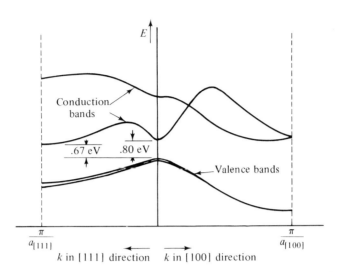

Fig. 6-13 E vs. k diagram for germanium at room temperature in the [100] and [111] directions.

are those of electrons in a square well potential, which are not degenerate. Overlapping occurs in real crystals because the uncoupled wave functions are atomic ones similar to those of the hydrogen atom in which there are many eigenfunctions with the same energy for the higher states.

Next, the effective mass in each direction may be different. That is, m^* is anisotropic and should be represented by a matrix array of values rather than by one value.

Because of the overlapping bands, the possibility exists that two or more bands may coincide in some part of **k**-space. In this case there may be two or more states with the same **k** and E, but the electrons in these states may have different effective masses. Furthermore, the effective mass of an electron in a specific band may depend strongly on which state of the band it occupies. Evidently the effective mass can become quite a complicated concept.

Fortunately, the curvature of $E(k)$ is relatively well defined at the bottom and top of each band. In semiconductor electronics, the states of prime importance are those near the top of the valence band and the bottom of the conduction band. When it is possible to approximate the actual $E(k)$ dependence with a parabolic form, $E = \alpha k^2$, then the effective mass is a constant. In most of what follows in this book we shall assume that the band structure is parabolic at the top of the valence band and at the bottom of the conduction band and therefore that the effective masses at the band edges are independent of k.

Another feature of the $E(k_i)$ diagrams for real crystals is the possibility of two types of energy gaps between bands, direct and indirect gaps. A *direct*

Sec. 6-4 Real Crystals
189

gap is one for which the highest energy of the valence band and the lowest energy of the conduction band occur for states with the same k. At room temperature the direct gap in germanium has a width of 0.80 eV, as pictured in Fig. 6-13. An *indirect gap* is one for which the energies on the opposite sides of the gap occur for states with different k. The indirect gap in germanium is 0.67 eV wide.

Real energy bands may have more than one minimum and one maximum, as with the lowest conduction band in germanium. Furthermore, the maxima and minima may occur at values of k other than $k = 0$ or at the zone edge.

Measuring Effective Mass by Cyclotron Resonance

One of the most important techniques of determining the band structure in solids is the measurement of the effective masses. Alternatively, measurement of the effective mass tensor itself may be an end of interest because of its relationship to the transport coefficients. Cyclotron resonance is the procedure most commonly employed to determine values of m^*.

A classical particle of charge $-e$ in a uniform static magnetic field **B** experiences the force

$$m\frac{d\mathbf{v}}{dt} = -e\mathbf{v} \times \mathbf{B} \tag{6-38}$$

It is a well-known result from electromagnetics that the orbit of the particle as a result of this force is a circle in a plane perpendicular to **B**. The particle gyrates at constant energy in its orbit at the *cyclotron frequency* ω_c:

$$\omega_c = \frac{eB}{m} \tag{6-39}$$

For an electron in a solid, Eq. (6-33) of the pseudoclassical description implies

$$\hbar\frac{d\mathbf{k}}{dt} = -e\mathbf{v}_g \times \mathbf{B} \tag{6-40}$$

where $\mathbf{v}_g$ is the expected or group velocity of the wave packet. Equation (6-40) holds provided the magnetic field is not sufficiently strong as to alter the band structure present in the absence of **B**. Equation (6-40) can be converted into a relation for the **k**-space behavior of the electron by employing the three-dimensional generalization of Eq. (6-24) for $\mathbf{v}_g$ with the result

$$\hbar\frac{d\mathbf{k}}{dt} = -\frac{e}{\hbar}\frac{\partial E}{\partial \mathbf{k}} \times \mathbf{B} \tag{6-41}$$

In Eq. (6-41) the cross-product is between the vector gradient in **k**-space and **B**. Therefore as the particle moves from state to state in **k**-space it does so on a plane perpendicular both to **B** and to the gradient in energy. That is, the energy is a constant of the motion as in the case of the classical particle.

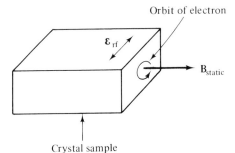

Fig. 6-14 Physical arrangement of a cyclotron resonance experiment to measure effective mass.

Furthermore, it can be shown that the particle gyrates in k-space and in coordinate space at the frequency

$$\omega_c = \frac{eB}{m_t^*} \tag{6-42}$$

where m_t^* is an effective mass in the plane perpendicular to the magnetic field.

One method of measuring the transverse effective mass utilizing the cyclotronic motion is illustrated in Fig. 6-14. The crystal is placed in a uniform static magnetic field. A high-frequency electric field is applied in a direction transverse to **B** but parallel to a surface of the crystal. How far the electric field penetrates into the crystal depends on the electromagnetic skin depth for the particular material and frequency. In semiconductors the rf field ordinarily penetrates the crystal completely so that $\mathcal{E}$ is spatially uniform over the circular orbit of an electron. When the magnitude of **B** is properly adjusted so that the electron gyrates at the same frequency as the applied electric field, the electron experiences a constant field in its frame of reference. The electron is accelerated by the field and gains energy from it. If the cyclotron frequency differs from the applied rf frequency, the electron experiences a time-varying field, being alternately accelerated and decelerated by it and gaining no net energy from it. The experimental effect is a resonant absorption of energy when the cyclotron frequency matches the rf frequency.

In metals the rf frequency field does not penetrate the solid. Therefore the field is spatially nonuniform. Only when the electron comes close to the surface does it experience the electric field. Resonant absorption occurs if the particle returns to the surface during each orbit at the time when the field is at the same phase. Resonance occurs not only when $\omega_{rf} = \omega_c$ but also at harmonics of the cyclotron frequency.

Measured values of effective mass vary from as little as 1% to as much as or more than the mass of a free electron.

BIBLIOGRAPHY

Introductory treatments of the quantum behavior of electrons in solids may be found in

KITTEL, C., *Elementary Solid State Physics.* New York: John Wiley & Sons, Inc., 1962.

ROSE, R. M., L. A. SHEPARD, and J. WULFF, *Structure and Properties of Materials,* Vol. IV, *Electronic Properties.* New York: John Wiley & Sons, Inc., 1966.

LONGINI, R. L., *Introductory Quantum Mechanics for the Solid State.* New York: John Wiley & Sons, Inc. (Interscience Division), 1970.

More advanced texts are

DEKKER, A. J., *Solid State Physics.* Englewood Cliffs, N.J.: Prentice-Hall, Inc., 1957.

KITTEL, C., *Introduction to Solid State Physics* (4th ed.). New York: John Wiley & Sons, Inc., 1971.

LEVINE, S. N., *Quantum Physics of Electronics.* New York: The Macmillan Company, 1965.

WANG, S., *Solid State Electronics.* New York: McGraw-Hill Book Company, 1966.

PROBLEMS

6-1. Suppose that all states but the state ψ_s in the *n*th energy band of a solid are occupied by electrons. What is the current that these $2N - 1$ electrons carry? *Hint:* Use Eq. (6-28).

6-2. The velocity of electrons in the states of the simplified Kronig-Penney model at the top and bottom of each band is zero. Show that the wave functions with $k = \pm \pi/d$ at the zone boundaries satisfy a Bragg reflection condition for waves in a one-dimensional lattice.

6-3. Sketch m^* vs. k for energy bands $n = 2$ and $n = 3$ of Fig. 6-12 for the simplified Kronig-Penney model.

6-4. What is the effective mass m^* for a free electron with $E = \hbar^2 k^2 / 2m_e$?

6-5. Show that Eq. (6-41) has an orbital solution in k-space at the cyclotron frequency. *Hint:* Take the time-derivative of Eq. (6-41) and make an analogy to Eq. (6-38) for a classical particle.

6-6. Both Eq. (6-41) and the discussion of cyclotron resonance that follows it neglect the effects of collisions of the electrons with crystal imperfections. Suppose that the electrons have a mean collision time τ with imperfections. How do you suppose that τ must be restricted so that the results of experiments described in Section 6-4 would be unaffected by these collisions?

6-7. A cyclotron resonance experiment shows strong absorption of rf energy at 24 GHz in a magnetic field $B = 8.60 \times 10^{-2}$ T. Find m_t^*.

7

Metals

A metal is a crystalline material whose electronic configuration and bonding properties are such that the Fermi energy lies in the conduction band. Because $\kappa T \ll E_F$, there are very many states in the conduction band that are occupied by electrons and very many that are not as dictated by the Fermi-Dirac probability function P_{FD}. In fact, for most metals there is about one electron in the conduction band for each atom of the crystal, about 10^{22} conduction electrons/cm^3.

A model for the electronic properties of a metallic crystal of length L based on the energy band configuration is shown in Fig. 7-1. The conduction band electrons are bound within a finite potential well of depth $-E_c$, where E_c is the lowest allowed energy in the conduction band. An electron outside the solid is not bound to the crystal and is therefore free. A free electron may have any energy E greater than its potential energy U, which for reference we shall take to be zero. The Fermi level is somewhere in the conduction band. The minimum difference in energy between a free electron and one with $E = E_F$ is called the *work function* W. We shall assume that the density of allowed states in the conduction band, (dN_s/dE), is the same as that for electrons of an appropriate effective mass in a three-dimensional infinite square well. Equation (5-80) must be modified to reflect the model that takes E_c as the minimum energy. The density of states in the conduction band is

$$\left(\frac{dN_s}{dE}\right)_{\text{conduction band}} = \frac{V}{2\pi^2}\left(\frac{2m^*}{\hbar^2}\right)^{3/2}(E - E_c)^{1/2} \qquad (7\text{-}1)$$

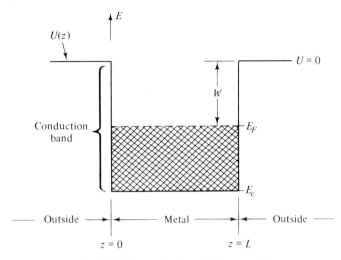

Fig. 7-1 Energy band model for a metal.

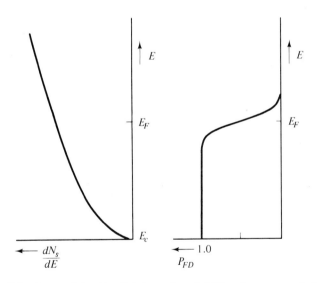

Fig. 7-2 Assumed density of states and probability of occupancy in the conduction band of a metal.

Figure 7-2 shows the density of states and the occupation probability P_{FD} plotted against energy for $\kappa T \ll E_F$. Notice that most of the states below E_F are occupied. There are, however, many available empty states within a few κT of E_F that are not occupied. Effectively an electron can assume almost any energy in this range. Because of the high density of electrons and the availability of states, a metal may be considered to be a plasma of highly mobile pseudoclassical electrons in a fixed background of the ionic charges of the crystal. Some authors refer to this model as an electron gas, the electrons being free to move about in the metal as molecules move in a gas. Other authors use the term "free electron theory of metals." We shall avoid the latter terminology so as not to confuse conduction band electrons with electrons that have escaped the metal and are truly free.

The electronic properties of metals such as conductivity arise from the transport of charge. In principle the transport theory of Chapter 3 is applicable to metals with appropriate modifications to account for quantum mechanical effects. For example, the classical distribution function $f(\mathbf{r}, \mathbf{p})$ must be replaced by the product $(dN_s/dE)P_{FD}$.

Suppose as an example that we wish to calculate the mobility of the conduction electrons in a metal. We could do so by determining $(dN_s/dE)P_{FD}$ in the nonequilibrium case of an applied electric field by solving the kinetic equation (3-30). Having obtained the nonequilibrium distribution of electrons among the allowed states, we could then calculate the average velocity of the system and relate that to the electric field through the mobility. In practice this is a difficult task with pitfalls at every step.

One problem encountered in attempting to solve the kinetic equation is how to handle the collision term. It should be remembered that the assumed band structure of the crystal already accounts for the interaction of each electron with all the atoms and all the other electrons in the lattice structure. Yet collisional effects occur because every disturbed system approaches equilibrium when the disturbance is removed. The collisional term in the kinetic equation may be the only one available to drive the system toward equilibrium. The relaxation approximation might be employed if the disturbance is small and if the relaxation time τ has meaning. In Chapter 3 we discovered that τ was effectively the mean collision time. The question is, "collisions with what?" The answer is, "with imperfections in the regular array of the crystalline lattice upon which the band theory is based." It is impossible for a real crystal not to have some imperfections, as we shall discover when we discuss this problem in more detail in the next chapter. In solids at room temperature, τ ranges over values from 10^{-14} to 10^{-12} sec.

Another problem is the actual solution of the kinetic equation in a quantum mechanical formulation and the subsequent calculation of the average velocity. In the pseudoclassical description, the kinetic equation is formulated in $\mathbf{k}$-space since the effective momentum is $\hbar \mathbf{k}$ and the velocity of an electron is the group velocity. Assuming that the kinetic equation in $\mathbf{k}$-space can be solved, the subsequent calculation of the average velocity of the electrons is mathematically difficult to do, involving P_{FD}, (dN_s/dE), $(\partial E/\partial k)$, and τ.

While calculation of the transport coefficients of metals from the kinetic theory is conceivably possible, it is not always necessary. An easier approach would be to use a pseudoclassical mean electron model like the one discussed at the beginning of Section 3.4. Suppose that the average electron of charge

$-e$ and mass m^* loses all of its momentum at each collision with an imperfection. The pseudoclassical equation of motion for this electron in an electric field $\mathcal{E}$ is

$$m^* \frac{d\mathbf{v}}{dt} = -e\mathcal{E} - \frac{m^*\mathbf{v}}{\tau} \qquad (7\text{-}2)$$

For time variations of the form $\exp\{j\omega t\}$, Eq. (7-2) becomes

$$m^*(j\omega\tau + 1)\mathbf{v} = -e\mathcal{E}\tau \qquad (7\text{-}3)$$

or

$$\mathbf{v} = \frac{-e\tau}{m^*(1 + j\omega\tau)}\mathcal{E} \qquad (7\text{-}4)$$

From Eq. (7-4) the low-frequency mobility for the electron plasma in a metal is

$$\mu = \frac{e\tau}{m^*} \qquad (7\text{-}5)$$

where τ is the mean collision time and m^* is the average effective mass. This pseudoclassical approach is very useful not only because it is simple but also because the average effective mass is a measurable quantity.

The primary effect of an increase in temperature on the mobility of the electrons in a metal is a decrease in the mean collision time τ. As we shall see in Chapter 8, the effective number of a certain class of imperfection increases with temperature so that interactions occur more frequently at higher temperatures. Generally metallic conductors have negative temperature coefficients of conductivity, meaning that as temperature increases conductivity decreases. The change in E_F or P_{FD} due to a small temperature change has little effect on mobility or conductivity because E_F lies in the band where a large number of states are always unoccupied.

7.1 ELECTRON EMISSION

An important electronic property of metals is their ability to emit electrons through a variety of physical processes. In this section we shall discuss photoemission and secondary emission, which are due to the impact on a metal of optical photons and particles, respectively, emission due to the application of high electric fields, and thermal emission from a heated cathode. All of these processes depend on the behavior of electrons at the potential barriers represented by the surfaces at $z = 0$ and $z = L$ in Fig. 7-1. Thermal emission is particularly important not only in vacuum tube devices but also in semiconductor junction devices where the flow of charge carriers from one semiconductor to another may be impeded by the natural potential barrier that appears between two dissimilar materials.

Photoemission

The phenomenon of photoemission was introduced in Chapter 5. It has been observed that optical radiation of an appropriate frequency v incident upon a cold photocathode causes electrons to be emitted and that the maximum kinetic energy of the photoelectrons depends on v and the cathode material according to

$$(\tfrac{1}{2}mv^2)_{\max} = hv - W \tag{7-6}$$

This experimental observation is easy to understand using the energy band model of metals. Suppose we have a metal at a low temperature that for practical purpose we can take as $T = 0$. Then all the electrons will occupy states with $E < E_F$. Figure 7-3 shows the energy diagram of such a metal and

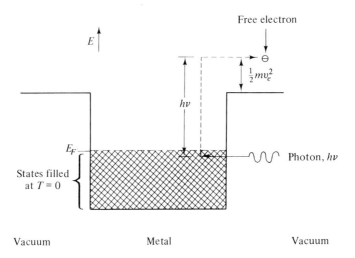

Fig. 7-3 Energy diagram for photoemission.

illustrates the energy considerations that explain photoemission. A single photon hv is incident upon the metal. If the photon is absorbed, it must give up all of its energy according to the quantization hypothesis. If an electron can absorb this energy, it will move up in the energy diagram by an amount hv. If the electron originally occupied a state within hv joules of the exterior potential and if the electron after absorbing the quantum of energy moves toward the surface, there is a finite probability that the electron will overcome the potential barrier represented by the surface of the solid and escape. Photoemission is a three-step process: the electron-photon interaction, transport of the electron to the surface, and emission over the potential barrier. The most energetic of the emitted electrons come from states at or near E_F so that their energy when free is given by Eq. (7-6).

Sec. 7-1 Electron Emission 197

The work function W is equivalent to an energy threshold for the incident photons. Typical values for useful photocathode materials range between 1 and 5 eV depending on the material, its crystalline structure, and the condition of its surfaces. It is common for the surfaces of solids to be coated with impurities such as oxides formed by the absorption of oxygen from the atmosphere. Surface impurities may increase or decrease the effective work function of a material, making it more or less useful as a photoemitter.

Secondary Emission

Electron emission from the surface of a solid that has been bombarded by charged particles, electrons or ions, is called secondary emission. There are three stages in the secondary emission process: The primary particle enters the solid and delivers energy to an electron, and the energetic electron diffuses to the surface and depending on conditions escapes. Ordinarily emission of secondary electrons occurs less efficiently when the incident primaries are heavy ions than when they are electrons, all other things being equal, because electron-electron collisions are more effective in the transfer of energy.

The number of emitted secondaries per primary is γ, the secondary yield coefficient. Figure 7-4 shows a typical plot of γ as a function of the energy of primary electrons. Notice that the yield rises to a maximum value that may be greater than 1 and then falls off at higher energy. For a potential secondary electron to escape it must gain an energy of about W from the primary, de-

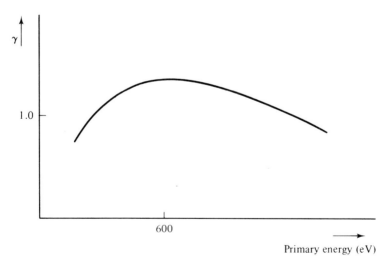

Fig. 7-4 Secondary yield coefficient vs. energy of the primary electrons for tungsten.

pending on its initial state. A single high-energy primary may give this energy to a number of electrons as it penetrates the solid. However, at higher energies the primaries penetrate far into the solid. The potential secondaries that are created far from the surface have little probability of escaping because they tend to lose their excess energy while diffusing toward the surface.

Figure 7-5 illustrates schematically the essential features of a *photomultiplier tube* utilizing both photoemissive materials and secondary emission.

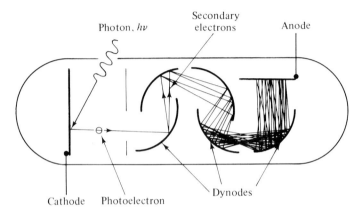

Fig. 7-5   Photomultiplier tube.

The purpose of such a device is to detect very low-level light signals. Light strikes a photocathode, which emits a photoelectron. The photoelectron is accelerated by an electric field to the first of a series of *dynodes*. The dynodes are made of material with high secondary yield. The secondary electrons are then accelerated into the second dynode, causing even more electrons to be emitted from it. The amplification process is repeated at each subsequent dynode until a sufficiently large signal is obtained. Practical tubes may have as many as 13 stages with current gains as much as 10^8. Alkali metals and their compounds are the materials most frequently used to coat the electrodes for photoemission and secondary emission in these devices.

Thermal Emission

Another very important emission process is thermal emission. Because the electrons are distributed in energy with the probability function P_{FD}, at nonzero temperatures some electrons have energy greater than E_F. A small fraction of these may even have energy sufficient to overcome the work function potential barrier at the surface and escape. Figure 7-6 shows an energy diagram at the surface of a metal and $P_{FD}(E)$ for an elevated temperature.

Let us calculate the thermally emitted current density using a statistical approach. We shall assume that only those electrons with energy $E > E_F +$

Sec. 7-1 Electron Emission

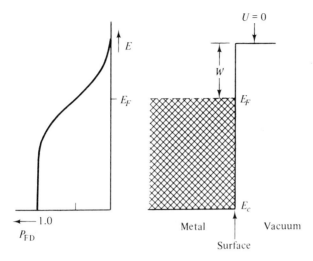

Fig. 7-6 Energy diagram with $P_{FD}(E)$ illustrating thermal emission from metals.

W directed toward the surface are emitted. The bottom of the conduction band E_c is a convenient reference for energy. The current density of the dn particles with velocity in the range dv_z about v_z is

$$dJ_z = -ev_z \, dn \tag{7-7}$$

The required number of particles dn expressed in terms of a phase-space distribution function is

$$dn = \int_{p_x=-\infty}^{\infty} \int_{p_y=-\infty}^{\infty} f(\mathbf{r}, \mathbf{p}) \, dp_x \, dp_y \, dp_z \tag{7-8}$$

For electrons, $f(\mathbf{r}, \mathbf{p})$ is just $(dN_s/d\mathbf{p})P_{FD}(\mathbf{p})$, where $(dN_s/d\mathbf{p})$ is the density of states in momentum space and $P_{FD}(\mathbf{p})$ is the Fermi-Dirac probability expressed in terms of momentum. To obtain the total emitted current density we need only substitute Eq. (7-8) into (7-7) and integrate over p_z from p_{zW} to infinity where

$$\frac{p_{zW}^2}{2m^*} = E_F + W \tag{7-9}$$

That is,

$$J_z = -e \int_{-\infty}^{\infty} \int_{-\infty}^{\infty} \int_{p_{zW}}^{\infty} \frac{p_z}{m^*} \left(\frac{dN_s}{d\mathbf{p}}\right) \left\{ \frac{1}{\epsilon^{[(p^2/2m^*)-E_F]/\kappa T} + 1} \right\} dp_x \, dp_y \, dp_z \tag{7-10}$$

To carry out the integral of Eq. (7-10) we need the density of states. For electrons in a three-dimensional square well of volume V, the density of states discounting spin was obtained in Eq. (5-77):

$$\left(\frac{dN_s}{d\mathbf{p}}\right) = \frac{V}{\pi^3 h^3} = \frac{8V}{h^3} \tag{7-11}$$

In the derivation of Section 5.6 that led up to this result, all the states were counted without regard to the sign of the momentum. However, the integral of Eq. (7-10) requires the explicit integration over each state whether its momentum is positive or negative. In order not to count the contributions of each state twice we must multiply Eq. (7-11) by $(\frac{1}{2})^3$ since half of the particles moving in each direction have positively directed momentum and half negatively directed momentum. On the other hand, we must account for the two possible values of spin so the appropriate density of states per unit volume is

$$\left(\frac{dN_s}{d\mathbf{p}}\right) = \frac{2}{h^3} \qquad (7\text{-}12)$$

The actual integration of Eq. (7-10) using Eq. (7-12) for the density of states is facilitated by recognizing that for the particles that escape $[(p^2/2m^*) - E_F]/\kappa T \gg 1$ so that the factor of unity in the denominator of P_{FD} can be neglected compared with the exponential. Equation (7-10) then becomes

$$J_z = \frac{-2e}{m^* h^3} \epsilon^{E_F/\kappa T} \int_{-\infty}^{\infty} \int_{-\infty}^{\infty} \int_{p_{z,W}}^{\infty} p_z \epsilon^{-p^2/2m^*\kappa T} dp_x \, dp_y \, dp_z \qquad (7\text{-}13)$$

The integrations over p_x and p_y each yield a factor of $(2\pi m \kappa T)^{1/2}$, leaving

$$J_z = \frac{-2e}{m^* h^3} (2\pi m^* \kappa T) \epsilon^{E_F/\kappa T} \int_{p_{z,W}}^{\infty} p_z \epsilon^{-p_z^2/2m^*\kappa T} dp_z$$

or

$$J_z = \frac{-2e}{m^* h^3} (2\pi m^* \kappa T) \epsilon^{E_F/\kappa T} (m^* \kappa T) \epsilon^{-p_{z,W}^2/2m^*\kappa T} \qquad (7\text{-}14)$$

By combining terms and using Eq. (7-9) for p_{zW}^2 we can transform Eq. (7-14) to

$$J_{z\,\text{thermal}} = \frac{-4\pi e m^*}{h^3} (\kappa T)^2 \epsilon^{-W/\kappa T} \qquad (7\text{-}15)$$

The minus sign in Eq. (7-15) accounts for the sign convention on current relative to the direction of motion of electrons. It is customary to denote the constants in Eq. (7-15) by A so that in magnitude

$$\boxed{J_{\text{thermal}} = AT^2 \epsilon^{-W/\kappa T}} \qquad (7\text{-}16)$$

Equation (7-16) is known as the Richardson-Dushman relation and shows the dependence of the thermally emitted current density on the barrier height above E_F as well as the temperature.

The constant A has a theoretical value of 1.20×10^6 A/(m°K)^2 for the free electron mass. It, along with the thermionic work function W, is amenable to measurement. Table 7-1 lists characteristic values of the measured parameters of commonly used thermal emitters. One reason A may be less than the value determined by the constants in Eq. (7-15) is that we have neglected the possibility of reflection of the electron back into the solid when it

Table 7-1. Thermal Emission Characteristics
of Common Cathode Materials

	Melting point (°K)	Average operating temperature (°K)	Work function (eV)	A [10^6 A/(m-°K)^2]
Tungsten	3683	2500	4.5	0.60
Tantalum	3271	2300	4.1	0.4–0.6
Cesium	303	293	1.93	1.62
Nickel coated with oxides of barium and strontium	—	1100	1.0	10^{-4}–10^{-5}

arrives at the surface. Quantum mechanically a particle with energy greater than the potential barrier that it is approaching has a finite nonzero probability of being reflected by the barrier as well as a finite probability of passing beyond the barrier.

Listed in Table 7-1 is a commonly used thermionic cathode made of nickel coated with a thin layer of barium oxide and strongium oxide. The oxide coating substantially reduces the work function for the cathode and lowers its operating temperature, but also greatly decreases the current density that can be drawn from such a cathode. The work function of nickel without a coating is 5.03 eV.

Field Emission

The emission of electrons from metallic surfaces is affected by the presence of an applied electric field. The effect of the field is to alter the potential energy of an electron outside the metal from that shown in Fig. 7-1. An electric field does not ordinarily alter the band structure of a metal because the electron screening length λ_e is very short and the field does not penetrate beyond the surface. Two changes occur in the surface barrier when a field is applied with the proper polarity to draw electrons away from the surface: The barrier height is reduced, allowing more thermal electrons to escape, and the thickness of the barrier is reduced so that electrons can tunnel out of the crystal.

Reduction of the barrier height in an applied electric field is called the *Schottky effect*. Figure 7-7 illustrates the effect on the electronic potential of an electric field applied to the surface of a realistic metallic solid. In Fig. 7-7(a) a realistic potential near the surface is sketched and compared with the idealization that we have heretofore employed.

It is sometimes useful to assume the following model to develop such a potential. An electron a distance z in front of a metallic surface induces net positive charge on the surface. The field of this induced charge is such as to

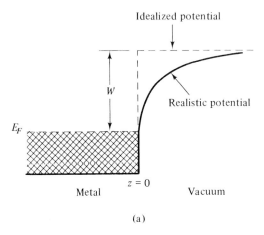

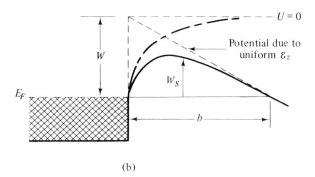

Fig. 7-7 Electronic potentials near the surface of a metal: (a) realistic potential, (b) with an applied electric field.

attract the electron to the surface. The field of the induced charge is the same as that of a hypothetical image charge $+e$ located a distance z behind the surface, $e/4\pi\varepsilon_0(2z)^2$. The potential of the electron in the image field is $U_{\text{image}} = -e^2/16\pi\varepsilon_0 z$. The potential of Fig. 7-7(a) has approximately this form for $z > 0$ except near $z = 0$.

Suppose now that we apply a uniform electric field ε_z at the surface of the metal. The total potential W_s of an electron outside the solid written with respect to the Fermi energy is

$$W_s = W - \frac{e^2}{16\pi\varepsilon_0 z} - e\varepsilon_z z \qquad (7\text{-}17)$$

It is easy to show that the maximum value of the potential barrier represented by the potential of Eq. (7-17) is

$$W_{s\,\max} = W - \frac{e}{2}\sqrt{\frac{e\mathcal{E}_z}{\pi\varepsilon_0}} \qquad (7\text{-}18)$$

The effect of the electric field is to lower the barrier height in proportion to $(\mathcal{E})^{1/2}$. The thermionic current density from a properly biased and heated cathode would be

$$J_{\text{thermal}} = AT^2\epsilon^{-W/\kappa T}\epsilon^{-\Delta W/\kappa T} \qquad (7\text{-}19)$$

where $\Delta W = W_{s\,\max} - W$, as determined from Eq. (7-18).

The current-voltage characteristic of a representative thermionic vacuum diode is sketched in Fig. 7-8. The regime in which the diode current was

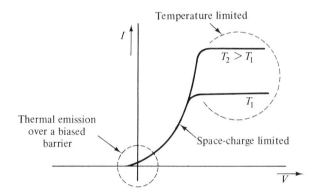

Fig. 7-8 *I-V* characteristic of a thermionic diode for two emission temperatures.

limited by the presence of space-charge was discussed in Section 4.2. The region of the characteristic near the origin exhibits behavior typical of thermal emission over a biased barrier. At high-current levels the characteristic may saturate, the current being limited by the emission temperature of the electrons. Figure 7-8 also illustrates the behavior of the *I-V* curve as a function of temperature.

If a very high electric field is applied to the surface of a metal the potential barrier of Fig. 7-7(b) may become thin. The barrier thickness b at the Fermi energy is approximately

$$b \simeq \frac{W}{e\mathcal{E}_z} \qquad (7\text{-}20)$$

Electrons may tunnel through this barrier with a probability that depends on both $W_{s\,\max}$ and b. While the tunneling probability may be small, there are very many electrons with energies near E_F that might tunnel out. It can be shown quantum mechanically that the current density due to tunneling electrons has the form

$$J_{\text{field}} = B\mathcal{E}_z^2 \epsilon^{-\Phi/\mathcal{E}_z} \qquad (7\text{-}21)$$

where B and Φ are constants that depend on W and E_F. Equation (7-21) has been verified experimentally. Notice the similarity in form between Eq. (7-21) for field emission current and Eq. (7-16) for the thermal emission current.

Field emission occurs for electric fields of the order of 10^9 V/m. Such fields are not uncommon in front of sharp corners and points on a conductor in a high-voltage circuit. The field-emitted electrons in these circumstances may initiate a corona discharge that is visible as a glow in the atmosphere surrounding the emitting region. Electron microscopes operate on a similar principle, electrons being field-emitted from a sharp, pointed electrode and accelerated toward a fluorescent screen. For an appropriate field pattern between the point and the screen, the pattern illuminated on the screen reproduces the emission pattern of the material of the electrode. The emission pattern is related to the crystalline structure. Thus, electron microscopes are used to investigate the structure of solids.

7.2 CONTACT POTENTIALS

When two dissimilar metals or semiconductors are brought in contact with one another, an electric field is developed in a transition region between them in much the same manner as in the sheath region formed between a plasma and a foreign body. The potential difference of this field is called the *contact potential*. Observations of the effects of the contact potential between metals were first reported by Galvani in 1780 in his famous experiments on quivering frogs legs and by Volta in 1800 with his voltaic pile. The lead-zinc wet cell battery is the modern form of the voltaic pile. More importantly for the purposes of this study, the operation of electronic devices made of junctions between a metal and a semiconductor or between two semiconductors depends on the existence of the sheath region and the electric field that exists in it.

Suppose that we have two dissimilar metals at low temperature. The energy diagrams for the two individual metals are sketched in Fig. 7-9. Each has its own work function and its own Fermi energy. When the two metals are brought into contact, electrons may flow from one to another either by thermal emission over the barrier between them or by tunneling through that barrier. For the metals pictured in Fig. 7-9 more electrons will flow from metal 2 into metal 1 as the two are brought together because there are more electrons in higher-energy states in metal 2 and there are more empty available states at lower energy in metal 1. Thus metal 1 will build up a net negative

Sec. 7-2 Contact Potentials

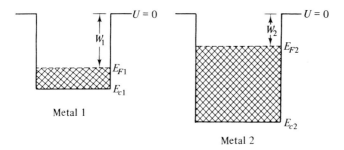

Fig. 7-9 Energy diagram for two dissimilar metals at $T = 0$.

charge. An electric field will arise between the metals with such polarity as to make it more difficult for additional electrons to leave metal 2. In the steady state electrons may flow in both directions, but there can be no net flow of particles from one metal to the other. At the absolute zero of temperature, $T = 0$, no net flow occurs only when the two Fermi levels are at the same energy, since all states below E_F are filled at $T = 0$. It can be demonstrated from thermodynamic considerations that at any temperature the equilibrium condition occurs when the two Fermi levels line up. In fact, in equilibrium there can be only one Fermi energy, the chemical potential for electrons in the system composed of the two metals.

Figure 7-10 shows the energy diagram of the two-metal system in equilibrium. Because the Fermi levels are at the same energy, the potentials just

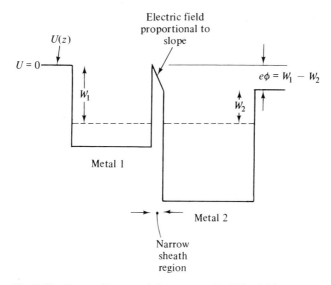

Fig. 7-10 Energy diagram of the two metals of Fig. 7-9 in contact.

outside the surfaces of the two metals differ by the difference in the work functions of the metals. The contact potential ϕ is defined as this difference expressed in volts:

$$e\phi = W_1 - W_2 \qquad (7\text{-}22)$$

The electric field that creates the potential difference appears in a sheath region between the two metals. The width of this transition region is of the order of the plasma screening length, which is a very small distance in metals.

BIBLIOGRAPHY

General references for the material in this chapter on electron emission are

ROSE, R. M., L. A. SHEPARD, and J. WULFF, *The Structure and Properties of Materials*, Vol. IV, *Electronic Properties*. New York: John Wiley & Sons, Inc., 1966.

RYDER, J. D., *Electronic Fundamentals and Applications* (4th ed.). Englewood Cliffs, N.J.: Prentice-Hall, Inc., 1970.

More detailed treatments are found in

LEVINE, S. N., *Quantum Physics of Electronics*. New York: The Macmillan Company, 1965.

VAN DER ZIEL, A., *Solid State Physical Electronics* (2nd ed.). Englewood Cliffs, N.J.: Prentice-Hall, Inc., 1968.

The book by van der Ziel also contains a brief treatment of contact potentials.

PROBLEMS

7-1. Find the mean collision time τ in copper if its conductivity is 6×10^7 mhos/m, the electron density is 8.5×10^{28} m^{-3}, and the effective electron mass $m_e^* = 1.012 m_e$.

7-2. Estimate the photoelectric threshold frequency $\nu_{\min}$ for the material that yielded the data of Fig. 5-1(b).

7-3. Use the *Handbook of Physics and Chemistry* or a similar reference to compare the work functions of tungsten, tantalum, cesium, and germanium as obtained from experiments on photoemission, thermionic emission, and contact potentials.

7-4. Plot the thermionic current density from tungsten as a function of temperature by calculating J_{thermal} at several temperatures using Eq. (7-16) and the data of Table 7-1.

7-5. Show that the maximum surface potential $W_{s\,max}$ for a metal in the presence of an applied field $\mathcal{E}_z$ occurs at a distance from the surface of

$$z_{max} = \left[\frac{e}{16\pi\varepsilon_0 \mathcal{E}_z}\right]^{1/2}$$

Verify Eq. (7-18).

7-6. Use Eq. (7-20) to calculate the electric field required to reduce the surface barrier width b to 2×10^{-7} m for aluminum. Take the work function of aluminum to be 4.08 eV.

8

Semiconductors

Elemental semiconductors are materials with relatively small energy gaps between their valence and conduction bands and with their Fermi levels located somewhere in the gap. If we denote the highest energy in the valence band by E_v and the lowest in the conduction band by E_c, the gap width is $E_g = E_c - E_v$ and $E_v < E_F < E_c$. By a relatively small E_g we mean that while it may be many times κT, the gap energy is still small enough that electrons may be found in conduction band states with nonzero probability. Figure 8-1 shows an energy diagram for a semiconductor with E_F in the middle of the gap and the occupation probability P_{FD} at a finite temperature. Notice that the tail of P_{FD} is small but not zero at energies lying above E_c. Furthermore, the probability that a state is not occupied, $1 - P_{FD}$, is not zero at energies below E_v. In the low-temperature limit all allowed states below E_F would be occupied and those above E_F would be empty.

In elemental semiconductors E_F always lies above the top of the valence band but below the bottom of the conduction band. The Fermi energy in insulators also lies in the gap; however, E_g for insulators is much larger than for semiconductors. Therefore the probability that an electron would vacate a valence state of the insulator and occupy a conduction state is remote. Diamond, for example, has a band gap of 5.4 eV, while silicon and germanium, semiconductors with diamond-like crystalline structure, have gaps of 1.11 and 0.67 eV, respectively, at room temperature. In principle both the conduction band and valence band electrons in both semiconductors and insulators can

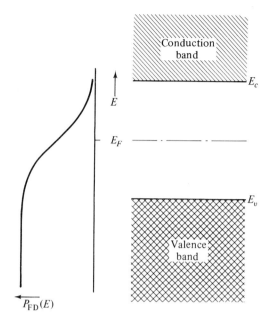

Fig. 8-1 Energy diagram of an intrinsic semiconductor with $P_{FD}(E)$.

support electrical current since these bands are either partially filled or partially empty at room temperature. The probabilities for insulators are so small, however, that practically no carriers exist in the conduction band nor are there unoccupied states in the valence band. The probabilities for semiconductors are greater, and electrons occupy the conduction band, leaving vacancies in the valence band. The carrier densities in semiconductors depend on E_g and are sensitive to the temperature through P_{FD}. Recall that in metals E_F lies in the conduction band so that there are many available carriers. The range of typical values of conductivity in metals, semiconductors, and insulators indicated in Table 3-1 is due primarily to the range of charge carrier densities in the respective materials.

Many semiconductors exist, both in the form of elemental materials such as silicon (Si), germanium (Ge), selenium (Se), and tellurium (Te) and as compounds such as indium antimonide (InSb), gallium arsenide (GaAs), and lead telluride (PbTe).

The most prevalent form in which semiconducting materials are used in electronics is not, however, the pure or intrinsic form. Semiconductors doped with impurities, so-called *extrinsic semiconductors*, have electronic properties that depend strongly on the impurities designed into them. As such, engineers have been able to design and manufacture these materials to perform according to need. Extrinsic semiconductors are the subject of most of this chapter, and their applications to electronics are the subject of the rest of this book. In this chapter we shall study impurities in Section 8.1. Then in Sections 8.2 and 8.3 we shall consider the charge densities in semiconductors and the current they carry. Section 8.4 is devoted to examples of bulk semiconductor devices and in Section 8.5 we shall investigate contacts between metals and semiconductors.

8.1 IMPURITIES AND EXTRINSIC SEMICONDUCTORS

Up to now our development of the behavior of electrons in solids has been based on the idealized model of a perfectly regular crystalline structure. This assumption is useful because it provides a clear physical picture that is relatively easy to describe mathematically. In reality, of course, this picture is an oversimplification of the truth. Solids are materials that are crystalline in form but not perfectly regular. The irregularities are called imperfections, defects, or faults. Imperfections are extremely important for the electrical and other transport properties as well as for the structural properties of solids. For example, the idealized model of an electron in a solid described in Chapter 6 presumed to account for all of the interactions of that electron with the regular lattice of atoms and all of the other electrons. A consequence of this model was the pseudoclassical description which, by itself, does not predict an Ohm's law relationship between an applied electric field and the resultant current density in the material. It is necessary to invoke "collisions" of some kind to achieve the correct observable relationship. Interactions with the regular crystalline atoms and all of the other electrons have already been taken into account. It is the interaction of the charge carriers with the irregularities of the structure that plays the primary scattering role of collisions.

If there are not too many of these irregularities we can base a model of a real solid on that of a perfect solid assuming that the small number of imperfections perturb the idealized model only slightly. In this section we shall examine various types of imperfections and how they alter the energy band picture of Fig. 8-1, with particular emphasis on the development of p-type and n-type extrinsic semiconductors.

Imperfections

Basically there are two general classes of irregularities by which real crystals differ from their idealized models, essential and nonessential imperfections. An *essential imperfection* is one that always exists regardless of how ideally the solid is prepared. *Nonessential imperfections* exist because the material cannot be prepared with perfect regularity by imperfect men and machines. Conceptually, however, the nonessential imperfections need not

Sec. 8-1 Impurities and Extrinsic Semiconductors

exist if techniques could be perfected. The essential imperfections, as we shall see, must always exist independently of manufacturing skills.

One type of essential imperfection that perturbs the periodic potential of a uniform regular array of lattice atoms or ions occurs because real crystals are of finite size whereas the mathematical lattice is infinite in extent. Surfaces represent essential imperfections not only because of the end of the periodicity of the crystal but also because of their physical or chemical condition. Mechanical polishing or chemical treatment of a solid alters its surface. When a solid is exposed to an atmosphere, atoms or molecules may be absorbed on its surfaces. Because these effects alter the configuration of the energy bands as do nonessential irregularities, we shall postpone our discussion momentarily until we have completed a categorization of the other irregularities.

A second essential imperfection is the random vibrational motion of the atoms, ions, or molecules of the crystal about the lattice sites. The crystal is bound together by forces with potential functions of the form sketched in Fig. 6-1. While neighboring ions are bound in the potential minimum, they are not motionless. They have thermal energy measured macroscopically by the crystal temperature. This energy is exhibited in an oscillatory motion about the nominal location of the potential minimum. Quantum mechanically the energy of this thermal motion is quantized since the atom is bound in a potential well. The quanta of energy are called *phonons* and their energy is proportional to the vibrational frequency. Even at the absolute zero of temperature, crystal atoms perform oscillatory motion, the energy of which is called the zero point energy. The uncertainty principle prohibits the atoms from being at rest at their nominal separation at $T = 0$ since neither position nor momentum can be known precisely.

When a solid is heated, phonons are created. The vibration of crystal atoms about their lattice sites is chiefly responsible for the specific heat of solids. Electrons may also interact with the vibrations. Such an interaction may transfer energy either from the electron to the vibrational energy, thus creating phonons, or from the lattice to the electron, thus destroying phonons. Naturally during an electron-phonon interaction the electron must change state so the interaction can occur only if an unoccupied allowed state is available into which the electron may transfer. Electron-phonon "collisions" are the principal effects that cause a disturbed electron distribution to relax toward equilibrium in the sense of nonequilibrium statistical mechanics. These are the interactions principally responsible for Ohm's law in solids.

The second general class of imperfections is the nonessential of which there are three types: point, line, and plane imperfections. An ordinary solid is not a single crystal but is composed of many small crystallites packed together. The orientation of adjacent crystallites may differ, giving rise to an irregularity in the potential along their interface. The interface is called a grain boundary and the defect a plane imperfection. Most semiconductor crystals prepared for solid-state devices are in the form of single crystals. Among the line imperfections is a variety of dislocations due, for example, to slip or to shear in the crystal.

By far the most significant type of nonessential defect for electronics is the point imperfection. Point defects occur when an atom is not in its customary position in the lattice. For example, an atom may be missing from its lattice site, causing a *vacancy*. Or it may be situated between regular lattice sites provided it does not cause too much distortion in the lattice. The latter defect is called an *interstitial*. Most importantly, an atom may be replaced by a foreign atom with different atomic radius and electronic structure, an *impurity*. Replacement defects are the ones used to create p-type and n-type semiconductors for device applications by introducing appropriate impurity atoms into the host crystal. The properties of these doped semiconductors are said to be extrinsic since they were externally controlled.

The Effects of Imperfections on Electronic States

Recall that the solution of Schrödinger's equation for the perfectly periodic potential of an ideal crystal yields eigenenergies and wave functions for a single electron in the entire system of coupled atoms. The allowed energies fall into bands. If the density of defects is so large as to destroy the fundamental periodicity of the lattice, then the solution for the allowed states is very difficult. If, on the other hand, the density of defects is low and does not alter the crystal periodicity, the solution for the allowed states can be obtained by considering the imperfections as a perturbation on the perfect crystal.

Consider, for example, replacement defects occurring at a frequency of 100 ppm of the host crystal. We can develop the following physical picture. An impurity replaces an original atom at a lattice site. The original atom contributed two states (counting spin) of each allowed energy band for the electrons of the solid. The replacement atom may have a totally different electronic structure. The allowed states in the solid due to the impurity may lie either in the original bands or in the previously forbidden gaps. Effectively a replacement removes two states from each band and replaces them with two new impurity states that may or may not be in the bands.

Unlike the original wave functions that extend throughout the solid, the wave functions of the impurity states are localized to the neighborhood of the imperfection in the otherwise periodic potential. An electron in one of these states is locally bound near the imperfection. What is meant by "near the imperfection" depends on the atoms involved. For example, an impurity-

Sec. 8-1 Impurities and Extrinsic Semiconductors

state wave function may extend over many lattice spacings and still be associated with the localized impurity.

For the case of interstitial defects, states may be added to the system without the removal of any original states from the bands depending on whether or not vacancies also occur.

If impurities of the same atoms are present with sufficient density that the impurity wave functions overlap, the defect states will split in energy, forming an allowed energy band of impurity states. The wave functions of impurity band states are not localized.

Classification of Bound States

The bound states present in solids due to localized imperfections are classified according to their location in the energy band diagram and their function. Their presence effectively decreases the mobility of the carriers in the bands. States in the originally forbidden gap between the valence and conduction bands are called donors, acceptors, traps, or recombination centers. These states also serve as scattering centers, effectively decreasing the mobility of the charge carriers.

The *donor* states lie close to the bottom of the conduction band, as illustrated in Fig. 8-2(a). In a host lattice of a valence IV semiconductor (Si,

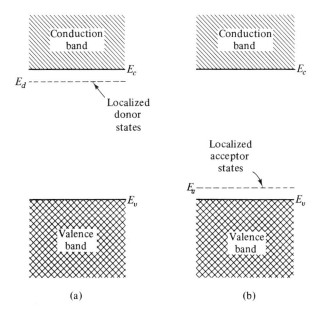

Fig. 8-2 Energy diagram of extrinsic semiconductors showing localized (a) donor and (b) acceptor states.

Ge), donor states arise from the replacement of an atom of valence IV with one from the valence V group in the periodic table of elements. Typically, phosphorus, arsenic, or antimony is used, replacing 1 to 1000 atoms per million of silicon or germanium. Physically the bond between the valence V impurity and its valence IV neighbors requires only four electrons. Thus the fifth electron of the impurity is not required for the bonding and is only loosely bound to the impurity. Because an electron in this donor state lies so close in energy to the conduction band, it may easily absorb thermal energy from phonons and be excited into a conduction band state. Thus the impurity donates electrons to the conduction band. Typical separations between the bottom of the conduction band E_c and the donor level E_d are between 0.01 and 0.05 eV. The effectiveness of supplying donor electrons to the conduction band depends on the temperature. The presence of donor impurities creates a partially filled conduction band that can support a current. When an electron is excited out of a donor state into the conduction band, the immobile impurity atom is left with a positive charge.

The *acceptor* states lie close to the top of the valence band, as illustrated in Fig. 8-2(b). In a host lattice of a valence IV semiconductor impurities from the valence III series of the periodic table create acceptor states. Boron, aluminum, gallium, and indium are examples. Bonding between the trivalent impurity atom and its neighbors is deficient by one electron. The energy state of an electron that would complete the bond is E_a, the acceptor level. Because typical separations $E_a - E_v$ are also between 0.01 and 0.05 eV, there is considerable probability depending on the temperature that an electron may be thermally excited from a valence band state to an acceptor state. The impurity state accepts electrons from the valence band into its incomplete bond. Once in an acceptor state the electron is localized and immobilized. However, the remaining valence band states are no longer completely filled so that they can now support a current.

The location of other types of bound states, traps and recombination centers, are shown in an energy diagram in Fig. 8-3. The distinction between shallow and deep traps and recombination centers depends on the location of the state within the gap and on what happens to electrons in them. For instance, electron traps are localized states into which conduction electrons may fall. Once in a trap an electron has a probability of being thermally excited back into the conduction band. The excitation probability depends on the difference in energy between the trap energy and E_c; the greater the difference, the less likely the excitation. A recombination center is a localized state in the gap from which the electron has a dominant probability of falling down to an unoccupied valence band state.

Any given impurity state may perform more than one function. The designation of states is not as important as the behavior of the electrons in them in relation to the states of the bands.

Sec. 8-1 Impurities and Extrinsic Semiconductors 215

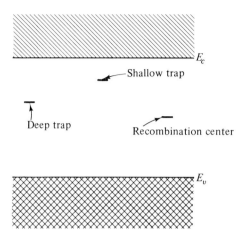

Fig. 8-3 Energy diagram showing traps and a recombination center.

The Fermi Level in Extrinsic Semiconductors

In *intrinsic* semiconductors the Fermi energy lies in or near the middle of the energy gap. Actually if the band structure were symmetrical with respect to the middle of the gap, E_F would be exactly in the middle for the following reason. The occupation probability $P_{FD}(E)$ has the symmetry property $P_{FD}(E_F + \Delta E) = 1 - P_{FD}(E_F - \Delta E)$. The number of electrons occupying conduction states must equal the number of vacant valence states since the electrons are thermally excited across the gap at nonzero temperatures. The number of occupied states in a range of energy is given by the product of P_{FD} and the density of available states. Since the number in the conduction band equals the number having vacated the valence band and because of the symmetry of P_{FD}, E_F lies in the middle of the gap if the band structures as described by the density of states are also symmetric. As we shall see in the next section, the Fermi energies in intrinsic germanium and silicon usually lie slightly toward the valence band from the middle of the gap because the band structures are not symmetric.

In *extrinsic* semiconductors with *donor* impurities the Fermi level at $T = 0$ lies between E_d and E_c since all of the bound states due to valence V impurities will be occupied at the absolute zero of temperature. As the temperature is increased, electrons will be excited out of the donor states into the conduction band. Since P_{FD} is a monotonically decreasing function of energy, the probability of occupancy for a conduction state is always less than for a donor state lying at lower energy. However, the density of conduction states is so much greater than that of the bound states that almost all of the donor electrons will be excited into the conduction band at moderate (room) temperatures, greatly increasing the conductivity of the material. At higher temperatures appreciable numbers of electrons may be excited thermally from the valence to the conduction band. Depending on the temperature and the impurity density, the electrons elevated from the valence band may far outnumber the donor electrons. In the latter case the semiconductor behaves like an intrinsic one and E_F is near the middle of the gap. Thus the Fermi level depends on the temperature moving from at or above E_d toward the middle of the gap with increasing temperature.

In *extrinsic* semiconductors with *acceptor* impurities, the acceptor states are all empty at $T = 0$ so the Fermi level lies between E_c and E_a. At high temperatures intrinsic behavior may again overwhelm the properties due to the presence of the impurities, in which case E_F lies near the middle of the gap. Thus the Fermi level moves from at or below E_a toward the middle of the gap with increasing temperature for this type of extrinsic semiconductor.

In the next section we shall study the thermal excitation of electrons and the location of the Fermi level quantitatively. Before proceeding, however, we shall anticipate the subject matter of Chapter 9, junctions between unlike extrinsic semiconductors, and state that the proper functioning of junction devices depends on maintaining a difference in the location of the Fermi level in relation to the band edges on either side of the junction. That is, the two materials must be extrinsic and different. At high temperatures doped semiconductors behave intrinsically. Junction devices therefore will not operate at arbitrarily high temperatures.

8.2 ELECTRONS AND HOLES

To examine quantitatively the density of charge carriers in semiconductors it is convenient to introduce the concept of a hole as an effective mobile positive charge carrier.

The Concept of Holes

Holes are a way of describing the behavior of partially empty energy bands. Recall that in Chapter 6 we discovered that only the electrons in a partially filled or partially empty band could support an electrical current because the net velocity of electrons in a completely filled band is

$$\sum_{j=1}^{2N} \mathbf{v}_j = 0, \quad \text{all states occupied} \quad (8\text{-}1)$$

where the summation is over the allowed states in the band, all of which are filled. Equation (8-1) results from the symmetry of the $E(k_l)$ energy band

Sec. 8-2 Electrons and Holes

structure with the velocity of an electron in the jth state being

$$\mathbf{v}_j = \frac{1}{\hbar} \frac{\partial E}{\partial \mathbf{k}}\bigg|_{\mathbf{k}_j} \tag{8-2}$$

For each state $\mathbf{k}_j$ with velocity $\mathbf{v}_j$ there is a state $-\mathbf{k}_j$ with velocity $-\mathbf{v}_j$ so that Eq. (8-1) holds.

On the other hand, suppose that all but the sth state is occupied. The current density due to the $2N - 1$ electrons is

$$\mathbf{J} = \left(-\frac{e}{V}\right) \sum_{\substack{j=1 \\ j \neq s}}^{2N} \mathbf{v}_j \tag{8-3}$$

where $-(e/V)$ is the charge density due to a single electron in the crystal of volume V. The summation in Eq. (8-3) may be rewritten as

$$\mathbf{J} = -\frac{e}{V}\left[\left(\sum_{j=1}^{2N} \mathbf{v}_j\right) - \mathbf{v}_s\right] \tag{8-4}$$

which can be evaluated using Eq. (8-1), yielding

$$\mathbf{J} = \frac{+e}{V}\mathbf{v}_s \tag{8-5}$$

In other words the net effect of all of the electrons occupying all but the sth state is the same as that of a single positive charge with a velocity the same as that of an electron in the sth state. Rather than deal with the $2N - 1$ electrons it is more convenient to deal with the single effective positive charge carrier called a *hole* that "occupies" the empty state.

When an electron is excited from a state in the valence band to one in the conduction band a hole is generated in the valence band. Both the electron in the conduction band and the hole in the valence band may carry current. A hole moves when an electron occupies its state, leaving another state unoccupied. When donor impurities are ionized an electron is elevated to the conduction band, leaving an impurity ion that is positively charged but immobile. When electrons are excited from the valence band into acceptor states, holes are generated in the valence band. These holes in the valence band can carry current, but the electrons in the acceptor states are bound and not mobile. De-excitation of a conduction electron into an unoccupied state in the valence band is called *electron-hole recombination* and eliminates both carriers from their respective bands.

When the electrons in a partially empty band gain energy from some source they move upward in the energy diagram and lower-energy states are vacated. Therefore, holes that gain energy move down in the energy diagram as illustrated in Fig. 8-4.

The pseudoclassical effective mass $m_h^*(k_s)$ of a hole in state k_s can be determined using the following argument. We have already determined that the hole velocity $v_h(k_s)$ is equal to the velocity of an electron in that state $v_e(k_s)$.

Increasing
electron
energy

Increasing
hole
energy

Fig. 8-4 Energy diagram for holes and electrons.

The holes move with the valence band electrons. It follows that the acceleration of a hole, $dv_h(k_s)/dt$, is equal to the acceleration of an electron in that state, $dv_e(k_s)/dt$. In the pseudoclassical description the acceleration equals the external force divided by the effective mass. For an external electric field $\mathcal{E}$

$$\frac{+e\mathcal{E}}{m_h^*} = \frac{-e\mathcal{E}}{m_e^*} \qquad (8\text{-}6)$$

or

$$\boxed{m_h^*(k_s) = -m_e^*(k_s)} \qquad (8\text{-}7)$$

Equation (8-7) is not cause for alarm: Recall that the effective mass is proportional to the curvature of the $E(k)$ band structure and that the curvature may be positive or negative. It always turns out that the band curvature is positive at the bottom of each band and negative at the top. In other words, an electron at or near the bottom of the conduction band has positive effective mass. An electron at or near the top of the valence band has a negative effective mass. But according to Eq. (8-7), a hole at or near the top of the valence band has positive mass as well as positive charge. Such a hole behaves pseudoclassically like a real particle with positive charge and positive mass.

Thermal Generation of Hole-Electron Pairs

By the generation of a hole-electron pair we mean the excitation of an electron into a conduction band state, creating a hole in the valence band state. There are three processes by which electrons and holes are generated in pairs within the bulk of semiconductors, by absorption of either thermal or electromagnetic energy or by impact ionization. In impact ionization the required energy is transferred from the kinetic energy of an energetic particle that collides with the neutral lattice atom, ionizing it.

Sec. 8-2 Electrons and Holes 219

Thermal generation of hole-electron pairs occurs when energy is transferred from lattice vibrations to valence band electrons. In thermal equlibrium the rate at which hole-electron pairs are being generated equals the rate at which the carriers are recombining since there must be an equilibrium distribution for holes and for electrons. The equilibrium distribution of electrons is given by the occupation probability P_{FD} multiplied by the density of states in the conduction band. The equilibrium distribution of holes is given by the probability of vacancy, $1 - P_{FD}$, multipled by the density of hole states in the valence band.

Let us now turn to a quantitative description of the equilibrium densities of holes and electrons in their respective bands. The carrier densities depend on the location of the Fermi level relative to the band edges, the temperature, and the densities of states in the respective bands. The number of electrons in the conduction band is

$$nV = \int_{\text{conduction band}} \left(\frac{dN_s}{dE}\right)_c P_{FD}(E)\, dE \tag{8-8}$$

where n is the spatially uniform electron density and V is the volume of the crystal. The lower limit of the integration in Eq. (8-8) is E_c, the energy at the bottom of the conduction band. The upper limit can be taken as $E \to \infty$ since $P_{FD}(E)$ vanishes rapidly with increasing energy. For the density of states we shall use the model of electrons in a three-dimensional square well, Eq. (5-80),

$$\left(\frac{dN_s}{dE}\right)_c = \frac{V}{2\pi^2}\left(\frac{2m_e^*}{\hbar^2}\right)^{3/2}(E - E_c)^{1/2} \tag{8-9}$$

where the reference for electron energy is taken to be the bottom of the band. By using Eq. (5-93) for P_{FD} and Eq. (8-9) for the density of states, Eq. (8-8) can be solved for the density of the conduction band electrons:

$$n = \frac{1}{2\pi^2}\left(\frac{2m_e^*}{\hbar^2}\right)^{3/2} \int_{E_c}^{\infty} \frac{(E - E_c)^{1/2}}{[\epsilon^{(E-E_F)/\kappa T} + 1]}\, dE \tag{8-10}$$

assuming that m_e^* is a constant in the band over the range of energy where the integrand is nonzero. To perform the integration of Eq. (8-10) it is convenient to change variables to $E_e = E - E_c$, the electron energy. Then we have

$$n = \frac{1}{2\pi^2}\left(\frac{2m_e^*}{\hbar^2}\right)^{3/2} \int_0^{\infty} \frac{E_e^{1/2}}{[\epsilon^{E_e/\kappa T}\epsilon^{-(E_F-E_c)/\kappa T} + 1]}\, dE_e \tag{8-11}$$

To simplify the integral of Eq. (8-11) we shall assume that $|E_F - E_c| \gg E_{e\,\text{max}}$ and that $|E_F - E_c|/\kappa T > 1$. The first assumption means that E_F is farther below E_c than any of the electrons are above the bottom of the band. The second assumption restricts the temperature so that the Fermi level has several times the thermal energy κT below E_c. When a material satisfies these assumptions it is said to be *nondegenerate*. When the full form of Eq. (8-11) must be used, the material is *degenerate*. Metals, with E_F in the conduction band, are degenerate.

Utilizing the assumptions of the nondegenerate case and recognizing that $E_F < E_c$, we can neglect the factor of unity in the denominator of Eq. (8-11) as being small compared with the exponential. We have

$$n = \frac{1}{2\pi^2}\left(\frac{2m_e^*}{h^2}\right)^{3/2} \epsilon^{(E_F - E_c)/\kappa T} \int_0^\infty E_e^{1/2} \epsilon^{-E_e/\kappa T} \, dE_e \tag{8-12}$$

The integral in Eq. (8-12) is of the form

$$I = \int_0^\infty z^{1/2} \epsilon^{-z} \, dz = \frac{\sqrt{\pi}}{2}$$

Making the substitution $z = E_e/\kappa T$, we can evaluate Eq. (8-12) as

$$\boxed{n = N_c(T)\epsilon^{-(E_c - E_F)/\kappa T}} \tag{8-13}$$

where

$$N_c(T) = \frac{1}{4}\left(\frac{2m_e^*\kappa T}{\pi h^2}\right)^{3/2} \tag{8-14}$$

is an effective density of states near the bottom of the conduction band. The density of conduction band electrons depends on the location of E_F relative to E_c and on the temperature through the factor of $(\kappa T)^{3/2}$ and through the exponent. The actual density given by Eq. (8-13) is not yet known because the location of E_F has so far not been determined.

The number of holes in the valence band is

$$pV = \int_{\substack{\text{valence}\\\text{band}}} \left(\frac{dN_s}{dE}\right)_v [1 - P_{\text{FD}}(E)] \, dE \tag{8-15}$$

where p is the uniform hole density and $(dN_s/dE)_v$ is the density of hole states,

$$\left(\frac{dN_s}{dE}\right)_v = \frac{V}{2\pi^2}\left(\frac{2m_h^*}{h^2}\right)^{3/2} (E_v - E)^{1/2} \tag{8-16}$$

taking the zero for hole energy to be E_v with the hole energy increasing downward in the energy diagram. The limits of integration in Eq. (8-15) may be taken as $-\infty$ and E_v since $(1 - P_{\text{FD}})$ vanishes at energies well below E_F and E_v is the upper limit of the band. Using Eq. (8-16) for the density of states and Eq. (5-93) for P_{FD} we can solve Eq. (8-15) for the hole density p for the nondegenerate case with the result

$$\boxed{p = N_v(T)\epsilon^{-(E_F - E_v)/\kappa T}} \tag{8-17}$$

Sec. 8-2 Electrons and Holes

where

$$N_v(T) = \frac{1}{4}\left(\frac{2m_h^*\kappa T}{\pi h^2}\right)^{3/2} \quad (8\text{-}18)$$

The density of valence band holes depends on the location of E_F and on temperature.

To find the actual densities n and p of the charge carriers in a semiconductor we must know the location of E_F relative to the band edges. For intrinsic semiconductors with carrier densities n_i and p_i, we know that $n_i = p_i$ since the conduction electrons are excited from valence band states. By setting $n_i = p_i$ from Eqs. (8-13) and (8-17) we obtain for intrinsic semiconductors

$$(m_e^*)^{3/2}\epsilon^{-(E_c - E_{Fi})/\kappa T} = (m_h^*)^{3/2}\epsilon^{-(E_{Fi} - E_v)/\kappa T}$$

from which we can solve for the intrinsic Fermi energy E_{Fi}:

$$E_{Fi} = \frac{1}{2}(E_v + E_c) + \frac{3}{4}\kappa T \ln\left(\frac{m_h^*}{m_e^*}\right) \quad (8\text{-}19)$$

The first term on the right-band side of Eq. (8-19) is the energy at the middle of the gap. Therefore, for intrinsic materials E_{Fi} differs from the middle of the gap at nonzero temperatures by an amount related to the band structures through the effective masses. If the band curvatures were the same, except, of course, for sign, E_F would be in the center of the gap at any temperature. Table 8-1 lists the effective masses of holes and electrons in silicon and germanium.

Table 8-1. Effective Masses of Charge Carriers in Silicon and Germanium in Units of the Free Electron Mass

	$\frac{m_e^*}{m_e}$	$\frac{m_h^*}{m_e}$
Ge	0.55	0.37
Si	1.10	0.59

The densities of the charge carriers in intrinsic materials can be determined by using Eq. (8-19) for E_{Fi} in Eqs. (8-13) and (8-17) for n_i and p_i, respectively. It is easier, however, to form the product $n_i p_i = n_i^2 = p_i^2$ using Eqs. (8-13) and (8-17) since this procedure eliminates E_F directly. The result is

$$n_i^2 = N_c(T)N_v(T)\epsilon^{-E_g/\kappa T} \quad (8\text{-}20)$$

where $E_g = E_c - E_v$ is the width of the energy gap. The intrinsic carrier density depends only on the temperature, the gap width, and the band structure. An important method of determining E_g experimentally is to measure n_i as a function of temperature.

The intrinsic densities of holes and electrons in silicon and germanium at room temperature are on the order of 10^{10} and 10^{13} cm^{-3}, respectively. The factors N_v and N_c are on the order of 10^{19} cm^{-3} for both materials. The ratios n/N_c and p/N_v can be interpreted using Eqs. (8-13) and (8-17) as probabilities for the occupation of states in the appropriate bands by electrons and holes, respectively. Thus N_c and N_v can be thought of as effective densities of available states at the respective band edges. From the numbers it is clear that very few of these available states are occupied by intrinsic charge carriers at room temperature.

In extrinsic semiconductors the number of conduction electrons does not equal the number of valence holes because charge carriers in the bands are supplied primarily by the impurity atoms. Suppose, for example, that we consider a semiconductor with donor-type impurities with a density of N_d. Charge neutrality requires that the number of electrons in the conduction band equal the number of valence band holes plus the number of ionized donor atoms. If we denote the number of unionized neutral donor atoms per unit volume as N_{d0}, the density of ionized donors is $N_d - N_{d0}$ and charge neutrality is expressed by

$$n = p + (N_d - N_{d0}) \quad \text{donor impurities} \quad (8\text{-}21)$$

At the absolute zero of temperature none of the donor states are ionized and there is no hole-electron pair thermal generation. The Fermi level must therefore lie above E_d. At a low temperature many donors may be ionized but few hole-electron pairs generated across the gap. Figure 8-5 shows an energy diagram for a semiconductor with donor impurities at a low temperature with the Fermi-Dirac probability for occupation of conduction band states superposed*. Notice that there is little probability for a vacancy to occur in the valence band and that the occupation probability at the donor level is greater than that in the conduction band. However, the density of states is so much greater in the conduction band than at the donor level that almost all of the donors will be ionized at moderate temperatures. The electron density from Eq. (8-21) is

$$n = p + (N_d - N_{d0}) \simeq (N_d - N_{d0}), \quad \text{low } T \quad (8\text{-}22)$$

since there are relatively few intrinsic hole-electron pairs at low temperatures. Thus the conductivity of the semiconductor is greatly enhanced by the addition of the impurity that supplies electrons to the conduction band. An ex-

---

*It can be shown that the probabilities for electronic occupation of isolated donor and acceptor states differ slightly from the Fermi-Dirac function, although these differences do not affect the arguments of this chapter. See, for example, Wang, *Solid State Electronics*, pp. 142ff. (listed in the Bibliography) for an explanation based on electron spin.

Sec. 8-2 Electrons and Holes   223

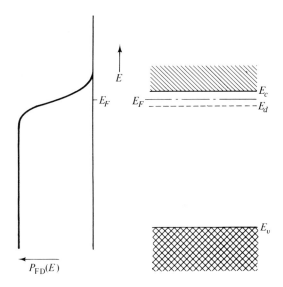

Fig. 8-5 Energy diagram of an *n*-type semiconductor at a low temperature.

trinsic semiconductor doped with donor impurities is called an *n-type semiconductor* because the density of negatively charged electrons is greater than that of valence band holes. In *n*-type semiconductors electrons are said to be the majority carriers and holes the minority carriers.

We can calculate the majority and minority carrier densities using Eqs. (8-13) and (8-17) provided we know the Fermi level. As before we can use the charge neutrality condition for *n*-type material to determine E_F. For example, the density of neutral donors N_{d0} is the number of donor states per unit volume that are occupied. That is,

$$N_{d0} = N_d P_{\text{FD}}(E_d) \qquad (8\text{-}23)$$

since all the donor states have the same energy. Substituting Eq. (8-23) into Eq. (8-22), we obtain, after some manipulation,

$$n = N_d \left[\frac{1}{1 + \epsilon^{(E_F - E_d)/\kappa T}} \right], \quad \text{low } T \qquad (8\text{-}24)$$

which for $\kappa T < E_F - E_d$, is

$$n = N_d \epsilon^{-(E_F - E_d)/\kappa T}, \quad \text{low } T \qquad (8\text{-}25)$$

At low temperatures $n \leq N_d$ according to Eq. (8-22) so E_F must be at least E_d or between E_d and E_c. We can solve for E_F by combining Eqs. (8-25) and (8-13):

$$N_d \epsilon^{-(E_F - E_d)/\kappa T} = N_c \epsilon^{-(E_c - E_F)/\kappa T}, \quad \text{low } T$$

from which we find

$$E_F = \frac{1}{2}(E_c + E_d) + \frac{1}{2}\kappa T \ln\left(\frac{N_d}{N_c}\right), \quad \text{low } T \quad (8\text{-}26)$$

Notice that at $T = 0$, E_F is midway between E_c and E_d and that for increasing temperature E_F falls since $N_d < N_c$.

At an intermediate temperature essentially all of the donors will be ionized owing to the high density of states in the conduction band. Then

$$n = N_d \quad \text{intermediate } T \quad (8\text{-}27)$$

assuming that $N_d \gg n_i$. Combining Eqs. (8-27) and (8-13) to find E_F we get

$$N_d = N_c \epsilon^{-(E_c - E_F)/\kappa T}, \quad \text{intermediate } T$$

from which we can solve for E_F:

$$E_F = E_c - \kappa T \ln\left(\frac{N_c}{N_d}\right), \quad \text{intermediate } T \quad (8\text{-}28)$$

At higher temperatures generation of hole-electron pairs across the gap becomes important. When the temperature is sufficiently high that $n_i > N_d$, the material behaves for all intents and purposes as if it were an intrinsic semiconductor.

An extrinsic semiconductor with acceptor impurities has more holes in

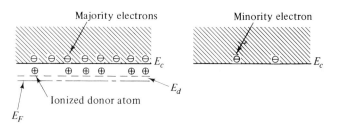

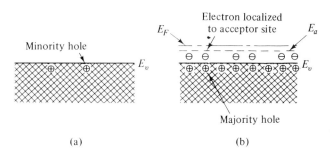

(a)   (b)

Fig. 8-6 Energy diagrams indicating the location of charges in (a) n-type and (b) p-type semiconductors.

the valence band than electrons in the conduction band at low and intermediate temperatures. Under such conditions it is called a *p*-type semiconductor because of its higher density of mobile positive charge. The holes are supplied thermally both by excitation of electrons into acceptor states and by generation of hole-electron pairs across the gap. The conduction electrons result from thermal generation. In the extrinsic regime the holes are the majority carrier, and the electrons the minority carrier. Determination of the carrier densities and the Fermi level under different temperature conditions proceeds in a manner analogous to the analysis of *n*-type semiconductors with analogous results. For example, the Fermi level at $T = 0$ is between E_v and E_a. At higher temperatures it moves toward the middle of the gap. When hole-electron pair generation creates carriers in numbers comparable to the extrinsic holes, the semiconductor behaves as if it were intrinsic.

Figure 8-6 illustrates the differences in the energy diagrams for *n*-type and *p*-type semiconductors. Also shown are the locations of the thermally generated charges.

Optical Generation of Holes and Electrons

Thermal generation is not the only process by which charge carriers can be created in semiconductors. Thermal generation provides charge densities in thermodynamic equilibrium. Carrier densities in excess of their equilibrium values may be generated by the application of an external source of energy. In particular, absorption of electromagnetic radiation can create excess charge carriers in their respective bands. The excess carriers increase the conductivity of the semiconductor sample over its value in the "dark," an effect known as *photoconductivity*.

A electron in an initial state E_i can absorb the energy of an optical photon $h\nu$ and make a transition to an unoccupied final state E_f provided both energy and momentum are conserved. If the initial and final states are in the valence and conduction bands, respectively, a hole-electron pair is generated. If E_i is at the donor level and E_f is in the conduction band, a conduction electron and an ionized impurity will be formed. Similarly, if E_i is in the valence band and $E_f = E_a$, a bound electron and a free hole will be created.

Absorption of a photon and transition from an initial to a final state must conserve both energy and momentum. The momentum of a photon is negligible in comparison with the effective momentum of an electron in a crystal. Therefore allowed transitions can be divided into two classes depending on whether the electron momentum $\hbar \mathbf{k}$ changes during the transition. Figure 8-7 illustrates the transitions on $E(k)$ diagrams. The first class, *direct transitions*, are those for which the electron momentum is unchanged. That is, the final state k_f equals the initial state k_i as pictured in Fig. 8-7(a). Conservation of

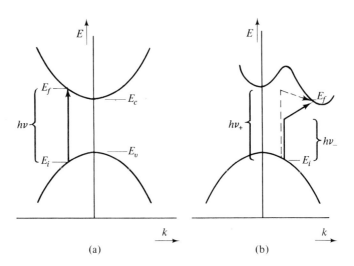

Fig. 8-7 Optical excitation of hole-electron pairs via (a) direct and (b) indirect transistions.

energy in direct transitions requires

$$E_f - E_i = h\nu = \frac{hc}{\lambda} \tag{8-29}$$

where c is the speed of light and λ is the wavelength of the radiation. In an *indirect transition*, illustrated in Fig. 8-7(b), the electron changes momentum. Momentum is conserved in this case by interaction with the lattice or with phonons. Phonons of energy E_p may be either created or destroyed. The figure illustrates two transitions, one creating and the other absorbing a single phonon. Conservation of energy in this case in expressed by

$$E_f - E_i \pm E_p = h\nu_\pm \tag{8-30}$$

The probabilities with which specific transitions occur can be calculated from a quantum mechanical theory of electrodynamics.

Both intrinsic and extrinsic semiconductors exhibit photoconductivity, as do insulators. Some commonly used photoconductive materials are lead sulfide (PbS, $E_g = 0.37$ eV at $T = 300°$ K), lead selenide (PbSe, $E_g = 0.27$ eV), lead telluride (PbTe, $E_g = 0.30$ eV), cadmium sulfide (CdS, $E_g = 2.42$ eV), silicon and germanium doped with either gold or mercury, and various III-V compounds.

Recombination and Carrier Lifetimes

Recombination is the annihilation of a hole-electron pair and can occur either in a band-to-band transition or through an intermediate state serving as a recombination center. Both radiative and nonradiative recombination can take place in semiconductors.

Radiative recombination is the inverse of the optical absorption process. A photon is emitted when an electron falls into a lower-energy state. When light is emitted the material is said to be *luminescent*. Both direct and indirect downward transitions can cause the emission of photons. In direct gap semiconductors direct transitions are more likely to occur, while in materials with indirect gaps such as silicon and germanium the indirect transitions involving intermediate states are much more likely. The equilibrium distribution of electrons in energy states according to the temperature of the semiconductor must be disturbed by an external influence to provide the energy that is eventually radiated. When an electric current provides the excess energy the process is called electroluminescence. When optical excitation creates the nonequilibrium condition the process is called photoluminescence. The luminescence may occur only during the application of the excitation or it may continue for some time after termination of the excitation. The former phenomenon is called fluorescence, and the latter phosphorescence.

Nonradiative transitions are the dominant recombination processes in most materials. These transitions occur both under equilibrium and nonequilibrium conditions. In equilibrium they balance the thermal generation of hole-electron pairs. Nonradiative recombination usually requires a third body to conserve momentum in addition to the electron and hole that are recombining. The third body may be one or more phonons, a lattice defect, or a surface imperfection.

For practical purposes it is convenient to describe recombination processes in terms of recombination rates and carrier lifetimes. Suppose, for example, that we consider the equation of continuity for one charge species in the presence of generation and recombination. If holes are being generated at a rate of g holes per unit time and are recombining at a rate of r holes per unit time, the equation of continuity is, from Eq. (3-17),

$$\frac{\partial p}{\partial t} + \nabla \cdot (p \langle \mathbf{v}_h \rangle) = g - r \qquad (8\text{-}31)$$

where $\langle \mathbf{v}_h \rangle$ is the average velocity of the holes. In equilibrium the generation and recombination rates are equal, $\langle \mathbf{v}_h \rangle = 0$, and $p = p_0$ is constant and uniform. In a nonequilibrium case there will be an excess or deficiency of holes, that is, $p \neq p_0$. To isolate recombination from other effects we shall assume that the system has been excited to a nonequilibrium configuration that is nonetheless spatially uniform. At time $t = 0$ we shall turn off the excess generation rate so that for $t > 0$ the hole density is governed by

$$\frac{dp}{dt} = g_0 - r \qquad (8\text{-}32)$$

Now the recombination rate r must depend on the density of holes available for recombination. We shall therefore write that dependence phenomenologically as

$$r = \frac{p}{\tau_h} \tag{8-33}$$

where τ_h hides a good deal of ignorance. The equilibrium generation rate g_0 equals the equilibrium recombination rate $r_0 = -p_0/\tau_h$. Substitution of Eq. (8-33) into Eq. (8-32) permits us to interpret τ_h as the time scale for recombination. Equation (8-32) has the solution

$$p - p_0 = (p - p_0)_{t=0} \epsilon^{-t/\tau_h} \tag{8-34}$$

The carrier density in excess of its equilibrium value decays exponentially toward zero from its initial value with a time constant τ_h. Thus τ_h is referred to as the hole *lifetime* or *recombination time*. In effect τ_h is an average time that an excess carrier exists between its generation and annihilation. Lifetimes or recombination times depend on the imperfections in the material and in typical cases are many orders of magnitude longer than ordinary collision times. In other words, an average carrier experiences very many collisions with lattice vibrations and other imperfections before suffering an interaction that leads to recombination.

As we shall see in the next chapter, carrier lifetimes are an important consideration in the design and fabrication of semiconductor junction devices. In these devices an excess carrier concentration may be injected into the semiconductor by an electric field at a contact. It is of importance to know how far these carriers travel by diffusion before they recombine. The diffusive flow of carriers with nonuniform densities is an important mechanism that results in current flow. Consider Eq. (8-31) for the continuity of holes under the condition of a steady-state flow of holes from a surface across which they are somehow injected. With the hole particle current density $\Gamma = p\langle \mathbf{v}_h \rangle$ given by the diffusion law, we have

$$\nabla \cdot \Gamma = \nabla \cdot (-D_h \nabla p) = g - r$$

or

$$D_h \nabla^2 p = \frac{-(p - p_0)}{\tau_h} \tag{8-35}$$

It is easy to demonstrate that in one dimension Eq. (8-35) has the solution for $z > 0$

$$[p(z) - p_0] = [p(z=0) - p_0]\epsilon^{-z/L_h} \tag{8-36}$$

where

$$L_h = (D_h \tau_h)^{1/2} \tag{8-37}$$

is called the *diffusion length* for holes. Equation (8-36) indicates that the density of the excess carriers decays exponentially into the material with a scale length given by L_h for holes. Thus the excess holes will not arrive at a distance several times L_h from the surface, having recombined along the way.

A similar treatment applies to the continuity of electrons leading analogously to τ_e and L_e, the lifetime and diffusion length for electrons.

Semimetals

Semimetals are materials in which the Fermi level lies in a region of overlap of the conduction and valence bands in the energy diagram as indicated in Fig. 6-7. Bismuth, arsenic, antimony, selenium, and tellurium are examples of semimetals. Because of the overlapping of the bands, valence electrons easily enter conduction band states. The intrinsic densities of electrons and holes in semimetals range from 3×10^{17} cm^{-3} in bismuth to 2×10^{20} cm^{-3} in arsenic.

8.3 CURRENT IN SEMICONDUCTORS

Electric current in semiconductors may be carried by both electrons in the conduction band and holes in the valence band. The electron current density is

$$\mathbf{J}_e = -ne\langle \mathbf{v}_e \rangle \tag{8-38}$$

The average electron drift velocity $\langle \mathbf{v}_e \rangle$ is related to the applied electric field $\mathcal{E}$ that produces the current by the electron mobility μ_e:

$$\langle \mathbf{v}_e \rangle = -\mu_e \mathcal{E} \tag{8-39}$$

for isotropic media. Similarly, the hole current density is

$$\mathbf{J}_h = pe\langle \mathbf{v}_h \rangle \tag{8-40}$$

where

$$\langle \mathbf{v}_h \rangle = \mu_h \mathcal{E} \tag{8-41}$$

for isotropic media. The net current density is the sum of $\mathbf{J}_e$ and $\mathbf{J}_h$ or

$$\mathbf{J} = \sigma \mathcal{E} \tag{8-42}$$

in which σ is the isotropic conductivity:

$$\boxed{\sigma = ne\mu_e + pe\mu_h} \tag{8-43}$$

Equation (8-42) is the intensive form of Ohm's law for linear isotropic media. So that the linear relationship between $\mathbf{J}$ and $\mathcal{E}$ will be valid the mobilities defined by Eqs. (8-39) and (8-41) must not depend on the electric field strength $\mathcal{E}$. It is not only possible but more often than not true, however, that the mobilities depend on the vector direction of $\mathcal{E}$: Most semiconductors are not isotropic. The various components of $\langle \mathbf{v} \rangle$ are related to the components of $\mathcal{E}$ by differing values of mobility. For anisotropic materials each of Eqs. (8-39) and (8-41) is replaced by three equations for the components of $\langle \mathbf{v} \rangle$ in terms of the components of $\mathcal{E}$. The two sets of three equations may be written in a compact matrix form in which the mobility is a three-by-three tensor. The anisotropic conductivity is therefore also a tensor. For the most part we shall ignore the complications arising from anisotropies by considering only one-dimensional problems. However, the effects of the anisotropies should be kept in mind, especially in connection with the operation of integrated circuits in which the current flow is in many cases two-dimensional.

In the previous section we studied the carrier densities n and p in both intrinsic and extrinsic materials as functions of the Fermi level and the temperature. It is also possible to develop a model for the mobilities from the transport theory provided the relaxation mechanisms are understood. For our purposes we shall briefly discuss the collision mechanisms and the results. There are two primary scattering mechanisms for the charge carriers, one being interactions with lattice vibrations and the other being collisions with ionized and neutral impurities or defects. Scattering by impurities is the dominant mechanism determining the mobility in heavily doped extrinsic materials at low temperatures. Impurity scattering yields a functional dependence of mobility with $T^{3/2}$, mobility increasing with increasing temperature. On the other hand, collisions with phonons dominate mobility under intrinsic conditions and limit it to a $T^{-3/2}$ dependence. Table 8-2 lists room-

Table 8-2. Intrinsic Carrier Mobilities at Room Temperature in cm^2/V-sec.

	μ_e	μ_h
Si	1,500	600
Ge	3,900	1,900
InSb	77,000	750
InAs	33,000	460
PbS	550	600

temperature mobilities in several intrinsic semiconductors and Figure 8-8 shows typical data of μ vs. T for various doping levels in n-type germanium. Table 8-2 indicates that in general $\mu_e \neq \mu_h$ as might be expected at least since $m_e^* \neq m_h^*$. Furthermore, relaxation times for the electrons and holes may differ. Notice in Fig. 8-8 that at higher temperatures for the extrinsic materials the phonon scattering begins to dominate. Figure 8-8 also indicates that the mobility depends on the doping concentration. Figure 8-9 shows the dependence of electron mobility in n-type germanium at 300°K as a function of the density of the conduction band electrons. The intrinsic density in germanium is about 10^{13} cm^{-3}. The carrier densities higher than the intrinsic value are due to the ionization of donor impurities of varying densities.

In addition to the dependence of the mobilities on temperature and im-

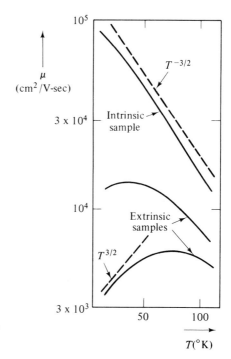

Fig. 8-8 Temperature dependence of μ_e in n-type germanium.

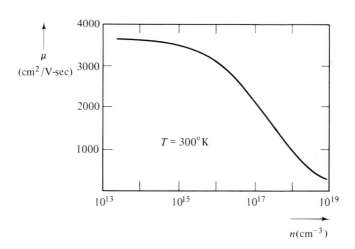

Fig. 8-9 Electron mobility vs. electron density in germanium at room temperature.

purity concentrations, μ may depend on the field strength $\mathcal{E}$ for high fields. That is, at high fields $\langle \mathbf{v} \rangle$ is not linearly related to $\mathcal{E}$ and hence the semiconductor is not ohmic. We anticipated this difficulty in Section 3.4 when we derived the mobility as a transport coefficient from the kinetic equation. Recall that the linear solution was predicated on the assumption that the field $\mathcal{E}$ is small so that the deviation of the distribution function from its equilibrium configuration is small and directly proportional to $\mathcal{E}$. We should therefore not be surprised that the experimental evidence indicates that $\langle \mathbf{v} \rangle$ varies nonlinearly with $\mathcal{E}$ at high fields. Data to this effect are plotted in Fig. 8-10. Physically, the deviation can be understood in terms of phonon interactions. At high fields the charge carriers gain significant energy from the

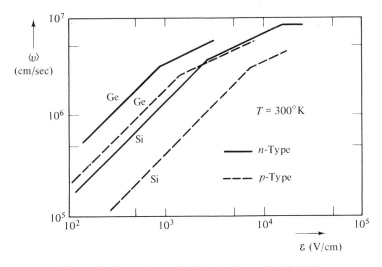

Fig. 8-10 Dependence of drift velocity on electric field.

source field between collisions and subsequently have a higher thermodynamic temperature than the lattice. During collisional interactions there is a net flow of energy from the hot carriers to the lattice, creating phonons. Dissipation of the hot carrier energy reduces their drift velocity over what it would have been if the charges and the lattice were at the same temperature. Actually there are two separate mechanisms for the generation of phonons by hot electrons. One is effective at high fields and results in a drift velocity proportional to $\mathcal{E}^{1/2}$. The second is effective at very high fields and yields a drift velocity that is independent of $\mathcal{E}$. The magnitude of this field-independent velocity, the so-called *scatter-limited velocity*, is about the 10^7 cm/sec, which is roughly the thermal velocity of an electron with free mass m_e at room temperature.

Diffusion

Charge carrier diffusion is also affected by temperature and impurities. According to the Einstein relation developed in Section 3.5, the diffusion coefficient for a carrier is related to its mobility by the absolute temperature:

$$D = \frac{\kappa T}{e}\mu \qquad (8\text{-}44)$$

The Einstein relation has been verified over a significant range of temperatures and electric fields by a simultaneous measurement of both D and μ. One version of the technique, known as the Shockley-Haynes experiment after the original investigators, is illustrated in Fig. 8-11. Excess minority carriers

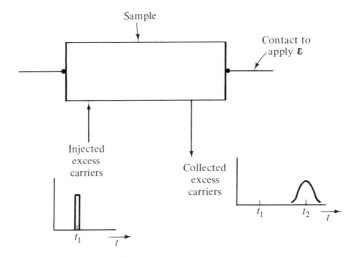

Fig. 8-11 Shockley-Haynes experiment to measure mobility and diffusion.

are injected near one end of a sample upon which a field $\mathcal{E}$ is imposed. The narrow pulse of injected carriers must be formed in a small region of the sample. As the carriers drift down the sample in the electric field they also diffuse spatially. The mobility is obtained from the drift velocity, which is determined from the time of flight of the carrier pulse between the point of injection and a collecting probe. The width of the collected current pulse is related to the diffusion coefficient.

An important feature of carrier diffusion in semiconductors is its ambipolar characteristic. Two species of charged particles, electrons and holes, may be distributed nonuniformly in the solid and hence will diffuse. Recall that in Chapter 4 we discovered that the internal electric field due to the nonequilibrium charge densities strongly influences the diffusion. For equal densities, the charges diffuse together with a diffusion coefficient that is intermediate between the two but closer to that of the less mobile of the two species. Such is also the case for intrinsic semiconductors. For extrinsic semiconductors, on the other hand, the mobile carrier densities are not equal. However, beginning with Eq. (4-28) it is not difficult to show that extrinsic ambipolar diffusion is determined by the minority carrier diffusion coefficient. For example, if $p\mu_p \gg n\mu_n$, it turns out that $D_a \simeq D_n$, the diffusion coefficient of the less numerous electrons. For p-type silicon the hole and electron mobilities are of the same order of magnitude but $p \gg n$ so that the effect of the mobilities in the inequality is not important. The Shockley-Haynes experiment measures minority carrier diffusion and mobility. That the minority carrier dominates ambipolar diffusion in extrinsic semiconductors is important for device applications that we shall study in the next chapter.

The Hall Effect

In extrinsic semiconductors the majority charge densities are the principal agents that carry ohmic electric current. That current can in fact be carried by the positive holes has been demonstrated by measurement of the Hall effect.

Demonstration of the Hall effect may be obtained from the experimental configuration illustrated in Fig. 8-12. The semiconducting sample under test

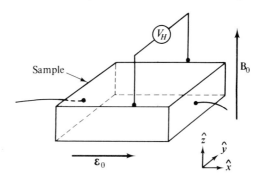

Fig. 8-12 Hall effect experiment.

is placed in a uniform constant magnetic field $\mathbf{B}_0 = B_0 \hat{z}$. An external electric field $\mathbf{\mathcal{E}}_0 = \mathcal{E}_0 \hat{x}$ is applied in a direction perpendicular to $\mathbf{B}_0$. As carriers of charge q drift in the $\hat{x}$-direction due to the influence of $\mathbf{\mathcal{E}}_0$ with velocity

$$v_x = \pm \mu \mathcal{E}_0 \qquad (8\text{-}45)$$

they experience a magnetic force $q\mathbf{v} \times \mathbf{B}$. The sign in Eq. (8-45) depends on the sign of the charge, positive sign for positive charge and negative sign for negative charge. The magnetic force impels the charges in a direction transverse to both $\mathcal{E}_0$ and $\mathbf{B}_0$ until a charge imbalance creates a transverse electric field $\mathcal{E}_H$ that depends on the sign and density of the majority carrier. A voltage V_H proportional to the Hall field $\mathcal{E}_H$ is developed across the sample.

Consider, for example, the behavior of electrons in an n-type semiconductor illustrated in Fig. 8-13(a). The electrons drift in $\mathcal{E}_0$ in the negative $\hat{x}$-direction so the magnetic force $-e(v_x\hat{x} \times \mathbf{B}_0)$ causes them to move in the

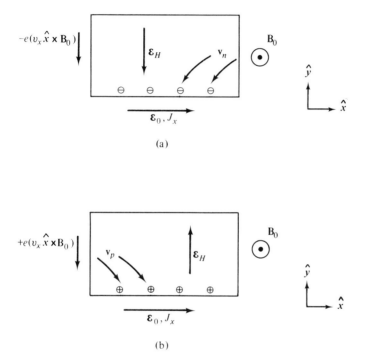

Fig. 8-13 Carrier flow and fields in the Hall effect experiment with (a) electrons and (b) holes as majority carriers.

negative $\hat{y}$-direction. In the steady state the electrons accumulate near the surface indicated until the force of the transverse Hall field balances the magnetic force. Note that $\mathcal{E}_H < 0$ for n-type semiconductors in the configuration shown. Holes in a p-type semiconductor cause the opposite transverse field, as illustrated in Fig. 8-13(b). The holes drift in the $+\hat{x}$-direction in $\mathcal{E}_0$ so that the magnetic force impels them in the negative $\hat{y}$-direction until the transverse field stops the transverse flow. Notice that $\mathcal{E}_H$ due to the holes is oppositely directed from that due to negative carriers.

Quantitatively in either case the transverse field balances the magnetic force due to the drifting carriers so that

$$\mathcal{E}_H = (v_x \hat{\mathbf{x}} \times \mathbf{B}_0) \tag{8-46}$$

We can manipulate Eq. (8-46) for $\mathcal{E}_H$ into easily measurable quantities by using Eqs. (8-45) and (8-42),

$$\mathcal{E}_H = \pm \frac{\mu}{\sigma} J_x B_0 \tag{8-47}$$

since J_x and B_0 are perpendicular. For an extrinsic semiconductor the conductivity σ is $ne\mu_e$ for n-type material and $pe\mu_h$ for p-type material. Then

$$\mathcal{E}_H = \begin{cases} \dfrac{+J_x B_0}{pe}, & p\text{-type} \\ \dfrac{-J_x B_0}{ne}, & n\text{-type} \end{cases} \tag{8-48}$$

The *Hall coefficient* R_H is defined by

$$\boxed{R_H \equiv \frac{\mathcal{E}_H}{J_x B_0} = \begin{cases} +\dfrac{1}{pe}, & p\text{-type} \\ -\dfrac{1}{ne}, & n\text{-type} \end{cases}} \tag{8-49}$$

The Hall coefficient has the sign of the majority carrier and is inversely proportional to the majority carrier density. Thus measuring R_H determines not only the density but also the majority carrier.

Monovalent metals have negative Hall coefficients, while beryllium, zinc, and cadmium have $R_H > 0$. Among the semimetals in which the electron and hole concentrations are equal, bismuth has a negative coefficient, while arsenic and antimony have positive coefficients. When the carrier densities are of comparable magnitude the principal contribution to the current is from the carrier with the greater mobility.

From Eq. (8-47) the Hall coefficient of an extrinsic semiconductor is related to the mobility by

$$\mu = \pm \sigma R_H \tag{8-50}$$

The conductivity σ is easily measured in the same experiment, being $\sigma = \mathcal{E}_0 / J_x$ so that the carrier mobility can also be obtained from the data of a Hall effect experiment.

The Hall angle θ_H is the angle that the total electric field $\mathcal{E}_0 + \mathcal{E}_H$ makes with the applied field $\mathcal{E}_0$. In other words,

$$\tan \theta_H = \frac{\mathcal{E}_H}{\mathcal{E}_0} \tag{8-51}$$

Using Eq. (8-47) for $\mathcal{E}_H$ and subsequently Eq. (8-42) for $\mathcal{E}_0$, we can rewrite

Eq. (8-51) as

$$\tan \theta_H = \mu B_0 \qquad (8\text{-}52)$$

Measurement of θ_H and B_0 also yields the mobility μ. Equation (8-52) can be manipulated even further using a pseudoclassical model for μ in terms of the collisional relaxation time τ and the effective mass m^*. If we take

$$\mu = \frac{e\tau}{m^*}$$

as in Eq. (3-62), we have, for small θ_H,

$$\theta_H \simeq \left(\frac{eB}{m^*}\right)\tau = \omega_c \tau \qquad (8\text{-}53)$$

where ω_c is the cyclotron frequency. Since $\omega_c = 2\pi/T_c$, where T_c is the period of the cyclotron orbit, θ_H is the average fractional number of 2π radians traversed by a particle between collisions in its cyclotron orbit.

8.4 BULK SEMICONDUCTOR DEVICES

The properties of bulk semiconductors studied in this chapter suggest two classes of devices that utilize these characteristics, radiation detectors and temperature-sensitive resistors. In addition, the characteristic behavior known as negative differential conductivity of certain III-V semiconducting compounds can be understood in terms of their unique energy band structure. So-called Gunn-effect devices utilize negative differential conductivity. In this section we shall study these three applications of bulk semiconductors to electronics.

Radiation Detectors

The absorption of electromagnetic photons and the subsequent generation of hole and electron densities in excess of their equilibrium values increases the conductivity of semiconductors. This excess conductivity above its value "in the dark" is called *photoconductivity*. Materials sensitive to a spectrum of wavelengths or frequencies find application in the detection of radiation evidenced by a change in bulk resistivity when illuminated.

Cadmium sulfide is a material commonly used to detect optical radiation in the visible spectrum. Commercial CdS photoconductors can have a "dark" resistance as high as $2 \times 10^6 \, \Omega$ and an "illuminated" resistance as low as $10 \, \Omega$.

Semiconductor radiation detectors are also available for X-ray and γ-ray energies as well as for high-energy nuclear particles. Particle detectors operate on the same principles as optical detectors: The incident particle creates holes and electrons that produce a measurable current proportional to the incident flux. While a photon can generate only one hole-electron pair, a high-energy particle may produce very many hole-electron pairs as it penetrates a crystal and its energy is absorbed. Particle detectors can therefore be very sensitive instruments, especially when operated at low temperatures to reduce the background current of the thermally generated carriers.

Thermistors

A *thermistor* is a device whose resistance is temperature-sensitive, usually decreasing with increasing temperature. Conventional wire wound metallic resistors have positive temperature coefficients of resistivity; that is, their resistance increases with increasing temperature because of decreased mobility. Intrinsic semiconductors such as high-resistivity germanium have negative temperature coefficients of resistivity owing to the strong dependence on temperature of the generation of hole-electron pairs. Because the carrier density increases approximately exponentially with temperature, the resistivity decreases markedly for only small increases in temperature.

Thermistors can be used as temperature sensors for electronic thermometers. They are also used in electronic circuits to compensate for the change in resistance with temperature of ordinary components where variation of component values cannot be tolerated.

Many of the applications of thermistors require that the power dissipated in the device not be high enough to alter its temperature, thus changing its resistance. When a thermistor does dissipate sufficient power to change its temperature its terminal characteristic exhibits negative differential resistivity. That is, the increase in temperature due to dissipation in the device causes the current to increase while the voltage decreases. In this regime the I-V curve has a negative slope.

Devices that exhibit negative differential conductivity are useful for making oscillators, amplifiers, and switching circuits. Unfortunately thermistors are not useful in these applications because their response characteristics are too slow. However, certain bulk semiconducting compounds have negative resistance characteristics over a limited range of operating parameters utilizing mechanisms unrelated to the temperature sensitivity of the resistivity. They are very useful as oscillators and amplifiers at microwave frequencies. These materials are used to make Gunn-effect devices.

Gunn-Effect Devices

Gallium arsenide is one of several semiconductors whose conduction band is comprised of several overlapping $E(k)$ bands two of which are of the

Sec. 8-4 Bulk Semiconductor Devices

form shown in Fig. 8-14. The two subbands have different minimum energies, E_L for the lower- and E_U for the higher-energy minimum. Notice that the curvatures of the two bands are also different so that an electron in the valley of the lower band has smaller effective mass than one in the upper valley. The different effective masses are reflected in different mobilities for these electrons. In GaAs μ_L is about 60 times the value of μ_U. A further feature of the conduction bands is the fact that the upper subband has a very high density of states compared with the $k = 0$ minimum.

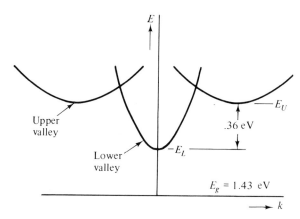

Fig. 8-14 Conduction band $E(k)$ in GaAs.

Let us now see how these features of the conduction band explain the negative differential conductivity first observed in n-type GaAs by J. B. Gunn in 1963. At low electric fields conduction electrons occupy states in the lower valley and carry ohmic current $J = \sigma \mathcal{E}$ with conductivity characteristic of the extrinsic density and the mobility μ_L. As the applied field is increased the electrons gain energy from it and move upward in the lower subband. When the electric field is raised even higher the hot electrons may gain enough energy to transfer into the upper valley band. Actually the probability of this intervalley transfer is very good if the field is beyond a threshold value because there are many available states in the upper valley. As the electrons transfer to the upper valley, their mobility is decreased, thus decreasing the differential conductivity of the sample. The sketch of Fig. 8-15 represents the average electron drift velocity in a medium that exhibits this transferred electron mechanism plotted as a function of applied field.

The region of the figure where $\langle v_e \rangle$ decreases with increasing $\mathcal{E}$ is one of negative differential conductivity. The slope of the current-voltage characteristic is proportional to

$$\frac{dJ}{d\mathcal{E}} = \sigma + \mathcal{E}\frac{d\sigma}{d\mathcal{E}} \qquad (8\text{-}54)$$

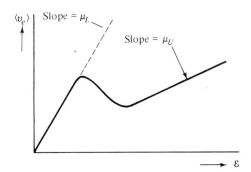

Fig. 8-15 Electron drift velocity vs. $\mathcal{E}$ in an n-type Gunn-effect semiconductor.

hence its being negative depends on $d\sigma/d\mathcal{E}$. There are two conceivable ways in which the conductivity can decrease with increasing electric field. One is the Gunn effect of negative differential mobility. The second is the possibility of the free carrier density decreasing with increasing field. The latter effect can occur if imperfections or impurities whose trapping efficiency increases with $\mathcal{E}$ are present. For the most part, however, and especially for extrinsic materials the mobility variation is the technically important process. So that the mobility variation will be effective in producing the negative differential conductivity the difference in energy between the valley minima must be large enough so that the upper valley is unoccupied at low fields but small enough so that intervalley transfer occurs at realistic fields. The threshold field should not raise the electron temperature so high as to cause significant generation of electrons from the valence band by impact ionization. Thus the energy difference $E_U - E_L$ must be less than E_g. In GaAs, $E_U - E_L = 0.36$ eV, while $E_g = 1.43$ eV at room temperature.

When a d.c. bias at or above the threshold field of about 3000 V/cm is applied to an n-type GaAs sample the charge densities and electric field within the sample become nonuniform, creating *domains*. Suppose, for example, that the field is just at the threshold value, the field is uniform, and the charge densities are drifting uniformly, carrying ohmic current. Now let us assume that for some reason a nonuniformity arises in either the field or the charge density. The cause might be a nonuniformity in the structure of the cathode end of the diode or in the impurity doping. Yet another possible cause for nonuniformity is noise, the thermal fluctuations in the carrier density. For the sake of argument we shall assume that fluctuations in the carrier density create a local excess of electrons. The electric field near this excess negative charge is composed of both the applied field $\mathcal{E}_0$ and the field of the charges, as indicated in Fig. 8-16. The net electric field on the anode side of the domain of excess negative charge is greater than that on the cathode side. If $\mathcal{E}_0$ is at threshold, the field at the anode side of the accumulation

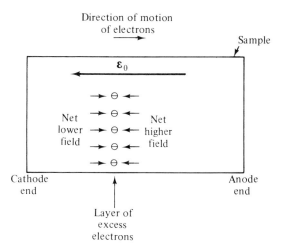

Fig. 8-16 Principle of formation of the space-charge accumulation layer in a Gunn-effect diode.

of negative charge exceeds threshold and the electrons in the domain move with the lesser mobility μ_U. The upstream electrons, however, drift with the low field mobility $\mu_L \gg \mu_U$. The upstream electrons, drifting faster than those in the domain, overtake the domain and accumulate in it. Downstream from the accumulated electrons the field is below threshold. The downstream electrons drift with the higher mobility, creating a region with a deficiency of electrons. The regions of excess and deficiency of electrons form a dipole layer. As the domain drifts toward the anode the accumulation of space charge continues, increasing the space-charge field, in turn causing more charge to accumulate in the dipole. The distributions of $\mathcal{E}$ and n in the accumulation mode are indicated in Fig. 8-17. As the accumulation layer grows and propagates toward the anode the voltage across the device remains constant due to the redistribution of $\mathcal{E}(z)$. The instability grows until the accumulated charge is collected at the anode circuit contact. As the domain is collected the electric field distribution changes to keep the voltage, $V = -\int \mathcal{E}\, dz$, a constant, $\mathcal{E}$ rising until the threshold field is reached. When the threshold field is reached again, a new domain will be formed, starting the process again.

The current through the device oscillates with a time period characteristic of the time required for the domain to drift between the cathode and the anode. The Gunn diode then is a transit-time device, its frequency of operation depending on its length. Devices with active regions 10 μm thick operate at frequencies near 10 GHz. The efficiency of such a device for delivering power at its fundamental frequency depends on its output wave form. Since for d.c. biasing the output consists of a series of current spikes, one with the

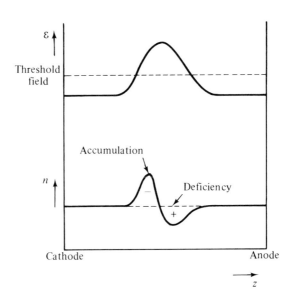

Fig. 8-17 Electric field and electron density variations in the space-charge accumulation mode of a Gunn-effect diode.

arrival of each domain at the anode, its frequency content is high in harmonics. It turns out that for maximum efficiency the spatial extent of the domain should be about half the length of the drift space. The size of the accumulation layer is generally inversely proportional to the extrinsic carrier density. Thus for a good output wave form a criterion for the doping concentrations and active length of the diode can be established. Ordinary Gunn diodes operate at low power levels, a watt or less, with low overall efficiencies, a few percent or less.

The frequency of oscillation of the Gunn diode operated in the Gunn domain mode is limited by the transit time of the charges across the device. Another mode of operation is possible. The so-called limited space-charge accumulation (L.S.A.) mode takes advantage of the negative differential conductivity with better frequency and power capabilities than the transit-time mode. For L.S.A. operation the diode is placed in a resonant microwave cavity that oscillates at the desired frequency. The voltage applied to the device is the sum of a d.c. bias near threshold and the a.c. voltage of the tuned circuit. The a.c. voltage must be large enough to swing the diode below threshold long enough during each cycle for the accumulated space charge to disperse and above threshold for a short time so that the space charge does not accumulate too much yet the negative differential conductivity is effective. The doping concentration is critical in determining the range of the frequency of operation. Operated at the proper frequency an L.S.A. diode acts like a negative resistance at that frequency supplying a.c. energy to the

Sec. 8-5 Metal-Semiconductor Contacts

tuned circuit. The energy comes from the d.c. bias supply. L.S.A. diodes operate at appreciable power levels at frequencies up to a few hundred gigahertz.

8.5 METAL-SEMICONDUCTOR CONTACTS

To utilize the electronic properties of a semiconductor the metallic conductors of a circuit must somehow be connected to it. Metallic contacts for semiconductors can be formed by soldering, by alloying, or by depositing a thin layer of atoms from a metallic vapor introduced into a vacuum chamber. Generally a potential difference appears between the metal and the semiconductor just as in the case of two dissimilar metals. An electric field in the transition region between the substances is responsible for the contact potential. The electric field at equilibrium arises from the redistribution of charge required thermodynamically by the establishment of a single Fermi energy, the chemical potential, for the whole system. The electric field in the contact region controls the flow of charge across the contact. The equilibrium field, of course, allows no net current to flow across the junction. By applying external biases to the contact this field can be altered, causing current to flow between the metal and the semiconductor. That is, the metal-semiconductor junction has a current-voltage relationship. Contact I-V characteristics fall into two classes, ohmic and rectifying. The I-V characteristic of an *ohmic contact* is linear and symmetric about the point $I = 0$, $V = 0$; an ohmic contact presents the effect of a small resistance. The I-V characteristic of a *rectifying contact* is nonlinear and asymmetric with respect to the externally applied bias. When a rectifying contact is forward-biased, a forward current flows, and when reverse-biased, the much smaller reverse current flows. The forward current is very much larger than the reverse current, and the voltage dependence of the two is different. A rectifying junction between a metal and a semiconductor is objectionable if the purpose of the contact is to connect a device to a linear circuit. On the other hand, the rectifying characteristics of a metal-semiconductor junction may be useful, for example, in nonlinear switching circuits.

The nature of any junction depends in large part on the internal contact potential. Recall from the discussion of Section 7.2 that the contact potential (in volts) is proportional to the difference in the work functions W_1 and W_2 of two metals. The work function is the minimum energy on the average that must be supplied to a charge carrier in order to cause its emission from the material. In a metal, the work function is the difference between the external potential energy of a free electron, usually taken as zero, and the Fermi level. In semiconductors the work function depends on impurity doping and temperature. For n-type semiconductors with ionized donors the true work function W_n is the difference between the external potential and the Fermi level, which is near the bottom of the conduction band E_c. For p-type semiconductors, the true work function W_p is the difference between the external potential and the Fermi level, which is near the top of the valence band E_v.

Rectifying Contacts

Rectifying contacts occur when the work function of the metal W_m is either greater than the work function of an n-type semiconductor W_n or less than the work function of a p-type semiconductor W_p. Figure 8-18 shows the energy diagrams of isolated metals and semiconductors that will form rectifying contacts when joined.

Let us first consider the contact between the metal and n-type semiconductor of Fig. 8-18(a). As contact is made between the two solids, electrons in states near E_{cn} flow into the metal where lower-energy, unoccupied states are available. The electron flow to the metal continues until a uniform Fermi

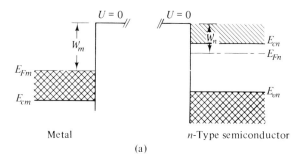

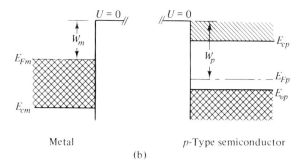

Fig. 8-18 Energy diagrams of isolated metals that form rectifying contacts with (a) n-type and (b) p-type semiconductors.

level is established. When equilibrium is established the metal will have net negative charge and the semiconductor net positive charge since electrons have moved from the semiconductor to the metal. The excess electronic charge in a metal effectively resides on its surface owing to the high density of mobile electrons in the metal. The excess positive charge in the semiconductor resides on ionized donor atoms that, because of their relative low density, lie in a distributed region of nonzero thickness near the interface with the metal. Much like a screening length, the thickness of this layer of positive charge is inversely proportional to the donor density. Due to these charges an electric field exists between the two solids, the spatial integral of the field being the contact potential $\phi = -\int \mathcal{E}\, dz$, where the $\hat{z}$-direction is assumed to be normal to the interface between the materials. Because of the nature of the distribution of the excess charges the field is nonzero on the semiconductor side of the interface but does not exist inside the metal. The charge distributions, the electric field, and the energy diagram of the contact are indicated in Fig. 8-19. The effect of the field is represented by the bending of the electronic energy band structure since $\mathcal{E}$ is proportional to the gradient of the potential energy. The contact energy $e\phi$ represents a potential barrier to the additional net flow of electrons in either direction across the barrier in equilibrium. For electrons in the semiconductor the electric field in the

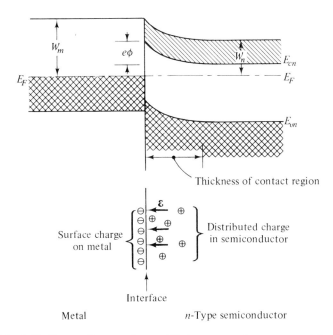

Fig. 8-19 Energy diagram of an equilibrium contact between materials of Fig. 8-18(a).

contact region repels the electrons. For electrons in the metal, the excess electrons at the surface repel the electrons from the interior. Of course, because of their thermal distribution, some electrons on each side have sufficient energy to overcome this potential barrier. In equilibrium the contact potential barrier is such that equal numbers of electrons flow in each direction, and hence no net current flows.

When a rectifying n-type semiconductor-metal contact is biased by an external supply as in Fig. 8-20(a) the balance of charge flow across the interface is upset, allowing net current to flow. We shall assume that the contact

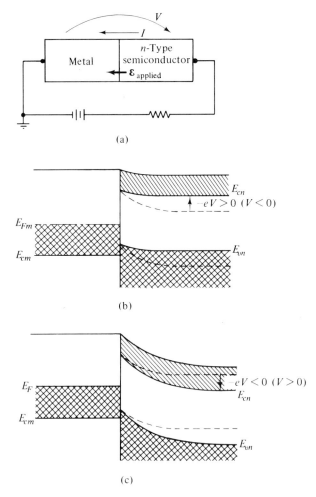

Fig. 8-20 Rectifying n-type semiconductor-metal contact (a) in a circuit with energy diagrams for (b) forward bias and (c) reverse bias voltages.

Sec. 8-5 Metal-Semiconductor Contacts

between the semiconductor and the biasing circuit is an ideal ohmic one. The applied potential appears as an additional electric field in the electron-depleted contact region. The high conductivity of the bulk of the metal and of the semiconductor keeps the electric fields in the bulk regions small so that most of the potential drop appears in the contact region. When the semiconductor is biased negatively with respect to the metal, the field in the contact region is reduced as indicated in Fig. 8-20(b). The barrier to electron flow from semiconductor to metal is reduced, but the barrier from metal to semiconductor is unchanged. Therefore a net electron flow occurs from semiconductor to metal, experienced as an electric current from metal to semiconductor. The current increases exponentially with $|V|$ because the emission of electrons from the semiconductor to the metal over the contact potential barrier depends exponentially on the barrier height. On the other hand, when the semicondctor is biased positively with respect to the metal the field in the contact is increased, reducing the flow of electrons from semiconductor to metal from its unbiased value. Figure 8-20(c) indicates that the barrier to electron flow from metal to semiconductor is not affected by the application of reverse bias so that net electric current flows from semiconductor to metal. This reverse current quickly saturates with reverse bias as the barrier to the semiconductor electrons builds up. The reverse current is much smaller than forward currents as indicated on the I-V characteristic sketched in Fig. 8-21. Notice how much this characteristic resembles that of an ideal diode.

Fig. 8-21 I-V characteristic of a rectifying n-type semiconductor-metal contact near the origin.

Next we shall consider a rectifying contact between a p-type semiconductor and a metal with $W_p > W_m$, materials illustrated in Fig. 8-18(b). The energy diagram for these materials in contact is shown in Fig. 8-22. The electric field in the contact region results because the metal is a large reservoir of electrons that move into the semiconductor to occupy previously empty states at lower energies. A positive surface charge is left on the metal, and a negative charge is distributed in a layer of the semiconductor near the interface. The negative charge layer is due to ionized acceptors, that is, acceptor states filled with electrons thereby being negatively charged. The potential barrier due to the contact field retards the flow of majority carrier holes

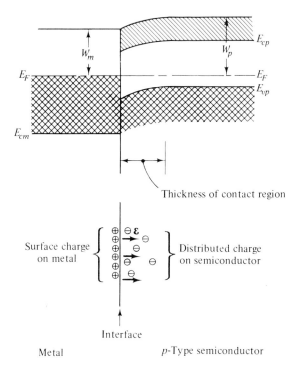

Fig. 8-22 Energy diagram of an equilibrium contact between materials of Fig. 8-18(b).

from the semiconductor to the metal. Particle current of holes from semiconductor to metal results in an electrical current flowing from the semiconductor to the metal. On the other hand, the contact field extracts the minority carrier electrons from the semiconductor. Particle current of electrons from semiconductor to metal results in an electrical current flowing from the metal to the semiconductor. In equilibrium the contact potential is such that these two currents balance and no net current flows across the interface.

When an external bias is applied to the p-type semiconductor as in the circuit of Fig. 8-23(a), the balance of the electron and hole currents across the junction is upset. When the semiconductor is biased positively, the field in the contact region is reduced, causing the bands to bend as indicated in Fig. 8-23(b). With positive bias the barrier to hole flow from semiconductor to metal is reduced, permitting an increase in hole current while the electron current across the interface is unaffected. Thus a net electrical current flows from the semiconductor to the metal for positive bias. The current increases exponentially with V as the barrier height for the majority holes is lowered. The reverse bias condition, $V < 0$, is sketched in Fig. 8-23(c). Hole flow from

Sec. 8-5 Metal-Semiconductor Contacts

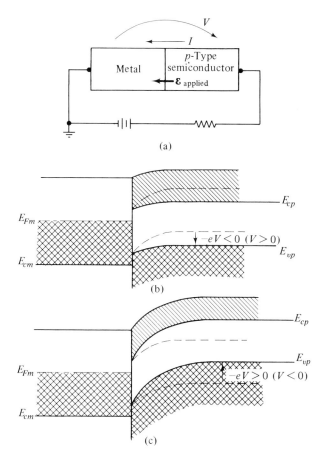

Fig. 8-23 Rectifying p-type semiconductor-metal contact (a) in a circuit with energy diagrams for (b) forward bias and (c) reverse bias voltages.

semiconductor to metal is reduced by the increased barrier. The minority carrier electron current is relatively unaffected by moderate reverse bias voltages so that a net electrical current flowing from the metal to the semiconductor results from reverse bias. This reverse current is very much smaller than the currents that flow under forward bias conditions because of the very low density of the minority carrier electrons. The I-V characteristic of a rectifying p-type semiconductor-metal contact is sketched in Fig. 8-24.

Rectifying metal-semiconductor contacts are called Schottky barriers. Their d.c. I-V characteristics are similar to those of p-n junctions, which we shall study in the next chapter. Schottky barrier diodes and p-n junction diodes have similar applications in electronics as well.

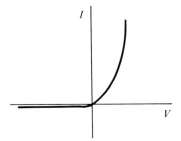

Fig. 8-24 I-V characteristic of a rectifying p-type semiconductor-metal contact near the origin.

Ohmic Contacts

Ohmic contacts occur when the metallic work function W_m is either less than the work function of an n-type semiconductor W_n or greater than the work function of a p-type semiconductor W_p. The energy diagrams for an appropriate n-type semiconductor and a metal before and after contact are sketched in Fig. 8-25. When the solids are brought into contact electrons flow from the metal, leaving a positive metallic surface charge, to the semiconductor. In the semiconductor the excess charges occupy conduction band states and lie very close to the surface since they are not bound elsewhere in the semiconductor. As indicated in Fig. 8-25(b) essentially no barrier exists for electron flow in either direction across the interface since the difference between E_{c_n} and E_F is generally very small. Bias with either polarity causes current to flow freely. An analogous diagram can be constructed for an ohmic contact between a p-type semiconductor and a metal.

Practical Problems with Contacts

In practice, producing contacts is not simply a matter of selecting materials based on tabulated values of their respective work functions and joining them together. Because the concepts we have discussed in this section depend critically on the distribution of available energy states near the surface of the semiconductor, surface imperfections play a dominant role in determining the characteristic of an actual junction. Imperfections tend to make the contacts ohmic. Ohmic contacts usually result when metallic leads are soldered or welded to semiconductors because of surface damage that results from the fabrication technique. Sometimes surfaces are mechanically processed such as by sand blasting to damage the crystalline structure of the surface deliberately in order to achieve ohmic contacts.

On the other hand, to produce a Schottky barrier the surfaces must be very carefully prepared such as by chemical removal of surface layers of con-

Sec. 8-5 Metal-Semiconductor Contacts

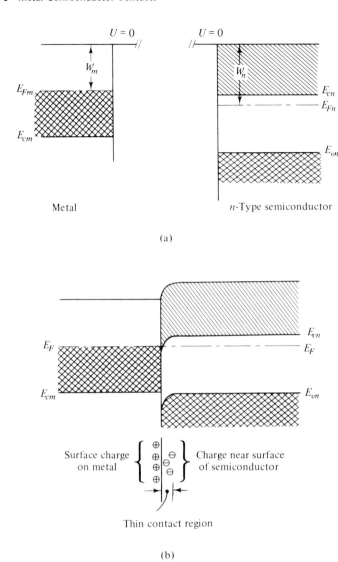

Fig. 8-25 Energy diagram for a metal and an n-type semiconductor, $W_m < W_n$: (a) isolated, (b) after contact.

taminants. Another preparation technique used on semiconductors in vacuum is to dislodge impurities by the impact of ions of inert gases, a process called sputtering or plasma etching. The metal is then deposited on the cleaned surface by condensation of a metallic vapor introduced into the vacuum chamber by evaporation from a heated source.

BIBLIOGRAPHY

For further reading on the electronic properties of semiconductors, see

ADLER, R. B., A. C. SMITH, and R. L. LONGINI, *Introduction to Semiconductor Physics*, SEEC Vol. 1. New York: John Wiley & Sons, Inc., 1964.

RYDER, J. D., *Electronic Fundamentals and Applications* (4th ed.). Englewood Cliffs, N.J.: Prentice-Hall, Inc., 1970.

DEKKER, A. J., *Solid State Physics*. Englewood Cliffs, N.J.: Prentice-Hall, Inc., 1957.

GIBBONS, J. F., *Semiconductor Electronics*. New York: McGraw-Hill Book Company, 1966.

More advanced treatments of the subjects of this chapter may be found in

VAN DER ZIEL, A., *Solid State Physical Electronics* (2nd ed.). Englewood Cliffs, N.J.: Prentice-Hall, Inc., 1968.

KITTEL, C., *Introduction to Solid State Physics* (4th ed.). New York: John Wiley & Sons, Inc., 1971.

WANG, S., *Solid State Electronics*. New York: McGraw-Hill Book Company, 1966.

SZE, S. M., *Physics of Semiconductor Devices*. New York: John Wiley & Sons, Inc. (Interscience Division), 1969.

Optical processes are treated thoroughly in

PANKOVE, J. I., *Optical Processes in Semiconductors*. Englewood Cliffs, N.J.: Prentice-Hall, Inc., 1971.

For additional reading on metal-semiconductor contacts see the books by van der Ziel or Sze listed above or

ANKRUM, P. D., *Semiconductor Electronics*. Englewood Cliffs, N.J.: Prentice-Hall, Inc., 1971.

The current state of the art of the application of bulk semiconductor effects to electronic devices is best followed by regular reading of the periodical literature, particularly in

IEEE Spectrum, a publication of the Institute of Electrical and Electronic Engineers, Inc., 345 East 47th Street, New York, 10017. Presenting review articles and those of broad interest to electrical engineering.

Electronics, McGraw-Hill Building, 330 West 42nd Street, New York, 10036. A trade magazine with articles related to electronics technology and the state of the commercial art.

Proceedings of the IEEE. Research and review papers in broad areas of electrical engineering.

PROBLEMS

8-1. Suppose that a semiconductor is doped with donor and acceptor impurities with densities N_d and N_a, respectively. At fixed temperature where in the energy diagram is the Fermi level for the cases $N_d > 2N_a$, $N_d = 2N_a$, $N_a < N_d < 2N_a$, $N_d = N_a$, $2N_d = N_a$? Use Fermi-Dirac statistics for the occupation probability of donor and acceptor states.

8-2. Derive Eq. (8-17) from Eq. (8-15).

8-3. Calculate the deviation of the Fermi level in intrinsic silicon from the middle of the energy gap at room temperature using the data of Table 8-1.

8-4. Calculate the effective density of states at the top of the valence band N_v and at the bottom of the conduction band N_c for both silicon and germanium at room temperature using the data of Table 8-1.

8-5. Show that the extrinsic carrier densities can be written as

$$n = n_i \epsilon^{(E_F - E_{Fi})/\kappa T}$$
$$p = p_i \epsilon^{-(E_F - E_{Fi})/\kappa T}$$

where n_i and p_i are the intrinsic densities ($n_i = p_i$) and E_{Fi} is the location of the Fermi level under instrinsic conditions. Then show that the product np equals the product $n_i p_i$.

8-6. Suppose that n-type germanium is doped with impurities of concentration $N_d = 10^{16}$ cm^{-3}. Find the location of the Fermi level relative to the bottom of the conduction band at room temperature if $N_c = 10^{19}$ cm^{-3}, the intrinsic density is $n_i = 10^{13}$ cm^{-3}, and all the donors are ionized. If the donor energy level is 0.02 eV below the bottom of the conduction band, find the probability that a donor state actually is ionized using the Fermi-Dirac function as an approximation to the occupation probability for donor states.

8-7. Impurity compensation is the reduction of free carrier densities and sample conductivity by the introduction of acceptor impurities in n-type semiconductors or of donors in p-type material. Explain this phenomenon by estimating the conduction electron density in an n-type semiconductor before and after adding acceptors with the same impurity concentration as the donors. For simplicity assume equal ionization energies for the impurities and equal effective masses for the charge carriers.

8-8. Calculate the maximum wavelengths of light absorbed by pure silicon and pure germanium. Silicon, germanium, and diamond have the same crystalline structure but different appearances. Explain.

8-9. Intrinsic photoconductivity results from optical generation of hole-electron pairs across the band gap, while extrinsic photoconductivity results from ionization of impurities. Why may intrinsic photodetectors be made so much thinner than extrinsic detectors?

8-10. Infrared sensors can be made from the compounds PbS, PbTe, and PbSe. Explain.

8-11. Calculate the intrinsic conductivity of silicon at room temperature. What donor concentration is required to obtain a conductivity of 10 mhos/m if all of the donors are ionized. What concentration of acceptors is required for p-type germanium for the same conductivity?

8-12. Estimate the diffusion lengths for holes and electrons in silicon and germanium using average recombination times of 10 μsec for silicon and 1 msec for germanium. Use the data from Table 8-2 and the Einstein relation to find the appropriate diffusion coefficients.

8-13. Calculate the upper limit on the electric field $\mathcal{E}$ for which silicon is ohmic by using

$$v_{\text{drift}} \simeq (0.1) v_{\text{scatter-limited}}$$

Compare your result with the data of Fig. 8-10.

8-14. Find the ambipolar diffusion coefficient D_a in semiconductors by rewriting Eq. (4-28) in terms of the gradients of the excess carriers $(n - n_0)$, $(p - p_0)$. Assuming that $\nabla(n - n_0) = \nabla(p - p_0)$ and that $p\mu_p \gg n\mu_n$, show that $D_a = D_n$.

8-15. The Hall coefficient for copper is -7.3×10^{-11} m^3/C and its conductivity may be taken as 6.7×10^7 mhos/m. Calculate the electron density and mobility.

8-16. A germanium sample has $R_H = -7.5 \times 10^{-3}$ m^3/C and conductivity of 10 mhos/m. Find the carrier density, mobility, and collisional relaxation time.

8-17. A semiconductor has both positive and negative charge carriers of densities p and n, respectively. Taking both carriers into account, show that the Hall coefficient is

$$R_H = \frac{-1}{e} \frac{(b^2 n - p)}{(bn + p)^2}$$

where $b = \mu_n/\mu_p$. *Hint:* Assume that the Hall field is made of two parts, one due to electrons, and the other to holes. Then use an expression for transverse current density to find the net $\mathcal{E}_H$ in terms of B_0, $\mathcal{E}_0$, the concentrations and mobilities. Last, relate $\mathcal{E}_0$ to J_x to find R_H.

8-18. Calculate the temperature sensitivity at room temperature of the resistance of a thermistor made of intrinsic germanium by finding the fractional change in resistivity $\partial \rho / \rho$ for a fractional change in temperature $\partial T/T$, where ρ is the inverse of the conductivity. Use Eq. (8-20) for $n = n_i$ neglecting the effect of T on N_v and N_c.

8-19. Calculate the drift velocity of a space-charge accumulation layer in a Gunn diode operating at 10 GHz with an active region 10^{-5} m thick. If the space charge drifts in an average field of about 3000 V/cm, estimate its mobility.

9

Semiconductor Junctions

While electronic effects occurring in bulk semiconductors have been applied to a variety of important devices, far and away the most prevalent use of solid-state materials in electronics is in configurations of two or more materials with different types of impurity dopants. When two different substances are in contact with one another, an electric potential appears between them, the electric field arising in the region of transition from one to the other. This field is responsible for the unique electronic properties of the junction region and of the devices made up of one or more of these junctions. The rectifying Schottky diode of the previous chapter is comprised of a metal-to-semiconductor contact. In this chapter we shall introduce an additional class of junctions based on the contact between two dissimilar extrinsic semiconductors. In its most common form this is the p-n junction. Junction diodes, junction transistors, and junction field-effect transistors are examples of the practical application of the p-n junction. Another configuration of dissimilar materials that we shall consider in Chapter 11 is the metal-insulator-semiconductor dual contact that in its most common form is a layer of an oxide of a semiconductor between a metal and the semiconductor, the so-called MOS structure. Insulated-gate field-effect transistors and charge storage devices are examples of the application of MOS structures in electronics.

Before proceeding with our study, a word about semiconductor electronics. By virtue of his interests the reader is already aware of the revolutionary impact that semiconductor devices have had on electronics technology.

Not only do solid-state devices have the advantages of smaller size, lighter weight, lower power consumption, lower cost, and higher reliability when compared with comparable vacuum tube devices: Some electronic phenomena such as the Gunn effect are unique to semiconductors, and the crowning glory of semiconductor technology is the integrated circuit. Semiconductor devices in both discrete and integrated forms are being used extensively in and in large measure are responsible for the modern development of computers, of both broadcast and telephonic communications, of power generation, distribution, and control, of electronic instrumentation, including medical electronics, and of consumer products and appliances.

The subject matter of this chapter is treated at an introductory level. Our purpose is to present the fundamental relationships between the physical principles we have studied so far and their applications to common devices. For our purposes it is necessary that our discussion be brief. After all, whole books have been written about almost every topic we shall discuss. Depending on the interest of the student, further study may be advisable in the physics of solid-state devices, their capabilities and limitations, or their equivalent circuit representation or modeling.

9.1 THE p-n JUNCTION

The transition between a p-type and an n-type extrinsic semiconductor is called a *p-n junction*. Ordinarily the transition occurs within the same host lattice and is the result of differential doping with predominantly acceptor impurities in the p-type region and predominantly donor impurities in the n-type region.

Techniques of Fabrication

A p-n junction can be manufactured in one of several ways. One method is to vary the doping concentrations as the crystal is being drawn from a liquid melt. Crystals can be grown by withdrawing a small seed from a molten bath of the semiconductor. If impurities are introduced into the bath, the resulting crystalline semiconductor will be extrinsic. For example, if acceptor impurities have been added to the melt, the semiconductor will be p-type. To make a grown p-n junction, donor impurities are added to the melt after a p-type region has been formed and the drawing resumed. So that an n-type region will be formed the donors must be added at a concentration greater than that required just to compensate for the acceptor impurities still present. When the drawing process is complete, a single crystal ingot of semiconductor will have been produced with spatially dependent doping concentrations. The transition between n- and p-type regions in grown junctions is smooth, such a transition being referred to as a *graded junction*.

Alloying techniques, on the other hand, produce a transition region that may be less than 10^{-6} cm thick, an *abrupt junction*. An alloyed or fused junction is formed in the following manner. A crystalline wafer of extrinsic semiconductor is obtained by slicing an ingot, for example. An impurity of a kind other than the one it already contains is placed on top of the wafer. For example, a pellet of aluminum, an acceptor impurity, might be placed on a wafer of n-type silicon. The combination is then heated to a temperature above the eutectic temperature for the mixture of the semiconductor and the impurity but below the melting point of the original semiconductor. The eutectic temperature is the lowest temperature at which a solid-liquid phase transition for the two-component system takes place. The molten zone is composed of a mixture of semiconductor and impurity atoms. Upon cooling, a heavily doped semiconductor recrystalizes on top of the wafer and under the impurity. An abrupt junction is formed at the boundary between the original crystal and the alloyed material.

Another method of forming a junction utilizes vapor epitaxial growth of crystals. In epitaxy a crystal is formed by the deposition of atoms from a vapor onto a heated crystalline surface. The deposited atoms will form a crystal if the binding forces and natural lattice spacings correspond to those of the original seed on which the crystal is to be grown. A junction is formed by varying the impurity introduced into the vapor.

Yet another method of forming p-n junctions utilizes the controlled diffusion of impurity atoms into the host crystal. A heated extrinsic crystal is exposed to a vapor of impurity atoms of the other kind, and these atoms diffuse into the solid. The impurity concentrations can be carefully controlled by adjusting the vapor pressure and the temperature. Figure 3-17 shows the temperature dependence of the diffusion coefficients of a variety of impurities into silicon. A p-n junction is formed if the diffused impurity is sufficiently concentrated in part of the crystal to overcompensate for the impurities originally present. A diffused junction may be relatively abrupt or graded. This vapor-diffusion process is utilized to fabricate p-n junctions in both discrete devices and integrated circuits. Impurities may also be diffused into a host lattice from liquid or solid forms placed on the surface of the semiconductor.

Qualitative Description of the Junction

Before proceeding to the quantitative analysis of the p-n junction that will develop our understanding of its electrical characteristics, its application to devices, and its modeling, we shall briefly describe it physically to emphasize its relationship to the concepts we have already developed.

The p-n junction is characterized by severe gradients in the charge carrier densities across the region of transition from n-type to p-type. On either side of the junction, the majority carrier density is determined by the doping concentration. The majority carriers are the extrinsic charges, electrons in the n-type and holes in the p-type material. The minority carrier densities, holes in the n-type and electrons in the p-type region, are determined in equilibrium by the energy band gap E_g and the temperature T. Because each carrier is the majority carrier on one side of the junction and the minority on the other, strong gradients in density exist across the transition region. These gradients drive diffusive particle currents across the transition region, each carrier diffusing from the region of its majority to the region in which it is the minority carrier.

This diffusive flow is impeded by an internal electric field that arises in the following manner. As electrons diffuse from the n-region across the junction they leave behind net positive charge residing on the ionized donor atoms. As holes diffuse from the p-region they leave behind net negative charge on the ionized acceptor atoms. Thus either side of the junction is charged, the n-side positively and the p-side negatively. That is, a dipole of charge exists in the transition region. The electric field of this dipole is in such direction as to retard the flow of majority carriers across the junction, opposing the diffusive flow. The vector field is directed from the n-region toward the p-region so that, for example, electrons diffusing in the gradient toward the p-region experience a retarding force. On the other hand, minority carriers are swept up by the dipole field and caused to cross the transition region. For instance, electrons in the p-region that wander by random thermal processes into the region of influence of the dipole field experience a force that pulls them across the junction to the n-region.

The region of transition in which the charge dipole and internal electric field exist is variously referred to as the depletion or the space-charge layer. The term *depletion layer* refers to the fact that the mobile carriers are swept through the region and many fewer carriers exist there than in the bulk regions. The term *space-charge layer* refers to the fact that net charge exists in this space while the n-type and p-type regions are charge neutral. Thus the dipole electric field can be calculated if the space-charge density is known.

The electric field in the space-charge layer causes a difference in electric potential to exist between the n- and p-regions. This potential difference is the negative of the line integral of the electric field across the transition region and in equilibrium is nothing more than the contact potential ϕ_0.

An Energy Band Diagram for the Junction

Based on these descriptive comments it is a relatively easy task to construct an energy band diagram for an unbiased p-n junction in equilibrium.

Sec. 9-1 The p-n Junction

Such a diagram is shown in Fig. 9-1(b), while Fig. 9-1(a) indicates the physical configuration of a one-dimensional junction, locations of the space charge, and the polarity of the internal contact field $\mathcal{E}_0$. In the n-region the Fermi level is close to the bottom of the conduction band, while it is near the top of the valence band in the p-region. The metallurgical junction is that plane in which the impurity concentrations exactly compensate one another. The space-charge layer extends from a depth of w_n in the n-side of the metallurgical junction to a depth of w_p on the p-side. The contact potential appears across the distance $w = w_n + w_p$, the width of the space-charge layer.

The band diagram of Fig. 9-1(b) indicates that the contact potential is a barrier to the flow of majority electrons from left to right as well as a barrier to the flow of majority holes from right to left. However, thermally generated minority carriers that appear at their respective edges of the space-charge layer fall down the potential hill and cross the layer. In equilibrium with no external bias applied, no net current can flow. If minority electrons, for example, flow easily from right to left, this current must be balanced by an equal

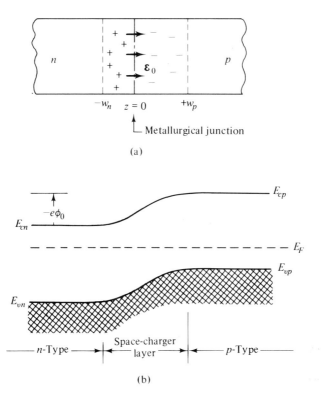

Fig. 9-1 Equilibrium p-n junction: (a) physical configuration, (b) energy band diagram.

flow of electrons from left to right. In other words, the minority electron current across the junction must be balanced by a majority electron current, the latter being due to diffusion of electrons from n to p against the retarding force represented by the potential barrier $e\phi_0$. Thermodynamically, a separate but similar detailed balance of hole currents must also exist in equilibrium.

When a bias voltage is applied between the n- and p-regions the potential barrier in the space-charge layer is altered, upsetting the detailed balance in the charge flow and resulting in a net current. If the bias is applied in a polarity to lower the potential barrier in the junction, the flow of majority carriers is greatly enhanced since injection over a barrier depends exponentially on the barrier height as well as on the temperature as described in Section 7.1. Minority carrier flow is unaffected by a decreased barrier. Therefore under this biasing condition, known as *forward bias*, the net current that flows across the junction is due to the injection of majority carriers against the barrier field. If bias of the opposite polarity is applied, the barrier height is increased, dramatically reducing the number of majority carriers that can overcome the enhanced repulsive force. Again the minority carrier flow is unaffected by an increased barrier provided the electric field does not become too large. In this *reverse bias* condition, minority carrier flow across the junction accounts for the current.

Because the forward and reverse currents are due to different mechanisms that balance only when no bias is applied and because of the internal electric field in the space-charge layer, the current-voltage characteristic of a p-n junction is asymmetric. Furthermore, it is nonlinear. Proper operation of a p-n junction depends on the controlling influence of the electric field on the charge flow. The field in turn must be alterable by the application of the external bias. So that the external bias will appear across the space-charge layer, the conductivities of the n- and p-regions should be relatively high and that of the depletion region low. Since the space-charge region contains relatively few mobile charges, its conductivity is low. So that the field in the space-charge layer will have the desired controlling effect, it should extend only over a well-localized region of space. A small depletion layer means that the contact potential results from a strong electric field. The electric field in practical junctions is often sufficiently high that the drift of the carriers is scatter-limited and the mobility is not independent of the field. Besides the strong electric field, there are also large gradients in the charge densities in the space-charge layer. Because of the strong field and gradients and the small width of the depletion region, carrier transit times are typically less than a recombination time. Therefore recombination may be assumed not to occur inside the junction.

In the next three sections we shall develop a quantitative analysis of the charge carrier densities and their behavior under equilibrium and under biased conditions.

9.2 THE p-n JUNCTION IN EQUILIBRIUM

The electron and hole densities on the p-side of a p-n junction will be designated n_p and p_p, respectively, the subscript p indicating values on the p-side. On the n-side the carrier densities are n_n and p_n. The majority carrier concentrations then are p_p and n_n, while n_p and p_n are the minority densities. The values of the respective concentrations under unbiased equilibrium conditions are further designated by n_{p0}, n_{n0}, p_{p0}, and p_{n0}.

From the discussion of Section 8.2 it is easy to show using Eqs. (8-13), (8-17), and (8-20) that in equilibrium the product of the charge densities in an extrinsic semiconductor depends only on the temperature and the band gap,

$$np = n_i^2(E_g, T) \quad (9\text{-}1)$$

where n_i^2 is the square of the intrinsic charge density given by Eq. (8-20). Equation (9-1) is independent of the doping concentrations and hence of the location in the energy diagram of the Fermi level E_F. In the p-region of the p-n junction of Fig. 9-1, E_F is near the top of the valence band E_{vp}, and almost all of the acceptor states are ionized. Therefore the majority hole density in the uniform p-region away from the space-charge layer is approximately given by

$$p_{p0} = N_a \quad (9\text{-}2)$$

while according to Eq. (9-1), the minority concentration is

$$n_{p0} = \frac{n_i^2}{N_a} \quad (9\text{-}3)$$

In the n-region away from the space-charge layer the carrier densities are

$$n_{n0} = N_d, \quad p_{n0} = \frac{n_i^2}{N_d} \quad (9\text{-}4)$$

since E_F is close to the bottom of the conduction band, E_{cn}, on that side of the junction. Because the extrinsic carrier densities are much greater than the intrinsic density, Eqs. (9-2)–(9-4) indicate that very strong concentration gradients exist in the transition region.

More precisely, we may examine the difference in the electron density across the junction with the help of Eq. (8-13), repeated here as

$$n = N_c \epsilon^{-(E_c - E_F)/\kappa T} \quad (9\text{-}5)$$

where N_c is the effective density of states in the conduction band given by Eq. (8-14). Equation (9-5) can be used to find both n_{n0} and n_{p0} if we recall that the internal field of the contact potential causes the bottom of the conduction band to be at different energy levels on either side of the junction. In Fig. 9-1 the bottom of the conduction band is labeled E_{cp} and E_{cn} in the p- and n-

regions, respectively. We therefore write

$$n_{p0} = N_c \epsilon^{-(E_{cp}-E_F)/\kappa T} \tag{9-6}$$

and

$$n_{n0} = N_c \epsilon^{-(E_{cn}-E_F)/\kappa T} \tag{9-7}$$

since in thermodynamic equilibrium E_F is a constant. The ratio n_{p0}/n_{n0} is then

$$\frac{n_{p0}}{n_{n0}} = \epsilon^{-(E_{cp}-E_{cn})/\kappa T} \tag{9-8}$$

The difference in electron energy $(E_{cp} - E_{cn})$ in equilibrium defines the contact potential ϕ_0 according to

$$(E_{cp} - E_{cn}) \equiv -e\phi_0 \quad \text{in equilibrium} \tag{9-9}$$

so that

$$\frac{n_{p0}}{n_{n0}} = \epsilon^{e\phi_0/\kappa T} \tag{9-10}$$

Notice that by Eq. (9-9) ϕ_0 is the equilibrium electric potential of the p-side of the junction relative to that of the n-side. It turns out that by this definition ϕ_0 is negative and $n_{p0} < n_{n0}$. By taking the logarithm of both sides of Eq. (9-10) we can obtain an alternative expression for ϕ_0:

$$\phi_0 = \frac{\kappa T}{e} \ln\left(\frac{n_{p0}}{n_{n0}}\right) \tag{9-11}$$

Using Eqs. (9-3) and (9-4), this result may be rewritten as

$$\boxed{\phi_0 = \frac{\kappa T}{e} \ln\left(\frac{n_i^2}{N_a N_d}\right)} \tag{9-12}$$

Typical values of ϕ_0 calculated from this equation lie in the range of 0.1 to 1.0 V depending on the material, the temperature, and the impurity concentrations. For room-temperature germanium with $N_a = 4.1 \times 10^{18}/\text{cm}^3$ and $N_d = 1.4 \times 10^{15}/\text{cm}^3$, Eq. (9-12) yields $\phi_0 = -0.42$ V. Since $\kappa T/e \simeq 0.025$ V at room temperature, the electron density falls by a relative factor of

$$\epsilon^{-16.8} = 5.05 \times 10^{-8}$$

across the transition region. The negative value of ϕ_0 indicates that in equilibrium the potential of the p-side of the junction is lower than that of the n-side. This result is as should be expected since the net charge residing on the p-side is negative while that on the n-side is positive.

Equation (9-12) for ϕ_0 may also be obtained by considering the ratio of hole concentrations, p_{n0}/p_{p0}.

An Abrupt Junction in the Depletion Approximation

Having obtained expressions for the charge densities on either side of a junction and the height of the contact potential barrier, we now turn to the

Sec. 9-2 The p-n Junction in Equilibrium

problem of the variation of the electric potential inside the space-charge layer itself. To study this problem quantitatively we shall have to make certain assumptions about the variations of both the doping concentration and the space-charge density across the transition region. Having made these assumptions we can calculate not only the spatial variation of the electric potential $\phi(z)$ in the junction but also the electric field, $\mathcal{E} = -\nabla\phi$, and the width of the space-charge layer. For the purposes of illustration we shall consider a uniformly doped abrupt junction with a space-charge layer completely depleted of mobile charge.

Let the n-side of an abrupt junction be uniformly doped with a *net* impurity concentration of N_d donors per cubic centimeter. The p-side is uniformly doped with a *net* N_a acceptors per cubic centimeter. We shall assume that to support the contact potential the space-charge layer within a distance of w_p from the metallurgical junction on the p-side and of w_n on the n-side is completely devoid of mobile charges. Thus the space charge in the depletion layer is assumed to be due to the presence of immobile ionized impurity atoms as illustrated in Fig. 9-2(a) and (b). This assumption of complete depletion in the space-charge layer is an extreme but surprisingly useful simplification of the true physical situation in which the mobile charge densities exist in the space-charge layer, varying uniformly across it from their values on either side as given in Eqs. (9-2)–(9-4). The depletion assumption is made to simplify the mathematics.

The electric potential $\phi(z)$ in the space-charge layer is related to the charge density $\rho(z)$ through Poisson's equation of electrostatics, which for variations in only one direction has the form

$$\frac{d^2\phi}{dz^2} = -\frac{\rho(z)}{\varepsilon} \quad (9\text{-}13)$$

For the assumption of complete depletion of the uniformly doped impurities, the charge density on either side of the metallurgical junction, $z = 0$, is given by

$$\rho(z) = eN_d, \quad -w_n < z < 0 \quad (9\text{-}14)$$

or

$$\rho(z) = -eN_a, \quad 0 < z < w_p \quad (9\text{-}15)$$

Equation (9-13) may be integrated once using Eq. (9-14) for $z < 0$ and Eq. (9-15) for $z > 0$, yielding, respectively,

$$\frac{d\phi}{dz} = -\frac{eN_d}{\varepsilon}z + C_n, \quad -w_n < z < 0 \quad (9\text{-}16)$$

and

$$\frac{d\phi}{dz} = +\frac{eN_a}{\varepsilon}z + C_p, \quad 0 < z < w_p \quad (9\text{-}17)$$

where C_n and C_p are constants of integration. These two equations can be

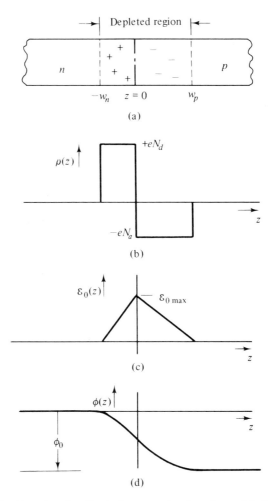

Fig. 9-2 Abrupt p-n junction in the approximation of complete depletion: (a) physical configuration, (b) space-charge density, (c) electric field, (d) electric potential.

rewritten in terms of the z-directed electric field $\mathcal{E}_z = -d\phi/dz$, resulting in

$$\mathcal{E}_{0z} = +\frac{eN_d}{\varepsilon}z - C_n, \quad -w_n < z < 0 \quad (9\text{-}18)$$

and

$$\mathcal{E}_{0z} = -\frac{eN_a}{\varepsilon}z - C_p, \quad 0 < z < w_p \quad (9\text{-}19)$$

To evaluate the constants of integration we note that since there is no net charge density outside the depleted region the electric field must be a constant

Sec. 9-2 The p-n Junction in Equilibrium

outside this region. Furthermore, since the field is confined to the depletion region, $\mathcal{E}_{0z} = 0$ at $z = -w_n$ and $z = w_p$. Since $\mathcal{E}_{0z} = 0$ at $z = -w_n$, Eq. (9-18) requires that $C_n = -eN_d w_n/\varepsilon$ and

$$\mathcal{E}_{0z} = \frac{eN_d}{\varepsilon}(z + w_n), \qquad -w_n < z < 0 \tag{9-20}$$

Since $\mathcal{E}_{0z} = 0$ at $z = w_p$, Eq. (9-19) requires that $C_p = -eN_a w_p/\varepsilon$ and

$$\mathcal{E}_{0z} = \frac{eN_a}{\varepsilon}(w_p - z), \qquad 0 < z < w_p \tag{9-21}$$

Equation (9-20) indicates that $\mathcal{E}_{0z}$ increases linearly with z from zero at $z = -w_n$ to its maximum value at $z = 0$, while Eq. (9-21) indicates that $\mathcal{E}_{0z}$ decreases linearly with z from its maximum value at $z = 0$ to zero at $z = w_p$. If the charge densities are assumed to be continuous functions of position, then ϕ and $\mathcal{E}_{0z}$ will also be continuous functions. Then the value of $\mathcal{E}_{0z}$ at $z = 0$ must be the same as determined either from Eq. (9-20) or Eq. (9-21). That is, at $z = 0$

$$\boxed{\mathcal{E}_{0\,\text{max}} = \frac{eN_d w_n}{\varepsilon} = \frac{eN_a w_p}{\varepsilon}} \tag{9-22}$$

As a consequence the total charge in the n-side of the depletion region, $+eN_d w_n$, is equal in magnitude to the total charge in the p-side, $-eN_a w_p$. In other words, the space charge in the depletion region forms a charge dipole. Furthermore, the width of the depletion region on each side of the metallurgical junction is inversely proportional to the net impurity concentration on that side:

$$\boxed{\frac{w_p}{w_n} = \frac{N_d}{N_a}} \tag{9-23}$$

The spatial variation of $\mathcal{E}_{0z}$ for the case under consideration is sketched in Fig. 9-2(c).

The spatial variation of the electric potential $\phi(z)$ may be obtained either by the second integration of Poisson's equation or equivalently by an integration of the electric field according to

$$\phi = -\int \mathcal{E}_z \, dz \tag{9-24}$$

On the n-side of the junction Eq. (9-20) may be used for $\mathcal{E}_{0z}$, yielding

$$\begin{aligned}\phi_n(z) &= -\int_{-w_n}^{z} \frac{eN_d}{\varepsilon}(z + w_n)\, dz \\ &= -\frac{eN_d}{\varepsilon}\left(\frac{z^2}{2} + w_n z\right) - \frac{eN_d w_n^2}{2\varepsilon} + K_1\end{aligned} \tag{9-25}$$

The constant of integration K_1 may be determined by assigning an arbitrary value for the potential at $z = -w_n$ because the absolute value of potential has no meaning. Only potential differences are important. For convenience we choose $\phi_n(z = -w_n) = 0$, which is equivalent to grounding the n-side of the junction. Then $K_1 = 0$. At the metallurgical junction

$$\phi_n(z = 0) = -\frac{eN_d w_n^2}{2\varepsilon} \tag{9-26}$$

On the p-side of the junction Eq. (9-21) may be used for $\mathcal{E}_{0z}$ in Eq. (9-24), yielding

$$\phi_p = -\int_0^z \frac{eN_a}{\varepsilon}(w_p - z)\, dz$$

$$= -\frac{eN_a}{\varepsilon}\left(w_p z - \frac{z^2}{2}\right) + K_2 \tag{9-27}$$

The constant of integration K_2 may be determined by evaluating ϕ_p at the metallurgical junction since the potential there is already determined by Eq. (9-26). That is,

$$K_2 = -\frac{eN_d w_n^2}{\varepsilon} \tag{9-28}$$

With the respective integration constants evaluated, Eqs. (9-25) and (9-27) give the potential variation across the space-charge layer for the assumed case of an abrupt junction with complete depletion in the region $-w_n < z < w_p$.

Because we have assumed that the potential in the n-region outside the depletion layer is zero, the value of ϕ_p at $z = w_p$ is the contact potential ϕ_0. Using Eq. (9-27) evaluated at $z = w_p$ and Eq. (9-28), we obtain

$$\boxed{\phi_0 = -\frac{e}{2\varepsilon}(N_d w_n^2 + N_a w_p^2)} \tag{9-29}$$

Equation (9-23) may be used to eliminate either w_n or w_p from Eq. (9-29), which can then be solved for either w_p or w_n, respectively. The results are

$$w_p = \left\{\frac{-2\varepsilon\phi_0}{eN_a[1 + (N_a/N_d)]}\right\}^{1/2} \tag{9-30}$$

$$w_n = \left\{\frac{-2\varepsilon\phi_0}{eN_d[1 + (N_d/N_a)]}\right\}^{1/2} \tag{9-31}$$

where $\phi_0 < 0$. Notice that the width of the depletion region in an unbiased abrupt junction depends on the square root of the equilibrium contact potential difference across the junction. By Eq. (9-22) the maximum electric field in the equilibrium space-charge layer also depends on $|\phi_0|^{1/2}$.

While we have been able to calculate the electric field, the potential, and the size of the depletion region, we are not able to calculate the drift and

Sec. 9-3 The Biased p-n Junction   267

diffusion currents in the space-charge region itself, because linear transport theory is not applicable in the large gradients of density and potential present in the depletion layer. Yet we know from the physical arguments advanced earlier in this section that in the unbiased junction a balance exists between the diffusive flow of majority carriers over the equilibrium potential barrier and the drift current of minority carriers swept across the junction by the internal field. The average currents flowing in either direction are enormous, as indicated in Problems 9-6 and 9-7, which use average gradients and fields and assumed effective values for the transport coefficients to estimate the currents. The point is that for typical currents drawn through *p-n* junction devices the balance between diffusion and drift is only slightly disturbed from the unbiased case. To accommodate the current flow in a biased junction only small changes are required in the charge carrier densities.

9.3 THE BIASED *p-n* JUNCTION

We now turn to the problem of the *p-n* junction to which an external bias voltage has been applied. Such a junction in the steady state is not in thermodynamic equilibrium. However, in most cases the deviations from the unbiased equilibrium case are small enough so that we can consider this problem as a perturbation about the equilibrium condition.

For the purpose of establishing a convention, let us assume that a bias voltage V is applied to a *p-n* junction using the circuit of Fig. 9-3. A variable voltage supply and a ballast resistor are connected in series with the *p-* and *n*-regions via ohmic contacts. By assuming that the *n*-region is grounded at zero potential we establish that as the reference for all voltages. Thus when the supply voltage V_{bb} is turned on and a current I flows, the *p*-region will be biased with a voltage $V = V_{bb} - IR$ with respect to the *n*-region.

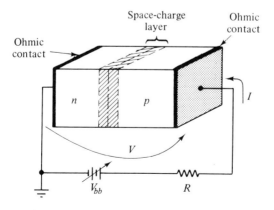

Fig. 9-3 Application of bias to a *p-n* junction.

The keystone to understanding the behavior of the carrier densities and currents in a biased junction is the effect of the applied bias on the potential barrier. If the conductivities of the n- and p-regions are moderately high, then essentially all of the applied bias voltage V appears across the depletion region. Because a voltage is the negative of the line integral of an electric field and because electric fields add vectorially, the applied potential and the built-in contact potential add. Figure 9-2(d) and Eq. (9-29) indicate that because of the contact potential the unbiased p-region is at a negative potential ϕ_0 with respect to the grounded n-region. To reduce this potential difference the applied bias V must make the p-region less negative. That is, a positive potential is applied to the p-region to reduce the potential difference across the space-charge layer. Figure 9-4(a) shows the spatial variations of both the

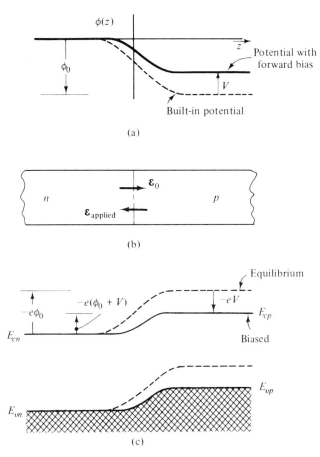

Fig. 9-4 Forward-biased p-n junction: (a) electric potential, (b) polarity of the applied field, (c) energy band diagram.

Sec. 9-3 The Biased p-n Junction

built-in potential and the potential with bias applied to reduce the junction barrier. Figure 9-4(b) indicates the polarity of the applied electric field that accomplishes this reduction. The effect of the applied bias on the electron energy band diagram is indicated in Fig. 9-4(c). Applying a positive potential lowers the electron energy levels because the potential energy of a charged particle in a potential ϕ is $q\phi$, where $q = -e$ for an electron.

Consideration of Fig. 9-4 leads to the conclusion that the net potential drop across a biased p-n junction is given by

$$\boxed{\phi = \phi_0 + V} \quad (9\text{-}32)$$

where ϕ_0 is the contact potential of Eq. (9-19) or Eq. (9-12). Since $\phi_0 < 0$, applying positive bias to the p-region reduces the potential drop across the junction. Applying negative bias to the p-region increases the magnitude of the potential drop across the junction. The former condition is called *forward bias* and the latter *reverse bias*.

A Biased Abrupt Junction in the Depletion Approximation

Before proceeding to a discussion of conditions of forward and reverse bias we must consider the effects of the applied bias on the space-charge layer and on the charge densities at either edge of the depletion region. The junction electric field and the width of the space-charge layer have already been determined for an unbiased abrupt junction in the approximation of complete depletion of the space-charge region. This calculation, which resulted in Eq. (9-22) for the maximum field and Eqs. (9-30) and (9-31) for the penetration of the depletion region into the p- and n-sides of the junction, respectively, was based on the fundamental relationships of Poisson's equation. The very same procedure applied to a biased junction under the same assumption leads to similar results, with the net potential drop ϕ replacing the equilibrium contact potential ϕ_0. Corresponding to Eqs. (9-30) and (9-31), the widths of the two parts of the depletion region are

$$\boxed{w_p = \left\{\frac{-2\varepsilon(\phi_0 + V)}{eN_a[1 + (N_a/N_d)]}\right\}^{1/2}} \quad (9\text{-}33)$$

and

$$\boxed{w_n = \left\{\frac{-2\varepsilon(\phi_0 + V)}{eN_d[1 + (N_d/N_a)]}\right\}^{1/2}} \quad (9\text{-}34)$$

while Eq. (9-22) remains valid for $\mathcal{E}_{max}$. Since $\mathcal{E}_{max}$ depends on w_n or w_p, both the width of the depletion layer and the maximum field in it depend on the applied voltage V. Recall that the reference for potential has been chosen so that ϕ_0 is negative.

Carrier Densities Due to the Bias

The carrier densities at either edge of the space-charge layer are also affected by the applied voltage. If we assume that conditions do not deviate greatly from the unbiased equilibrium case, we can write the ratio of electron densities at either edge of the depletion layer based on Eq. (9-10) as

$$\frac{n_p}{n_n} = \epsilon^{e(\phi_0+V)/\kappa T} \tag{9-35}$$

Using Eq. (9-10) in this equation yields the form

$$\frac{n_p}{n_n} = \frac{n_{p0}}{n_{n0}} \epsilon^{eV/\kappa T} \tag{9-36}$$

A similar ratio can be obtained for the hole densities:

$$\frac{p_n}{p_p} = \frac{p_{n0}}{p_{p0}} \epsilon^{eV/\kappa T} \tag{9-37}$$

Although the charge densities at the edges and the width of the depletion layer are altered by the application of the potential V, the charge neutrality of the n- and p-regions remains unchanged. Therefore in the n-region the excess of holes over the density under unbiased conditions must be balanced by an excess of electrons:

$$(p_n - p_{n0}) = (n_n - n_{n0}) \tag{9-38}$$

On the p-side

$$(n_p - n_{p0}) = (p_p - p_{p0}) \tag{9-39}$$

Recall, however, that the equilibrium minority carrier densities are very much smaller than their respective majority carrier densities. That is,

$$p_{n0} \ll n_{n0}, \qquad n_{p0} \ll p_{p0} \tag{9-40}$$

If the deviation from equilibrium is not too great, then the excess minority hole density in the n-region will be very much less than the equilibrium density of majority electrons,

$$(p_n - p_{n0}) \ll n_{n0} \tag{9-41}$$

so that from Eq. (9-38)

$$n_n \simeq n_{n0} \tag{9-42}$$

Similarly, the excess minority electron density in the p-region will be much less than the equilibrium density of majority holes,

$$(n_p - n_{p0}) \ll p_{p0} \tag{9-43}$$

so that from Eq. (9-39)

$$p_p \simeq p_{p0} \tag{9-44}$$

Sec. 9-3 The Biased p-n Junction

Equations (9-41)–(9-44) constitute conditions known as *low-level injection* since the carriers injected into the charge neutral region differ only slightly from equilibrium. In fact, the majority carrier densities are essentially unchanged. The minority densities are given by Eqs. (9-36) and (9-37), which, under conditions of low-level injection, take the form

$$n_p(z = w_p) = n_{p0}\epsilon^{eV/\kappa T} \qquad (9\text{-}45)$$

and

$$p_n(z = -w_n) = p_{n0}\epsilon^{eV/\kappa T} \qquad (9\text{-}46)$$

Even though the minority carrier densities at the edges of the space-charge layer depend exponentially on the applied potential, the percentage change in the majority carriers required to maintain charge neutrality is negligibly small since the equilibrium concentrations of the majority carriers is so much greater than that of the minority carriers.

Forward Bias

We have indicated in our discussion that the excess minority carriers given by Eqs. (9-45) and (9-46) appear at the edges of the space-charge layer as a result of the bias. In the case of forward bias, $V > 0$, the minority excess is positive—that is, the minority carrier densities exceed their equilibrium values—because under forward bias the potential barrier to majority carrier flow is reduced, permitting excess majority carriers to flow across the transition region into the other side where they are minority carriers. For example, for $V > 0$ more electrons are able to overcome the reduced barrier $\phi_0 + V$ and their thermal emission from the n-region is enhanced. Once in the p-region these electrons contribute to the excess minority population. But the excess minority carriers are lost by recombination as they diffuse into the neutral regions. Thus the bias controls the carrier densities at the edges of the neutral regions, and the processes of recombination and diffusion establish minority carrier gradients in the bulks of the two neutral regions. These gradients drive currents of charged particles that account for the net current through the junction under biased conditions. The electrical characteristics of a p-n junction are said to be minority carrier diffusion-limited since the current is determined by diffusion and recombination in the neutral regions while the bias voltage determines the minority carrier density at the edges of the neutral regions.

The minority carrier charge densities under conditions of forward bias are sketched in Fig. 9-5. We shall soon discover that the carrier densities decay exponentially moving away from the edges of the space-charge layer where their values are given by Eqs. (9-45) and (9-46) with $V > 0$. The scale

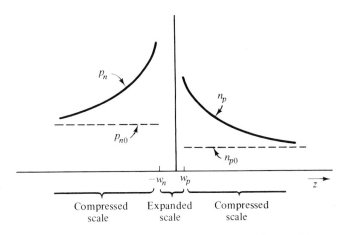

Fig. 9-5 Minority carrier densities under forward bias conditions.

lengths of the spatial decays of n_p and p_n are their respective diffusion lengths, which are usually many times the width of the space-charge layer. In Fig. 9-5 the width of the space-charge layer, $w_p + w_n$, is greatly exaggerated to indicate its presence. Qualitatively, the sketch indicates that both types of minority carriers diffuse away from the space-charge layer. Because we have neglected the potential drops inside the neutral n- and p-regions, the electric current densities carried by the minority electrons and holes, respectively, are

$$J_{ep} = eD_e \nabla n \qquad (9\text{-}47)$$

and

$$J_{hn} = -eD_h \nabla p \qquad (9\text{-}48)$$

due to diffusion. The notation of Eqs. (9-47) and (9-48) identifies the minority current densities, as for example J_{ep}, the electron current density in the p-region. The respective majority carrier currents are designated J_{hp} and J_{en}. Notice that although the minority charge carriers themselves each diffuse in opposite directions away from the junction, the net electric current flows from right to left in Fig. 9-5. Because the concentration gradients are functions of position, the diffusion currents are also functions of position. The minority carrier diffusion currents in the neutral regions are shown in Fig. 9-6(a). Because we have neglected recombination in the depletion region, the

Sec. 9-3 The Biased p-n Junction

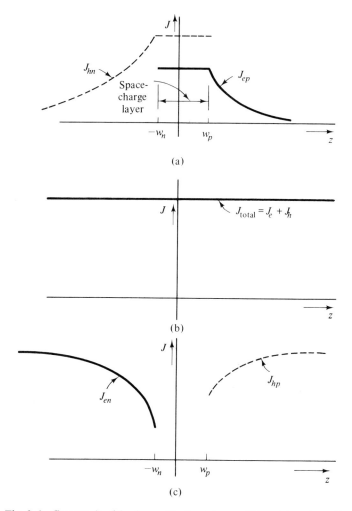

Fig. 9-6 Current densities in an idealized forward-biased junction: (a) minority carrier current densities, (b) total current density, (c) majority carrier current densities.

hole and electron currents in that region, also shown in Fig. 9-6(a), may be taken as constant.

The total current density is comprised of the flow of both electrons and holes. Under steady-state conditions the total current density must be independent of z, as dictated by the equation of continuity and indicated in Fig. 9-6(b). An expression for this total current density may be written at any position. For reasons that will become apparent shortly it is helpful to express the sum of the current densities for values at some position in the space-charge layer. Because J_e and J_h are constant throughout the space-charge layer, we can evaluate each at a convenient location. We can write, for example,

$$J_{\text{total}} = J_{hn}(z = -w_n) + J_{ep}(z = w_p) \qquad (9\text{-}49)$$

That is, we choose to evaluate the hole current density J_h at the n-edge of the depletion layer where the excess minority holes are injected into the n-region. Likewise it is convenient to evaluate J_e at $z = w_p$, the p-edge of the depletion layer where the excess minority electrons are injected into the p-region.

The difference between the total current density of Eq. (9-49) sketched in Fig. 9-6(b) and the minority carrier current densities of Fig. 9-6(a) is carried by the majority charges. The majority carrier currents are sketched in Fig. 9-6(c). The majority carriers flow from the ohmic contacts toward the junction to supply carriers for injection across the junction and for recombination with the excess minority carriers. Several diffusion lengths from the edges of the junction virtually all of the current is carried by majority carriers.

Reverse Bias

Under conditions of reverse bias, $V < 0$, the minority carrier concentrations at the edges of the space-charge layer are depressed below their equilibrium values as indicated by Eqs. (9-45) and (9-46) and as illustrated in Fig. 9-7. Again the width of the space-charge layer is exaggerated compared with the scale lengths of the density gradients. The depression of the minority concentrations near the space-charge layer under reverse-biased conditions results because the minority carriers continue to be extracted by the field in the space-charge layer but they are not resupplied by the current of majority carriers across the depletion region due to the heightened barrier. For example, the electrons that wander to the p-edge of the depletion region are swept across the junction by the electric field. In equilibrium these electrons were replaced by the injection across the transition region of electrons from the n-region where they were majority carriers. With the increased barrier of reverse bias this replacement process is less efficient, resulting in the depression of the electron density near w_p. The density profiles shown in Fig. 9-7 result from the extraction of minority carriers and their replacement by diffusion from the bulk regions toward the junction. The gradients drive the minority carriers toward the junction, each flow resulting in a current density from left to right in the figure. Thus minority carriers thermally generated within a few diffusion lengths of the edges of the space-charge layer carry the currents. The minority carrier current densities are sketched as functions of position in Fig. 9-8(a). Again the electron and hole currents are assumed

Sec. 9-3 The Biased p-n Junction

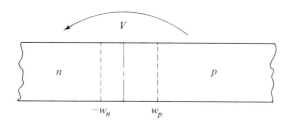

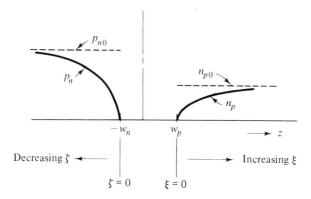

Fig. 9-7 Minority carrier concentrations under reverse bias conditions.

constant across the space-charge layer, no recombination occurring. Note that the current densities are negative, indicating flow in the direction opposite to that of the forward-biased condition. The total current is the sum of electron and hole currents, and Eq. (9-49) remains applicable. The majority carrier flow supplies the difference between the total and the minority carrier currents. The majority carriers flow away from the junction toward the contacts to remove the charges generated thermally near the space-charge layer and extracted from across the transition region by the electric field. The majority carrier current densities, shown in Fig. 9-8(c), dominate far from the depletion region.

It is important to note that under conditions of even small values of reverse bias the minority carrier concentrations near the junction are depressed well below their equilibrium values. Once the deficiencies in the minority concentrations have been established, greater reverse bias has little additional effect on the carrier distributions or current flow. This insensitivity to reverse bias results from the fact that the current is limited by the thermal generation and diffusion of the minority carriers near the edges of the space-charge layer.

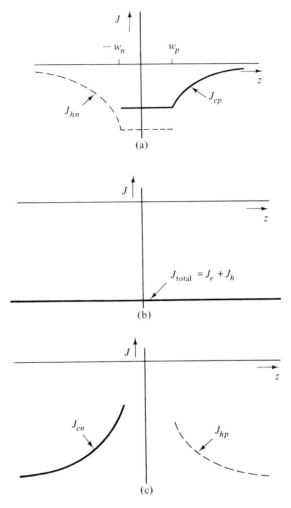

Fig. 9-8 Current densities in an idealized reverse-biased junction: (a) minority carrier current densities, (b) total current density, (c) majority carrier current densities.

9.4 THE *I-V* CHARACTERISTIC OF THE JUNCTION

As we remarked earlier, the current through a *p-n* junction under conditions of low-level injection is dominated by minority carrier diffusion. To determine the theoretical quantitative relationship between current and applied voltage we must be able to write and solve equations for the carrier densities and currents based on our physical picture of what is happening. We shall

Sec. 9-4 The I-V Characteristic of the Junction

therefore next consider quantitatively the problem of minority carrier diffusion leading to the *I-V* characteristic for the *p-n* junction.

Continuity of Minority Carriers

The minority carrier concentrations are governed by an equation of continuity similar to Eq. (3-17). Such a relation expresses the conservation of the particles. For excess minority carriers undergoing diffusion the continuity equation is called the *minority carrier diffusion equation*. For electrons in the *p*-region Eq. (3-17) in one dimension becomes

$$\frac{\partial n_p}{\partial t} - D_e \frac{\partial^2 n_p}{\partial \xi^2} = -\frac{(n_p - n_{p0})}{\tau_e} \quad (9\text{-}50)$$

where ξ is the coordinate direction and we have assumed that the net recombination rate on the right-hand side of Eq. (9-50) is proportional to the excess density through the carrier lifetime τ_e. The second term on the left of this equation is the divergence of the particle current driven by the density gradient. Solution of the minority carrier diffusion equation for the spatial dependence of n_p under steady-state conditions is facilitated by defining the excess concentration,

$$\delta n_p \equiv (n_p - n_{p0}) \quad (9\text{-}51)$$

and recognizing that n_{p0} is uniform. In terms of the excess concentration Eq. (9-50) in the steady state is

$$D_e \frac{d^2}{d\xi^2}(\delta n_p) = \frac{\delta n_p}{\tau_e} \quad (9\text{-}52)$$

It is easily verified that the general solution to this equation has the form

$$\delta n_p(\xi) = C_1 \epsilon^{-\xi/L_e} + C_2 \epsilon^{+\xi/L_e} \quad (9\text{-}53)$$

where C_1 and C_2 are constants of the integration that are determined by the values of δn_p at the boundaries of a specific problem and

$$L_e = (D_e \tau_e)^{1/2} \quad (9\text{-}54)$$

is the diffusion length introduced in Section 8.2, in this case for electrons in in the *p*-region.

To evaluate C_1 and C_2 we consider the *p*-region adjoining the depletion layer of a *p-n* junction extending, in the geometry of Fig. 9-1, from $z = w_p$ to $z \to \infty$. If we choose

$$\xi = z - w_p \quad (9\text{-}55)$$

then the origin of the coordinate system of Eq. (9-53) corresponds to the boundary of the *p*-region adjoining the space-charge layer. The electron density at $\xi = 0$ (or $z = w_p$) is given by Eq. (9-45) for low-level injection:

$$n_p(\xi = 0) = n_{p0} \epsilon^{eV/\kappa T} \quad (9\text{-}45)$$

The excess concentration at $\xi = 0$ using Eqs. (9-45) and (9-51) is

$$\delta n_p(\xi = 0) = n_{p0}[\epsilon^{eV/\kappa T} - 1] \qquad (9\text{-}56)$$

At the other boundary, $\xi \to \infty$, δn_p vanishes because of recombination. Thus conditions many diffusion lengths from the point of injection require that $C_2 = 0$. Then the boundary condition at $\xi = 0$ requires that $C_1 = \delta n_p(\xi = 0)$. Therefore Eq. (9-53) becomes

$$\delta n_p(\xi) = n_{p0}[\epsilon^{eV/\kappa T} - 1]\epsilon^{-\xi/L_e}, \qquad \xi \geq 0 \qquad (9\text{-}57)$$

which is the solution of the diffusion equation for the excess minority electron concentration.

Minority Electric Current Densities

The electric current density driven by this nonuniform electron concentration in the p-region is given by Eq. (9-47) in which the gradient of the density may be replaced by the gradient in the excess density:

$$J_{ep}(\xi) = eD_e \frac{d}{d\xi}(\delta n_p) \qquad (9\text{-}58)$$

Inserting the spatial gradient of δn_p from Eq. (9-57), this becomes

$$J_{ep}(\xi) = -\frac{eD_e n_{p0}}{L_e}[\epsilon^{eV/\kappa T} - 1]\epsilon^{-\xi/L_e}, \qquad \xi \geq 0 \qquad (9\text{-}59)$$

Analogous results can be obtained for the nonuniform minority hole concentration in the n-region, $z \leq -w_n$, and the electric current density driven by it. From the continuity equation, the excess hole density in the n-region is

$$\delta p_n(\zeta) = p_{n0}[\epsilon^{eV/\kappa T} - 1]\epsilon^{\zeta/L_h}, \qquad \zeta \leq 0 \qquad (9\text{-}60)$$

where the coordinate system has again been shifted according to

$$\zeta = z + w_n \qquad (9\text{-}61)$$

and L_h is the hole diffusion length in the n-region:

$$L_h = (D_h \tau_h)^{1/2} \qquad (9\text{-}62)$$

The minority hole electric current density using Eqs. (9-60) and (9-48) is

$$J_{hn}(\zeta) = -\frac{eD_h p_{n0}}{L_h}[\epsilon^{eV/\kappa T} - 1]\epsilon^{\zeta/L_h}, \qquad \zeta \leq 0 \qquad (9\text{-}63)$$

Junction Current

According to the argument developed in connection with Eq. (9-49), the total current through the junction is composed of the sum of electron and hole currents and is a constant independent of position. Furthermore, be-

Sec. 9-4 The *I-V* Characteristic of the Junction

cause the current densities do not vary in the space-charge layer, we may evaluate the currents at the edges of the *n*- and *p*-regions as a matter of convenience to utilize our solutions for the minority carrier current densities. If the cross-sectional area of the junction is A, then the total diode current may be expressed using Eqs. (9-49), (9-59), and (9-63) as

$$I = A[J_{ep}(\xi = 0) + J_{hn}(\zeta = 0)] \tag{9-64}$$

or

$$\boxed{I = I_s[e^{eV/\kappa T} - 1]} \tag{9-65}$$

where I_s is the *saturation current* given by

$$I_s = -eA\left[\frac{D_h p_{n0}}{L_h} + \frac{D_e n_{p0}}{L_e}\right] \tag{9-66}$$

The minus sign in Eq. (9-66) arises because of the definition of our coordinate system and the convention of positive bias, positive current flowing in the direction of decreasing z as indicated in Figs. 9-3, 9-5, and 9-6. Because we shall no longer need the coordinate system, we shall concentrate on the convention of applied bias of Fig. 9-3 and take I_s to be the magnitude of the expression in Eq. (9-66), i.e.,

$$\boxed{I_s = eA\left[\frac{D_h p_{n0}}{L_h} + \frac{D_e n_{p0}}{L_e}\right]} \tag{9-67}$$

An alternative expression for I_s is frequently used and can be obtained from Eq. (9-67) by substituting for the diffusion coefficients using Eqs. (9-54) and (9-62):

$$\boxed{I_s = eA\left[\frac{L_h p_{n0}}{\tau_h} + \frac{L_e n_{p0}}{\tau_e}\right]} \tag{9-68}$$

Equation (9-65) with either Eq. (9-67) or Eq. (9-68) is the idealized *I-V* characteristic of a *p-n* junction based on the physical picture that we have developed.

The functional dependence of the junction characteristic is shown in Fig. 9-9 in which the ratio I/I_s is plotted against $eV/\kappa T$, the applied bias normalized to the temperature. For forward bias, $V > 0$, the junction current depends strongly on the external voltage impressed upon the transition region. The dependence is exponential when V is a few times $\kappa T/e$. For instance, the error incurred by neglecting the factor of unity in Eq. (9-65) when the forward bias voltage is three times the thermal voltage is only 5%. Recall that at room temperature the thermal voltage is about 0.025 V. For reverse bias voltages, $V < 0$, that are more than a few times the magnitude of the thermal voltage the reverse current is essentially independent of the bias, being equal in magnitude to the saturation current. For example, the error incurred by neglecting the

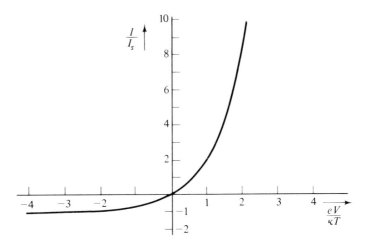

Fig. 9-9 Normalized plot of the *I-V* characteristic of the junction.

exponential factor in Eq. (9-65) for a reverse bias three times the thermal voltage is again only 5%.

Reverse Saturation Current

The reverse current through a *p-n* junction was interpreted in the previous section as being limited by the thermal generation of minority carriers in the regions adjacent to the depletion layer where the minority carrier densities are depressed below their equilibrium values. This interpretation is borne out by the mathematical expression for I_s given by Eq. (9-68). Notice that for the depressed minority carrier concentrations of reverse-biased conditions the term on the right-hand side of the diffusion equation, Eq. (9-50), represents net generation of carriers. The rate of thermal generation of minority electrons in the *p*-region is merely n_{p0}/τ_e. Similarly, the thermal generation rate of holes in the *n*-region is p_{n0}/τ_h. Thus the first term on the right-hand side of Eq. (9-68) is equivalent to a current due to *all* of the thermally generated holes in the *n*-region within a distance L_h from the edge of the space-charge layer $z = -w_n$. Similarly, the second term is equivalent to a current due to *all* of the thermally generated electrons in the *p*-region within a distance L_e from the edge of the space-charge layer $z = w_p$. Of course, not all of these carriers are collected, and some generated at distances of two and three diffusion lengths from the space-charge layer are collected. The actual current distributions are exponential, as given by Eqs. (9-59) and (9-63). However, the equivalent currents due to all the charges generated in the first diffusion length constitute a model that is frequently used. The "equivalence" is a mathematical consequence of the nature of the exponential function.

Sec. 9-4 The I-V Characteristic of the Junction

Because the saturation current is due to thermally generated carriers and because this process is a sensitive function of temperature, I_s depends strongly on temperature. To emphasize this dependence, Eq. (9-68) may be rewritten in terms of the intrinsic carrier density using Eqs. (9-3) and (9-4) for the equilibrium minority carrier densities. The result is

$$I_s = eAn_i^2\left[\frac{L_h}{N_d\tau_h} + \frac{L_e}{N_a\tau_e}\right] \quad (9\text{-}69)$$

While the diffusion lengths and lifetimes do depend on temperature, the principal dependence in Eq. (9-69) is through n_i^2. Recall that during the discussion of Section 8.2 we discovered that

$$n_i^2 = CT^3\epsilon^{-E_g/\kappa T} \quad (9\text{-}70)$$

where C is a constant depending on the effective masses and E_g is the energy gap.

An investigation of the temperature dependence of n_i^2 reveals just how sensitive the saturation current is, approximately doubling for an increase in temperature of about 10°K in germanium or 6°K in silicon near room temperature. However, saturation currents in silicon junctions are generally much smaller than those in germanium junctions so that silicon p-n diodes can operate at higher temperatures than germanium diodes and still have smaller reverse currents.

The actual saturation current under reverse-biased conditions in physical p-n junctions ordinarily exceeds that predicted by Eq. (9-69). Two effects may be responsible for this excess reverse current, leakage across the surface of the crystal and thermal generation of mobile carriers within the space-charge layer. Surface leakage currents are not well understood. Such currents increase with increased reverse bias. Because they depend on the condition of the surface of the crystal, careful fabrication of devices can minimize the deleterious effects of surface leakage.

Under moderate reverse bias voltages the space-charge layer is depleted of mobile charges except for those in transit across the junction. However, hole-electron pairs are always being produced, even in the depletion region. These carriers contribute to the reverse current, causing it to exceed the calculated value of I_s. The rate at which they are generated is determined by the temperature and the energy band gap, but the number of carriers so produced depends on the volume of the space-charge layer. Since the width of the depletion layer in an abrupt junction varies as the square root of the bias, the contribution of these carriers to the reverse current also has this dependence.

For larger values of reverse bias the reverse current increases dramatically, a condition known as *breakdown*. The mechanisms of reverse bias breakdown are the subject of the next section.

9.5 REVERSE BIAS BREAKDOWN

Reverse bias breakdown of a p-n junction is characterized by a very large increase in reverse current when the applied voltage reaches the breakdown voltage V_B, as illustrated on the I-V characteristic of Fig. 9-10. The breakdown voltage of typical junctions may be anywhere from less than a volt to

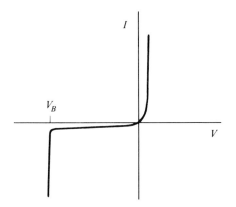

Fig. 9-10 I-V characteristic of a p-n junction showing reverse bias breakdown.

as much as 1 kV depending on the design of the junction. When breakdown occurs the junction current is limited by the external circuit that applies the bias. Because very large currents can flow under breakdown conditions if they are not otherwise limited externally, the semiconductor must be cooled effectively if irreversible thermal damage is to be avoided. In many instances the proximity of a heat sink is sufficient for temperature control. Semiconductor junction diodes designed to operate in the breakdown region are used as voltage references in applications related to the regulation and limiting of electronic signals.

Two mechanisms are responsible for reverse bias breakdown. The two processes are called avalanche and Zener breakdown.

Avalanche Breakdown

Of the two phenomena that result in reverse bias breakdown, the avalanche mechanism is more prevalent in devices. It occurs in the following way. An electric field exists in the depletion region around the metallurgical junction. As the minority carriers enter the depletion layer they are accelerated by this field, gaining energy from it. The mobile particles lose energy to the lattice when they suffer collisions with it. If the field is sufficiently high, the particles gain more energy during a mean free path than they can lose during collision, the amount of energy lost being determined by quantum mechanical considerations. Thus for a high field in the depletion region the minority

Sec. 9-5 Reverse Bias Breakdown   283

carriers may gain considerable energy over several mean free paths. It is therefore possible for an individual carrier to create a new hold-electron pair by impact ionization during its next collision with the lattice. This process continues for each new carrier created in the depletion region, leading to rapid multiplication or avalanching of the available carrier density and of the current.

The avalanche breakdown voltage V_{BA} depends on the impurity concentrations of the junction and is therefore amenable to design. The dependence arises from the fact that the maximum electric field and the width of the depletion layer vary with the doping levels. For example, let us consider an asymmetrically doped abrupt junction with $N_a \gg N_d$. For such a junction with low-level injection the maximum field and the extent of the space-charge layer into the p- and n-regions are given by Eqs. (9-22), (9-33), and (9-34), respectively:

$$\mathcal{E}_{\max} = \frac{eN_d w_n}{\varepsilon} = \frac{eN_a w_p}{\varepsilon} \quad (9\text{-}22)$$

$$w_p = \left\{ \frac{-2\varepsilon(\phi_0 + V)}{eN_a[1 + (N_a/N_d)]} \right\}^{1/2} \quad (9\text{-}33)$$

$$w_n = \left\{ \frac{-2\varepsilon(\phi_0 + V)}{eN_d[1 + (N_d/N_a)]} \right\}^{1/2} \quad (9\text{-}34)$$

Because $N_a \gg N_d$, most of the depletion layer lies on the n-side of the metallurgical junction, $w_n \gg w_p$. Furthermore, for reverse breakdown voltage, $|V| \gg |\phi_0|$. By combining Eqs. (9-34) and (9-22) under these conditions we obtain an expression for the maximum field at breakdown as a function of the reverse bias voltage and the impurity concentration on the more lightly doped side:

$$\mathcal{E}_{\max\,BA} = \left[\left(-\frac{2eN_d}{\varepsilon} \right) V_{BA} \right]^{1/2} \quad (9\text{-}71)$$

If avalanche breakdown depends only on the electric field, Eq. (9-71) predicts that the V_{BA} should vary inversely as the impurity concentration on the more lightly doped side of the junction. In reality avalanching also depends on the width of the depletion region and the mean free path, both of which are affected by the doping concentrations. Experimental observations indicate that V_{BA} varies inversely as the two-thirds power of the lesser of the impurity concentrations. For a silicon junction with $N_a \gg N_d$ the avalanche breakdown potential is about 60 V for $N_d = 10^{16}$ cm^{-3} and about 12 V for $N_d = 10^{17}$ cm^{-3}. Obviously for low breakdown voltages the lesser doping concentration must be quite high.

Zener Breakdown

To manufacture p-n junctions with low reverse bias breakdown voltages, the impurity concentrations should be high. High impurity concentrations produce very strong electric fields in narrow depletion regions. If the doping level is sufficiently high such as on the order of 10^{18} cm^{-3} or greater in silicon, the carriers in the depletion region may not be able to gain enough energy to sustain an avalanche because the depletion layer may be too narrow. However, breakdown may still occur through quantum mechanical tunneling. With a very high electric field in the depletion layer, conduction band states on the n-side may be at the same energy level as valance band states on the p-side of the junction. The probability then exists that an electron in a valance band state may tunnel directly into a conduction band state at the same energy, thus creating majority carriers on either side of the junction. The process, called the Zener mechanism, occurs in abrupt junction devices that have been heavily doped and depends on the field strength and the width of the energy gap. Because field-assisted tunneling is the process that underlies the operation of the tunnel diode, its detailed discussion will be postponed until Section 9.8.

Usually in a heavily doped junction both the avalanche and Zener mechanisms participate in breakdown. In a practical situation the major contribution of one mechanism rather than the other can be distinguished by the temperature dependence of the breakdown voltage. The Zener breakdown voltage decreases with increasing temperature because the increase in temperature actually decreases the energy gap between bands, making tunneling more probable. On the other hand, the avalanche breakdown voltage increases with increasing temperature. This effect results from an increase in the collision probability with temperature decreasing the mean free path and the energy gained between collisions.

9.6 NOISE IN p-n JUNCTIONS

The noise accompanying currents through a p-n junction can be considered to be due to the motion of the discrete charges across the depletion region. Under conditions of low-level injection the motions of the carriers are independent, there not being enough charge for their self-fields to influence their behavior significantly. Therefore the noise in a p-n junction is shot noise, a concept introduced in Section 3.7. Using Eq. (3-88), the mean square shot noise current per unit bandwidth is

$$\frac{\overline{I_{ns}^2}}{\Delta f} = 2eI \qquad (9\text{-}72)$$

Equation (9-72) is a valid representation of the low-frequency junction noise spectrum. It is independent of frequency for frequencies much less than the inverse of the transit time for carriers across the depletion region. For scatter-limited carriers traversing a depletion layer 10^{-4} cm wide the transit frequency

Sec. 9-6 Noise in p-n Junctions

is about 100 GHz so that Eq. (9-72) should be valid at least up to a few gigahertz.

The current though a p-n junction is given by Eq. (9-65),

$$I = I_s[\epsilon^{eV/\kappa T} - 1] \qquad (9\text{-}65)$$

in which the first term on the right is due to majority carrier diffusion against the field in the transition region and the second term is due to minority carrier drift in that field. The noise due to each of these two components of the junction current adds so that the spectral density of the shot noise current from Eq. (9-72) is

$$\frac{\overline{I_{ns}^2}}{\Delta f} = 2eI_s[\epsilon^{eV/\kappa T} + 1] \qquad (9\text{-}73)$$

Under conditions of reverse bias, $V < 0$, the spectral density is approximately

$$\frac{\overline{I_{ns}^2}}{\Delta f} \simeq 2eI_s, \qquad V < 0 \qquad (9\text{-}74)$$

due to the reverse saturation current only. For forward bias voltages greater than a few times the thermal voltage the factor of unity can be neglected in both Eq. (9-73) and the junction characteristic. Therefore

$$\frac{\overline{I_{ns}^2}}{\Delta f} \simeq 2eI, \qquad V > 0 \qquad (9\text{-}75)$$

For the equilibrium condition in which no bias is applied to the junction Eq. (9-73) predicts a noise spectrum of

$$\frac{\overline{I_{ns}^2}}{\Delta f} = 4eI_s, \qquad V = 0 \qquad (9\text{-}76)$$

While this result may appear to be in conflict with the concept of shot noise due to the flow of current, Eq. (9-76) is valid even though no net current flows when $V = 0$. Recall that in the equilibrium case both minority and majority carrier currents of magnitude I_s in fact do flow across the transition region. These currents are of opposite sign and hence their effects cancel but the noise accompanying them does not.

Equivalent Resistance

The shot noise in a p-n junction under equilibrium conditions, $V = 0$, can be related to thermal noise in a resistive medium. Thermal noise, it will be recalled from Section 3.7, is due to the random motion of charge carriers. In a one-dimensional system in thermodynamic equilibrium half of the charge carriers at one instant are moving in one direction and half in the other so that the net motion is zero. An analogous situation occurs in an unbiased depletion layer where there is a detailed balance of the flow of each type of carrier across the junction.

We shall now demonstrate that the shot noise in an unbiased junction given by Eq. (9-76) is equivalent to a thermal noise. To do so we shall eliminate the saturation current I_s from Eq. (9-76) in favor of an effective incremental resistance r in the following way.

Suppose that a p-n junction is d.c.-biased at a particular point of its operating characteristic $(I_{d.c.}, V_{d.c.})$, as indicated in Fig. 9-11(a). If a small low-frequency a.c. voltage is added to the d.c. bias, the instantaneous operating

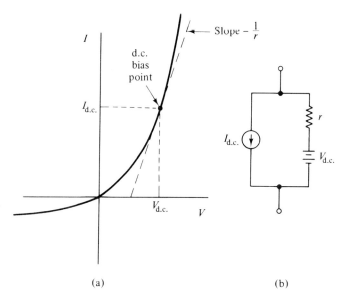

Fig. 9-11 Demonstration of the incremental resistance of a p-n junction: (a) I-V characteristic, (b) an equivalent circuit.

point moves back and forth on the I-V characteristic in the vicinity of $(I_{d.c.}, V_{d.c.})$. So long as the a.c. voltage is small, the a.c. operating point moves on a curve that can be approximated by the tangent to the characteristic at $(I_{d.c.}, V_{d.c.})$. This straight-line approximation to the path is the same as the I-V characteristic of a resistor whose value r is the inverse of the slope of the tangent. Insofar as the external circuit is concerned, the diode behaves like the equivalent circuit of Fig. 9-11(b). The slope of the I-V characteristic is the inverse of the incremental resistance and is called the incremental conductance $g = 1/r$.

From Eq. (9-65) the incremental conductance is

$$g = \frac{dI}{dV} = \frac{eI_s}{\kappa T}\epsilon^{eV/\kappa T} \tag{9-77}$$

Sec. 9-7 The p-n Junction Diode

The incremental resistance for the unbiased junction is therefore

$$\frac{1}{r_0} = g_0 = \frac{eI_s}{\kappa T} \qquad (9\text{-}78)$$

By substituting for I_s from Eq. (9-78) into Eq. (9-76) we obtain the result

$$\overline{I_{ns}^2} r_0 = 4\kappa T \Delta f \qquad (9\text{-}79)$$

Comparison of Eq. (9-79) with Eq. (3-86) for the thermal noise power of a resistive material in a bandwidth Δf indicates the equivalence of the shot noise with thermal noise of an equivalent resistance r_0 in equilibrium.

9.7 THE p-n JUNCTION DIODE

In the first part of this chapter we studied the electrical properties of the junction between n- and p-type semiconductors. Such a junction is the basic building block of a variety of electronic devices. The simplest of these is the junction diode, a device comprised of a single p-n junction. A schematic representation of one possible physical configuration of a discrete junction diode is shown in Fig. 9-12. Also in this figure are shown the circuit symbol

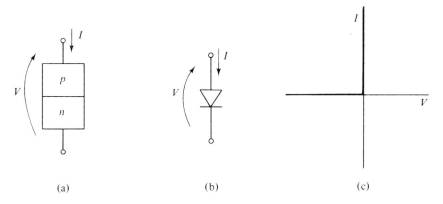

(a)   (b)   (c)

Fig. 9-12 p-n junction diode: (a) physical configuration, (b) circuit symbol of an idealized diode, (c) terminal characteristic of an idealized diode.

for an idealized two-terminal diode and its idealized I-V characteristic, concepts that were introduced in Chapter 1. The circuit symbol can be thought of as being composed of an arrowhead that indicates the direction of flow of forward current for forward bias conditions and a barrier to inhibit current flow in the reverse direction. For a p-n junction diode the arrowhead therefore always points from the p-region to the n-region.

The d.c. current-voltage characteristics of actual *p-n* junctions were studied in Section 9.4. Even a cursory examination of a typical d.c. characteristic such as that of Fig. 9-13 leads us to the conclusion that a junction diode in an electronic circuit cannot be represented by the idealization of Fig. 9-12(b) and (c). Furthermore, the d.c. current-voltage characteristic does not take into account dynamic effects.

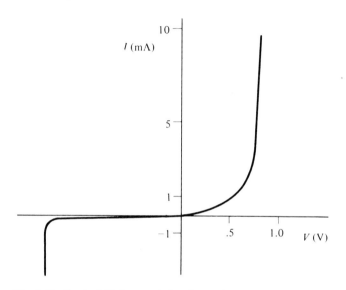

Fig. 9-13 Static *I-V* characteristic of a typical junction diode.

Equivalent Circuit Representation

The equivalent representation of a realistic diode in an electronic circuit utilizes combinations of idealized elements to simulate the actual physical behavior of the device for the purposes of analyzing and designing circuits. Just what combination of idealized resistances, capacitances, and voltage and current sources is required in addition to an ideal diode depends on the circuit application as well as on the physical mechanisms that produce the respective effects.

For example, in low-frequency small-signal circuits the d.c. analysis of the previous sections may describe the behavior of the junction with sufficient accuracy. In these circumstances the *I-V* characteristic of a diode biased at the point $(I_{d.c.}, V_{d.c.})$ can be approximated by a straight line tangent to the curve at the bias point, as indicated in Fig. 9-14(a). Because the *I-V* characteristic of a resistance is a straight line, this piecewise linear approximation of the diode characteristic is equivalent to a combination of an ideal diode, a voltage source, and a resistance. The equivalent circuit is shown in Fig.

Sec. 9-7 The p-n Junction Diode

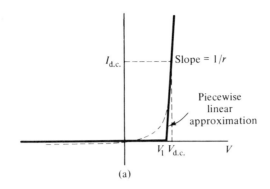

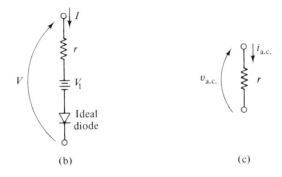

Fig. 9-14 Circuit representations of a junction diode: (a) piecewise linear approximation to the I-V characteristic, (b) a d.c. equivalent circuit, (c) equivalent circuit for small-amplitude a.c. signals.

9-14(b). Figure 9-14(c) shows the equivalent circuit for the small-signal a.c. components of current and voltage. Because this resistive effect arises with changes in the diode current and voltage, it is called the incremental resistance and is denoted by r. The incremental resistance is the inverse of the incremental conductance g introduced in Section 9.6. Repeating Eq. (9-77),

$$g = \frac{eI_s}{\kappa T}\epsilon^{eV/\kappa T} \qquad (9\text{-}77)$$

and combining it with the I-V characteristic of the junction, Eq. (9-65), we may write g as

$$g = \frac{e}{\kappa T}(I + I_s) \qquad (9\text{-}80)$$

where I is the total diode current. Then the incremental resistance is given by

$$r = \frac{\kappa T}{e(I + I_s)} \qquad (9\text{-}81)$$

This result and the equivalent circuits of Fig. 9-14 are valid for low forward currents and for very small reverse currents. For high forward currents the incremental resistance of Eq. 9-81 is small. When it is comparable with the bulk resistance R_b of the neutral n- and p-regions the latter effect must be taken into account. That is, for forward bias conditions the net a.c. resistance of the diode is $R_b + r$. For reverse currents approaching I_s in magnitude the idealization represented by Eq. (9-81) predicts a very large effective resistance. In actual junctions, as discussed in Section 9.5, the reverse current differs from the ideal saturation current due to leakage effects and to creation of carriers in the depletion layer. Backward resistances of typical junction diodes lie in the range 10^5 to 10^6 Ω or greater.

Dynamic Effects

In high-frequency applications inherent delays in the response of the junction to a changing stimulus may be represented by capacitances. There are two such capacitive effects of importance in junction diodes. One is associated with the behavior of the dipole layer of immobile charge in the depletion region and is called the space-charge or depletion layer capacitance. The second is associated with the behavior of the excess minority carriers stored in the neutral regions under conditions of forward bias and is called the storage or diffusion capacitance.

The *depletion layer capacitance* arises from the fact that the net dipole charge Q on each side of the depletion layer depends on the potential drop across the junction. Recall from the development in Section 9.1 that, in the approximation of complete depletion of the mobile charges in the space-charge region, the ionized impurity atoms in the transition region form a dipole layer. The net charge on the n-side of the metallurgical junction is positive and may be obtained by integrating the charge density shown in Fig. 9-15 over the appropriate volume. The result is

$$Q_n = eN_dAw_n \qquad (9\text{-}82)$$

where A is the cross-sectional area of the planar junction. The net charge on on the p-side is

$$Q_p = -eN_aAw_p \qquad (9\text{-}83)$$

The magnitudes of Q_n and Q_p are, of course, equal. When the net potential drop across the junction changes because of a change in applied bias δV the dipole charges Q_n and Q_p also change because w_n and w_p depend on V. Suppose that the junction is reverse-biased, $V < 0$, and that the change in V increases the bias, $\delta V < 0$. Then w_n and w_p increase, increasing the magnitudes of Q_n and Q_p. If $\delta Q_n > 0$, then majority electrons near the edge of the n-side of the space-charge region move away from the junction to expose the ionized donors. These electrons flow through the ohmic contact to the n-region and the external circuit and are neutralized by holes flowing away

Sec. 9-7 The p-n Junction Diode

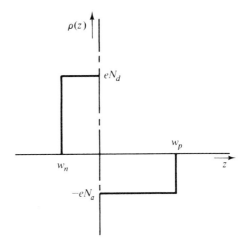

Fig. 9-15 Charge density in the space-charge layer of a junction diode.

from the junction in the p-region for similar reasons. Thus the current flowing in the external circuit is due to the motion of charge of sign opposite to that of the immobile charge in the depletion layer. That is,

$$i(t) = -\frac{dQ}{dt} \qquad (9\text{-}84)$$

where Q is the dipole charge and $i(t)$ is the a.c. component of the total junction current. The time dependence of Q in Eq. (9-84) is through its explicit dependence on the applied voltage V,

$$\frac{dQ}{dt} = \frac{dQ}{dV}\frac{dV}{dt} \qquad (9\text{-}85)$$

Combining Eqs. (9-84) and (9-85), we obtain

$$i = -\frac{dQ}{dV}\frac{dV}{dt} \qquad (9\text{-}86)$$

which has the form of the terminal characteristics of a capacitance. The equivalent depletion layer capacitance is therefore

$$C_{\text{depletion}} = -\frac{dQ}{dV} \qquad (9\text{-}87)$$

For the case of a uniformly doped abrupt junction the depletion capacitance may, for instance, be evaluated using Eq. (9-82) and Eq. (9-34) for w_n. The result is

$$\boxed{C_{\text{depletion}} = \frac{A}{2}\left\{\frac{-2e\varepsilon N_d}{(\phi_0 + V)[1 + (N_d/N_a)]}\right\}^{1/2}} \qquad (9\text{-}88)$$

For reverse bias voltages that are large compared with the contact potential, the capacitance of an abrupt junction varies inversely as the square root of the applied voltage. The functional variation of the depletion capacitance with voltage is determined by the impurity profiles and therefore is subject to design. The capacitance of a reverse-biased linearly graded junction depends on $V^{-1/3}$. Devices designed to take advantage of the voltage dependence of the junction capacitance are called varactors and are used as frequency multipliers and for electronic tuning. The depletion capacitance is relatively insensitive to forward bias voltages since $\phi_0 + V$ is always negative.

The *diffusion capacitance* is associated with the storage of excess minority carriers at the edges of the neutral regions under conditions of forward bias. For moderate values of forward bias this capacitive effect is much larger than the depletion layer capacitance. That the dynamic behavior of the excess minority carriers leads to a capacitive effect can be demonstrated by the following argument. The electric current density carried by the minority electrons in the p-region is due to diffusive flow in the gradient of the excess concentration. In Section 9.4 we wrote Eq. (9-58) as

$$J_{ep} = eD_e \frac{d}{d\xi}(\delta n_p) \qquad (9\text{-}58)$$

where $\xi = 0$ is the edge of the space-charge layer and δn_p is the excess concentration. From Eq. (9-58) and the similar one for hole current in the n-region it is evident that the current through a forward-biased diode depends on the gradient of the excess density. From Eq. (9-57) the excess electron density in the p-region is

$$\delta n_p = n_{p0}[\epsilon^{eV/\kappa T} - 1]\epsilon^{-\xi/L_e} \qquad (9\text{-}57)$$

assuming that this result calculated for the static case is also valid under dynamic conditions. Equation (9-57) shows that the minority carriers injected across the space-charge layer depend on the junction voltage. Thus the junction voltage can change only at the same rate at which the excess charge density can change. On the other hand, Eq. (9-58) indicates that the junction current changes with the gradient of the excess density. Thus the junction current can change more readily than the junction voltage. In other words, the voltage across the transition region cannot be changed rapidly without a large current flowing to supply the change in the minority carrier concentrations. The forward-biased junction is therefore a device that tends to maintain the voltage drop across it, as is a capacitor.

Before proceeding to a quantitative analysis of storage or diffusion capacitance we should consider the assumption that Eq. (9-57) is valid under dynamic as well as static conditions. The excess carrier concentrations represented by this equation are supplied by the injection of carriers over the potential barrier in the contact region. Equation (9-57) will be valid provided the junction voltage does not change on a time scale on the order of the lifetime of the minority carriers. This equation is the solution to a diffusion equation

Sec. 9-7 The p-n Junction Diode

including recombination. The excess minority carrier concentration of Eq. (9-57) is established on a time scale comparable with a recombination time. Depending on the material and the impurity concentrations, carrier lifetimes may be in the range of 10^{-8} to 10^{-5} sec. These lifetimes correspond to frequency limitations on the validity of Eq. (9-57) of 100 MHz and 100 kHz, respectively.

The diffusion capacitance may be analyzed quantitatively in much the same way as our previous analysis of depletion layer capacitance. We shall calculate the excess minority charge stored on either side of the junction and infer an incremental capacitance from the dependence of this charge on voltage. The total charge due to excess electrons in the p-region is

$$Q_{ep} = -\int_0^\infty eA\,\delta n_p(\xi)\,d\xi \qquad (9\text{-}89)$$

where we have assumed that the n-region is more than a few diffusion lengths thick. Using Eq. (9-57) for δn_p we can easily obtain

$$Q_{ep} = -eAn_{p0}L_e[\epsilon^{eV/\kappa T} - 1] \qquad (9\text{-}90)$$

The incremental capacitance associated with a change in this stored charge is

$$C_e = -\frac{dQ_e}{dV} = \frac{e^2 An_{p0}L_e}{\kappa T}\epsilon^{eV/\kappa T} \qquad (9\text{-}91)$$

Similarly, the incremental capacitance associated with a change in the excess hole density stored in the n-region is

$$C_h = \frac{e^2 Ap_{n0}L_h}{\kappa T}\epsilon^{eV/\kappa T} \qquad (9\text{-}92)$$

The net storage or diffusion capacitance is

$$C_{\text{diffusion}} = C_e + C_h = \frac{e^2 A}{\kappa T}(n_{p0}L_e + p_{n0}L_h)\epsilon^{eV/\kappa T} \qquad (9\text{-}93)$$

The diffusion capacitance is a strong function of the forward bias voltage.

Because the diode current has the same functional dependence on the junction voltage as the diffusion capacitance, it is convenient to recast Eq. (9-93) in terms of the current. Using Eq. (9-68) for I_s and assuming that the minority carrier lifetimes are comparable, Eq. (9-93) may be rewritten as

$$C_{\text{diffusion}} = \left(\frac{e\tau}{\kappa T}\right)I_s\epsilon^{eV/\kappa T} \qquad (9\text{-}94)$$

where τ is the common lifetime. Then using Eq. (9-65) this becomes, for $I \gg I_s$,

$$\boxed{C_{\text{diffusion}} = \left(\frac{e\tau}{\kappa T}\right)I} \qquad \text{forward bias} \qquad (9\text{-}95)$$

For carrier lifetimes on the order of 10 μsec and a forward current of a few milliamperes at room temperature, Eq. (9-95) predicts a diffusion capacitance on the order of microfarads. Notice that decreasing the carrier lifetime τ improves the high-frequency response of the junction. By contrast typical depletion layer capacitances are on the order of 100 pF.

Based on the foregoing discussion an equivalent circuit for a junction diode in response to a.c. signals can be devised. Such a circuit is shown in Fig. 9-16. The junction capacitance C_j represents the sum of the depletion layer and the diffusion capacitances. In addition to the elements shown, we might expect that at high frequencies the capacitance between the physical leads to the contacts becomes important. Furthermore, in some applications noise sources should be included.

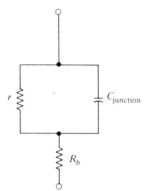

Fig. 9-16 Small-amplitude a.c. equivalent circuit of a junction diode.

Switching Times

In switching and other logic circuit applications the time interval required to change a junction diode from reverse to forward bias conditions, from "off" to "on," is of critical importance. Because the diode current depends on the profiles of the excess carrier concentrations, the switching times depend on how fast these densities can be changed. The minority carrier densities in a junction diode under forward and reverse bias conditions are sketched in Fig. 9-17. To turn a diode "on" implies to switch from a reverse to a forward bias condition between well-defined states. To turn it "off" means to switch from forward to reverse bias.

When a diode is turned on, majority carriers traverse the transition region quickly, and the minority carrier densities begin to build up at the edges of the space-charge layer. As the buildup continues, these carriers diffuse into the neutral regions and recombine. As the concentrations stored near the edges of the neutral regions approach their equilibrium values more and more of the injected carriers are consumed by recombination. Turn-on times in practical diodes can be very fast.

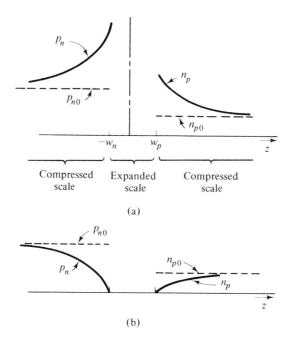

Fig. 9-17 Minority carrier densities in a junction diode: (a) forward bias, (b) reverse bias.

On the other hand, turn-off times present practical limitations to the applicability of p-n junction diodes as switches. When a forward-biased diode is suddenly switched off, two aspects delay the establishment of reverse-biased conditions. The excess minority charge densities of Fig. 9-17 must first be reduced to zero and then the reverse-biased profiles established. The excess concentrations are reduced to zero by recombination. The reverse bias profiles are established by the diffusion of the minority carriers within a few diffusion lengths of the edge of the space-charge layer. Therefore the turn-off time is limited by the recombination and diffusion processes.

In commercial junction diodes designed for switching small currents, the turn-off time may be as short as 10^{-9} sec. These devices usually have doping profiles that concentrate the stored excess minority charge in a narrow region close to the edge of the junction. Another technique for reducing switching times is to decrease the carrier recombination time by introducing gold atoms as impurities. Diodes capable of handling large currents may have turn-off times as long as 10^{-6} sec. Schottky barrier diodes, introduced in Chapter 8, are made of rectifying metal-semiconductor contacts and have I-V characteristics similar to those of semiconductor junctions. However, because they are majority carrier devices, there is no storage of minority charges. Therefore switching only requires charging of the depletion layer capacitance. Subsequently, Schottky diodes are faster than junction diodes when used as switching devices.

9.8 MICROWAVE DIODES

In Section 9.5 we studied two mechanisms of reverse bias breakdown, the avalanche and the Zener processes. Semiconductor junctions designed to operate in breakdown are used as voltage references or as limiters since the back bias voltage cannot exceed the breakdown voltage $-V_B$. The current through a diode in breakdown is determined by the external circuit. In most breakdown diodes the avalanche process accounts for the characteristic, while in the more heavily doped ones the Zener process also contributes. Breakdown voltages in conventional avalanche diodes lie in the range from about 5 V to as much as 1000 V. It will be recalled from the discussion of Section 9.5 that the lower breakdown voltages occur for more heavily doped junctions.

In this section we shall first continue our discussion of the Zener breakdown mechanism as it applies to devices known as tunnel diodes. We shall also study an application of the avalanche breakdown in so-called IMPATT devices. Both IMPATT and tunnel diodes exhibit negative differential conductivity at high frequencies as does the Gunn diode. The IMPATT diode is generally used as a microwave oscillator, while the tunnel diode is used as a microwave amplifier or as a fast switch.

Tunnel Diode

A *tunnel diode* consists of a p^+-n^+ junction, where the superscript $+$ indicates a high concentration of impurity atoms. For example, a silicon tunnel diode must have doping levels on both sides of an abrupt junction on the order of 10^{19} cm^{-3}. In such a junction the depletion layer is very narrow and the contact potential quite high. The electric field in the depletion region of a tunnel diode is high enough to cause a Zener-type "breakdown" with no bias applied.

Tunneling was introduced in Chapter 5 as a model to describe the nonzero probability that a quantum mechanical particle may be found in a region of space forbidden to it in classical physics. For example, a particle with energy E could be found "under" a potential barrier of energy $U > E$. Furthermore, a particle with energy E originally on one side of a thin potential barrier of height $U > E$ may at a later time by found on the other side.

The probability for tunneling "through" a thin potential barrier depends on the thickness and height of the barrier. When the barrier is the potential barrier in the thin depletion layer of a p^+-n^+ junction the tunneling probability also depends on the availability of allowed states on either side of the junction since the energy of the tunneling particle is the same before and after the process. For the simple examples of Chapter 5 this restriction on the tunneling process was not a problem since there were allowed states at all energies greater than zero on either side of the barrier. Electrons in solids, however, may occupy only discrete energy states so that to tunnel from one state to another the second state must be an unoccupied allowed state with the same energy as the first.

Let us now see how these quantum mechanical concepts explain the terminal characteristics of tunnel diodes. To develop the form of the I-V characteristic we shall consider the energy band diagram of the p^+-n^+ junction. The diagram for an unbiased junction in equilibrium is sketched in Fig. 9-18(a). Notice that the Fermi energy E_F lies within the conduction band on the n^+-side and within the valence band on the p^+-side. When the Fermi level lies in an allowed band a material is said to be degenerate. Heavily doped semiconductors may become degenerate when the impurity states interact to form impurity bands and these bands overlap the adjacent conduction or valence bands. The important point is that with E_F in the bands and the electrons distributed in the allowed states in equilibrium according to the Fermi-Dirac distribution, there are very many electrons in the conduction band on the n^+-side and very many holes or empty states in the valence band on the p^+-side.

In the unbiased equilibrium the drift and diffusion currents of holes and electrons across the space-charge layer are in balance just as in an ordinary junction. In addition to these carrier motions, electrons may tunnel in both directions across the transition region. Because tunneling occurs between allowed states at the same energy and because the occupation of states is determined by the statistics, the net tunneling current in equilibrium is also zero.

When a very small reverse bias voltage is applied to the junction a dramatic change occurs in the tunneling current. As indicated in Fig. 9-18(b) reverse bias raises the energy states on the p-side of the junction by the potential energy $|eV|$ relative to the states on the n-side. Tunneling continues to occur in both directions across the junction as in the equilibrium case but there is a net tunneling of electrons from p to n. This net tunneling current of electrons results from the greater availability of empty states in the conduction band on the n-side that are exposed to the elevated occupied states in the valence band on the p-side. That is, more electrons on the p-side see available empty states on the n-side. Furthermore, tunneling of electrons from the n- to the p-side is reduced because of the reduced availability of empty states on the p-side at energies where there are large numbers of electrons on the n-side. The resulting net tunneling of electrons from right to left in the diagram is equivalent to reverse electrical current in the convention of Fig.

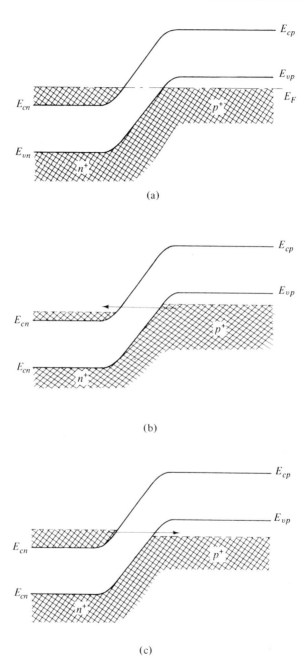

Fig. 9-18 Energy diagram of the p^+-n^+ junction of a tunnel diode: (a) unbiased, (b) reverse-biased, (c) forward-biased.

Sec. 9-8 Microwave Diodes

9-12. Because of the high density of electrons and available states, the tunneling current is very much greater than the ordinary reverse bias junction current.

When forward bias is applied to the junction the energy states on the p-side are lowered with respect to those on the n-side. Figure 9-18(c) indicates that the net tunneling for this case occurs from n to p since the electrons in the conduction band on the n-side are exposed to very many available empty states near the top of the valence band on the p-side. This net electron tunneling results in forward electrical current through the diode. The tunneling current does not continue to increase indefinitely with increasing forward bias, however. At some value of forward bias, depending on the degree of degeneracy, the band structures will have been shifted so that the bottom of the conduction band on the n-side lies just above the top of the valence band on the p-side. Under these circumstances electrons cannot tunnel across the junction because allowed states do not exist in the energy gap. Therefore no tunnèling current flows. However, net forward current does flow, governed by the field of the junction as in ordinary diodes. Figure 9-19 shows the theoretical tunneling current characteristic and the net I-V characteristic of a tunnel diode.

Indicated on Fig. 9-19(b) is a region of negative differential conductivity

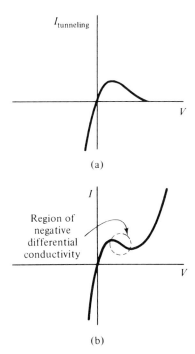

Fig. 9-19 Current in a tunnel diode: (a) tunnel current, (b) static I-V characteristic.

where the slope of the characteristic is negative. Ordinarily it is very difficult to obtain a stable d.c. operating condition on this part of the characteristic. Figure 9-20 shows a tunnel diode in a biasing circuit and two possible types

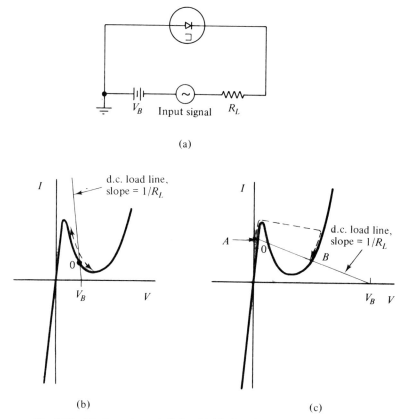

Fig. 9-20 Biasing a tunnel diode: (a) biasing circuit, (b) as an amplifier, (c) as a bistable switch.

of bias conditions. The slope of the load lines of Fig. 9-20(b) and (c) is the inverse of the bias resistor R_L. The point of intersection of the load line with the voltage axis is the instantaneous sum of the bias voltage and an a.c. input signal. The intersection of the load line and the characteristic of the tunnel diode indicates the instantaneous diode voltage and circuit current.

If the resistor and d.c. bias voltage are chosen so that the load line intersects the characteristic only in the region of negative differential conductivity as indicated in Fig. 9-20(b), then the diode acts as an amplifier of a small a.c. input signal. For very small changes in the sum of the bias voltage and the input signal, the instantaneous operating point makes large excursions along

Sec. 9-8 Microwave Diodes

the characteristic, as indicated by the dashed line in the figure. That is, for small variations in the input voltage signal there are large variations in the current through the diode. This current also flows through the series load resistor, causing a voltage to be developed across the load. The a.c. voltage drop across the load resistor is an amplified version of the input signal. The voltage gain, being the ratio of the a.c. output voltage to the a.c. input voltage signal, depends on the relative magnitudes of the load resistor and the differential resistance of the diode. If the magnitude of the incremental resistance of the diode in the region of negative differential conductivity is denoted by r_{td}, amplification occurs if $R_L < r_{td}$, the voltage gain being large if R_L is almost equal to r_{td}.

For the load line of Fig. 9-20(c) with $R_L > r_{td}$ the tunnel diode has two stable operating points, A and B, and operates as a bistable switch. The dashed line indicates the excursion of the instantaneous operating point when a positive input pulse is applied to a diode that originally is biased in operating condition A on the low-voltage section of the characteristic. If the input signal is of sufficient amplitude the operating point will be switched to the upper section of the characteristic with positive slope at higher voltage, finally arriving at point B when the input signal falls to zero. Applying a negative input pulse that exceeds a threshold value causes the operating condition to return to that of point A. The operating condition of the diode is therefore an indication of the polarity of the most recent pulse applied to it.

Tunnel diodes operating as amplifiers or switches are much faster than ordinary junction devices because the tunneling process does not depend on minority carrier diffusion and recombination. As microwave amplifiers they have operating frequencies in the range from a few to as much 100 GHz. Switching times in bistable circuits may be as short as 10^{-9} sec.

IMPATT Diode

The IMPATT diode combines the *imp*act *a*valanche ionization in a reverse-biased semiconductor junction with bulk *t*ransit-*t*ime properties to produce an overall high-frequency negative differential conductivity. In an IMPATT diode the a.c. current lags the a.c. voltage at high frequencies. When the terminal current is delayed by more than a quarter of a cycle, the component of the current *in phase* with the positive terminal voltage is negative. Under these conditions the device exhibits negative differential conductivity and when inserted in an appropriate resonant microwave cavity may oscillate spontaneously.

The phase delay of the current in an IMPATT device arises during two processes, the increase of the carrier density during avalanche breakdown in the high electric field present in a reverse-biased junction and the drift of the carriers across the device to the contact.

One form of the class of IMPATT devices is the Read diode, consisting of an n^+-p-i-p^+ structure, where i indicates a region of intrinsic characteristics. The physical configuration of a Read diode and the distribution of electric fields inside it are indicated in Fig. 9-21. A d.c. bias is applied to the diode so

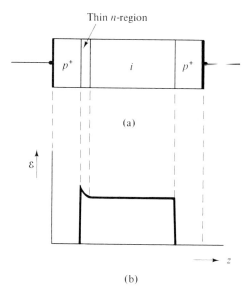

Fig. 9-21 IMPATT diode: (a) physical configuration, (b) spatial dependence of the electric field.

that the p^+-n junction is reverse-biased and a depletion layer extends throughout the narrow n-region and the intrinsic region. Because the p^+-regions are heavily doped, the electric fields extend only a negligibly short distance into these regions of the p^+-n and i-p^+ junctions. The highest field occurs on the n-side of the p^+-n junction.

The d.c. bias is adjusted so that during the positive half-cycle of an additional a.c. voltage impressed upon the device the maximum value of the total field is above the threshold value for which avalanche breakdown occurs. When the field is above this initial value the particle current increases exponentially with time. That is, the particle current continues to increase until the maximum value of the total field drops below the critical value. The peak value of the particle current therefore occurs not when the voltage across the device is a maximum but ideally a quarter-cycle later when the a.c. component of the field passes through zero. Thus the current generated by the avalanche mechanism has an inherent phase delay of 90° relative to the a.c. voltage driving the avalanche.

The avalanche process generates electrons that move to the p^+-region and holes that are injected into the intrinsic region. During the negative half-cycle of the a.c. voltage the peak current of holes drifts across the intrinsic region toward the p^+-contact, which acts as a collector. If the net field in the intrinsic region is always above the scatter-limited field, the holes drift with constant velocity independent of the field. The transit time for the drifting carriers is then determined only by the length L of the intrinsic region.

If the transit time is a quarter-period of the a.c. signal, another delay of 90° is introduced in drifting to the contact. The maximum positive a.c. current appears at the terminals when the a.c. voltage has its minimum value. The voltage and current are therefore 180° out of phase under optimal conditions. For a.c. signals the diode appears to be a negative resistance and energy is supplied to the a.c. field. The energy comes from the d.c. bias supply.

Other types of IMPATT devices operate on similar principles. Physical structures used include simple *p-n* junctions and *p-i-n* devices. IMPATT diodes have been produced to operate in a continuous wave mode with outputs of 100 mW at frequencies ranging from 40 to 110 GHz. Pulsed devices operate at frequencies up to 300 GHz. When placed in a suitable tunable resonant cavity, the oscillating frequency of some devices can be varied over a range of 10%.

9.9 OPTICAL DIODES

The absorption of optical energy by semiconducting materials creates hole-electron pairs that may contribute to the electrical properties of the material. Similarly, recombination of holes and electrons in certain circumstances may result in the emission of optical photons. The fundamental concepts of absorption and radiative recombination were introduced in Section 8.2. In this section we shall discuss the application of these effects in *p-n* junction devices.

Photodiode

Bulk photoconductivity is a term applied in general to processes in which optically generated charge carriers increase the conductivity of a bulk material or device. In contrast, a *photodiode* is a reverse-biased *p-n* junction designed to employ optical generation in and near the junction and used as a light detector. In operation it is the solid-state analog of the vacuum phototube.

The current in a "dark" photodiode (that is, without optical excitation of carriers) is the usual reverse bias saturation current I_s. Recall that this current is due to minority carriers thermally generated within a few diffusion lengths of either side of the space-charge layer. When this region of a diode is exposed to light of the proper frequency both minority and majority carriers are produced optically in equal numbers. However, the percentage changes in the majority carrier concentrations are much less than the percentage changes in the minority carrier densities. Therefore for practical purposes we may neglect the optically generated majority carriers as long as we remember that charge neutrality is maintained. The optically generated minority carriers, however, may exceed in number the thermally generated minority carriers. In this circumstance the optical generation rate for carriers near the junction determines the magnitude of the reverse current through the diode. The sensitivity of reverse current to light levels in practical photodiodes is in the range of 10 to 50 mA/lm.

Photovoltaic Cell

Photovoltaic effects refer to phenomena in which the absorption of optical energy generates an electric field in a medium resulting in a measurable voltage across it. The *solar cell* or *solar battery* is a p-n junction device utilizing a photovoltaic effect.

When a hole-electron pair is generated optically in an unbiased p-n junction the built-in field of the contact potential drives each of the charges to the side of the junction where it is the majority carrier. This motion of the optically generated carriers constitutes a reverse current across the junction. Once injected across the space-charge layer these charges are of the polarity to reduce the net dipole charge of the space-charge layer, thus reducing the potential drop across the junction. The change in the junction potential due to the optical excitation can be measured.

Two types of current flow in a photovoltaic cell. One is reverse current across the junction due to the minority carriers generated optically within a few diffusion lengths of the junction. The other is the ordinary diode forward current flowing in response to the reduced potential barrier. When operated as a photovoltaic cell, the optically generated reverse current is larger in magnitude than the component of the current that flows in response to the forward bias so that the voltage drop across the cell is positive while the current is negative. If the diode is short-circuited by an external circuit, only the reverse current of the optically generated minority carriers flows. As discussed previously, reverse current is due to some of the carriers generated within a few diffusion lengths but is equivalent to a current of all of the carriers generated within one diffusion length. If the generation rate due to optical absorption is denoted by g, the reverse photocurrent has magnitude

$$I_{sc} = eA(L_e + L_h)g \qquad (9\text{-}96)$$

where A is the cross-sectional area of the junction, L_e and L_h are the minority carrier diffusion lengths, and the subscript sc indicates short-circuit conditions. The generation rate g depends on the frequency and intensity of the light incident upon the junction. The ordinary current through the diode is given by Eq. (9-65), the I-V characteristic of the junction. The net current is therefore

$$I = I_s[e^{eV/\kappa T} - 1] - I_{sc} \qquad (9\text{-}97)$$

where I_s is the saturation current "in the dark." For back bias conditions Eq. (9-97) gives the current flowing in an excited photodiode. Under open circuit conditions, $I = 0$ and Eq. (9-97) may be solved for the open circuit voltage V_{oc} with the result

$$V_{oc} = \frac{\kappa T}{e} \ln\left[\frac{I_{sc}}{I_s} + 1\right] \qquad (9\text{-}98)$$

If the diode is connected across a finite load impedance, current will flow from n to p. The resulting potential across the load will be less than the open circuit voltage and is determined by the simultaneous satisfaction of Eq. (9-97) and the current-voltage characteristic of the load.

The spectral response of practical photocells depends on the material employed in their construction. Silicon cells are sensitive to the entire visible range. The peak sensitivity of indium antimonide is in the infrared. The open circuit voltage of silicon cells in full sunlight is about 0.6 V, while its output voltage under peak power conditions is about 0.4 V.

Light-Emitting Diodes

Recombination of an electron with a hole is the term used to describe the transition of an electron from a higher-energy state across an energy gap to an empty lower-energy state. The recombination may be radiative. If the transition occurs between states that themselves conserve momentum, that is, across a direct energy gap, a photon is emitted to conserve energy. If the transition occurs between states that do not conserve momentum, lattice vibrations, or photons, are generated and radiation may or may not be emitted. Many of the III-V semiconducting compounds such as indium antimonide and gallium arsenide have direct gaps, while silicon and germanium do not. Transitions in some junction diodes such as those made of gallium phosphide occur via impurity or defect levels.

Because recombination must occur for radiation to be emitted from a solid, a device intended as a light source must have an excess of recombinations over the equilibrium rate. That is, to induce emission, recombination must be induced. Because recombination rates are proportional to the excess in the minority carrier densities over their equilibrium values, a forward-biased p-n junction in a suitable semiconductor can be a good light source. The radiation from such a junction diode may emanate from the neutral regions adjoining the space-charge layer or from the space-charge layer itself. The former is the case for abrupt junctions since recombination in the depletion layer can usually be neglected. If, however, the impurity gradient is such that the transition region is wide, radiative recombination may occur within the space-charge layer.

When recombination occurs spontaneously between pairs of states differing in energy by the same amount, the electromagnetic fields radiated during different events have random phases, although they may all have the same frequency. The spontaneously emitted light from a diode is not strictly *monochromatic* but the spread in photon energy may be small. Because of the random emission times of the individual photons, the radiation is *incoherent*. The phenomenon in which incoherent light is emitted from a material under the influence of an electric current is called *electroluminescence*. An electroluminescent device converts electrical energy into light without employing heat. Practical electroluminescent diodes may be very small and operate at low voltages and currents depending on the material used. Typical values of operating parameters are about 10 mA with voltages less than 3 V. Furthermore, luminescent diodes may be switched rapidly, and their output is reasonably monochromatic. Such devices are useful in optical data transmission systems, in optical displays, and as indicator lights.

Schottky barrier diodes may also be used as light-emitting diodes.

Recombination may also occur in response to the presence of a photon. This process is called *stimulated emission*. A stimulated recombination occurs only if the stimulated photon has the same frequency as the initiating one. Furthermore, the radiated photon has the same phase and direction of propagation as the original one. Radiation composed of fields all of the same phase is said to be *coherent*. A device in which monochromatic coherent light is produced by stimulated emission is called a laser.

Junction Lasers

A p-n junction can act as a laser provided that radiative recombination occurs in the material and that the recombination can be enhanced. We have noted that recombinations in excess of equilibrium occur near the depletion layer of an abrupt junction under forward bias conditions. Light of the proper frequency traversing such a device is considerably amplified if very many emissions can be stimulated since as the recombinations increase in number more photons are stimulated, inducing even further recombinations. On the other hand, the photons may also be absorbed. Whether light is on balance emitted or absorbed depends on the relative populations of carriers originally in the initial and final states. In thermodynamic equilibrium, of course, more carriers are in lower-energy states than in higher-energy ones. However, the relative populations of the states may be altered under nonequilibrium conditions. When the number of particles in the higher-energy states is greater than the number in lower states, the populations are said to be inverted.

Sec. 9-9 Optical Diodes

Figure 9-22 illustrates as an example how forward bias applied to a degenerate p^+-n^+ junction can create a population inversion. When the unbiased junction of Fig. 9-22(a) is forward-biased the energy diagram of Fig. 9-22(b) results. The forward current of electrons across the reduced barrier creates an excess density of minority electrons on the p-side of the junction where there are also many holes with which to recombine. The region of population inversion occurs within a few diffusion lengths of the junction.

The overall efficiency of light amplification in the region of population inversion is increased by having the stimulating photons traverse the active region many times. This is accomplished by partially reflecting the emitted light by mirrors. In semiconductor lasers the mirrors are an integral part of the structure, usually being formed at the crystal surface.

Junction lasers are small, mechanically stable, and efficient. Several physical forms are available, including multiple-layer junction diodes, which are very efficient. Output power from a multijunction laser in continuous mode operation may be 40 mW at room temperature or higher for lower temperatures. Wavelengths range from 750 to 20,000 nm depending on the material.

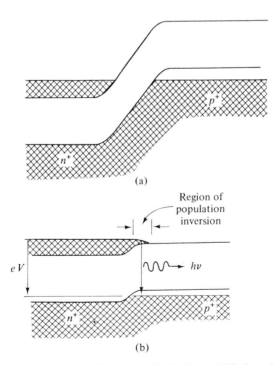

Fig. 9-22 Creation of population inversion by forward-biasing a junction laser: (a) unbiased, (b) forward-biased.

BIBLIOGRAPHY

For further readings on p-n junctions and their device applications, see

GRAY, P. E., et al., *Physical Electronics and Circuit Models of Transistors*, SEEC Vol. 2. New York: John Wiley & Sons, Inc., 1964.

ANKRUM, P. D., *Semiconductor Electronics*. Englewood Cliffs, N.J.: Prentice-Hall, Inc., 1971.

GIBBONS, J. F., *Semiconductor Electronics*. New York: McGraw-Hill Book Company, 1966.

NUSSBAUM, A., *Semiconductor Device Physics*. Englewood Cliffs, N.J.: Prentice-Hall, Inc., 1962.

VAN DER ZIEL, A., *Solid State Physical Electronics* (2nd ed.). Englewood Cliffs, N.J.: Prentice-Hall, Inc., 1968.

CHIRLIAN, P. M., *Electronic Circuits: Physical Principles, Analysis, and Design*. New York: McGraw-Hill Book Company, 1971.

More advanced treatments may be found in

WANG, S., *Solid State Electronics*, New York: McGraw-Hill Book Company, 1966.

SZE, S. M., *Physics of Semiconductor Devices*. New York: John Wiley & Sons, Inc. (Interscience Division), 1969.

Noise in p-n junctions is specifically treated in

THORNTON, R. D., et al., *Characteristics and Limitations of Transistors*, SEEC Vol. 4. New York: John Wiley & Sons, Inc., 1966.

VAN DER ZIEL, A., *Noise: Sources, Characterization, Measurement*. Englewood Cliffs, N.J.: Prentice-Hall, Inc., 1970.

Additional material on specialized topics may be found, for example, in

CHANG, K. K. N., *Parametric and Tunnel Diodes*. Englewood Cliffs, N.J.: Prentice-Hall, Inc., 1964.

TODD, C. D., *Zener and Avalanche Diodes*. New York: John Wiley & Sons, Inc. (Interscience Division), 1970.

SOOHOO, R. F., *Microwave Electronics*. Reading, Mass.: Addison-Wesley Publishing Company, Inc., 1971.

PROBLEMS

9-1. Obtain Eq. (9-12) by first calculating the ratio p_{n0}/p_{p0} using Eq. (8-17).

9-2. Calculate the equilibrium contact potential in a silicon p-n junction at room temperature if $N_a = 2.25 \times 10^{18}$ cm^{-3} and $N_d = 10^{15}$ cm^{-3}.

Chap. 9 Problems

9-3. The contact potential ϕ_0 cannot be measured by attaching voltmeter leads to either side of an equilibrium p-n junction. Explain.

9-4. Show that most of the depletion region in an unbiased abrupt junction lies on the side of lesser conductivity by demonstrating that $w_n \gg w_p$ if $N_a \gg N_d$.

9-5. Calculate w_n and $\mathcal{E}_{0\,\text{max}}$ for an unbiased junction in germanium if $N_a = 4.1 \times 10^{18}$ cm^{-3} and $N_d = 1.4 \times 10^{15}$ cm^{-3} so that $\phi_0 \simeq -0.42$ V.

9-6. Find the average hole density gradient across the depletion layer in Problem 9-5 and calculate the resulting electric diffusion current density *as if* we could write $J_h = -eD_h \Delta p$. Take $D_h = 49$ cm^2/sec.

9-7. Calculate the drifting hole electric current density in the depletion layer of Problem 9-5 *as if* we could write $J_h = pe\mu_h \mathcal{E}$. Use $\mathcal{E}_{0\,\text{max}}$ for $\mathcal{E}$ and assume that p is the geometric mean between p_{p0} and p_{n0}, i.e., $p = (p_{p0}p_{n0})^{1/2}$. Take $\mu_h = 1900$ cm^3/V-sec.

9-8. Calculate the excess minority carrier densities injected into the edges of the neutral regions by an applied forward bias of 0.4 V for a germanium junction at room temperature with $N_a = N_d = 10^{17}$ cm^{-3}. Calculate the percentage change in the majority carrier densities at the edges of the neutral regions.

9-9. The potential barrier in a p-n junction cannot be reduced to zero by application of sufficient forward bias voltage. What effect prevents this? Sketch an I-V characteristic that shows this effect for large forward bias voltages.

9-10. Derive Eqs. (9-60) and (9-63).

9-11. Calculate the electron diffusion length in silicon at room temperature assuming a recombination time of 3 μsec.

9-12. Calculate I_s for a germanium p-n junction of cross-sectional area 10^{-2} cm^2 for which $N_a = N_d = 10^{15}$ cm^{-3}. Assume that $\tau_e = \tau_h = 10$ μsec and that the diffusion coefficients are the same as those of intrinsic germanium at room temperature.

9-13. Estimate the temperature sensitivities of the saturation currents in silicon and germanium at room temperature by calculating the fractional change in n_i^2 per unit change in temperature, $(dn_i^2/n_i^2)/dT$. *Hint:* Take the natural logarithm of Eq. (9-70), and then differentiate with respect to temperature.

9-14. Calculate the equivalent noise resistance r_0 of the unbiased junction of Problem 9-12 at room temperature.

9-15. Estimate the incremental resistance r of a junction diode at room temperature biased for a forward current of 1 mA.

9-16. Show that for $N_a \gg N_d$ the space-charge layer capacitance of a p-n junction is equivalent to that of a parallel plane capacitor of area A, width w_n, and dielectric constant ε. Interpret this model on the basis of charging and discharging the capacitance.

9-17. Find the characteristic time required for charging or discharging the forward-biased p-n junction capacitance through the junction resistance r.

9-18. Calculate w_n, w_p, ϕ_0, and $\mathcal{E}_{0\,\text{max}}$ for a silicon tunnel diode at room temperature with $N_a = N_d = 10^{19}$ cm^{-3}.

9-19. Sketch the energy band diagram of a tunnel diode for forward bias conditions that just cut off the tunneling current.

9-20. Find the short-circuit photocurrent of a reverse-biased germanium photodiode in which the carrier lifetimes are 10 μsec and the area of the junction is 10^{-2} cm^2 if the junction is illuminated by a light source that causes the carrier generation rate to be 4.6×10^{18}/cm^3-sec. Compare the short-circuit current with the reverse-biased dark current. Calculate the open circuit voltage of the diode at room temperature. Assume that $N_a = N_d = 10^{15}$ cm^{-3}.

10

Multijunction Devices

In the previous chapter we studied the *p-n* junction and some applications of the properties of a single junction to common electronic devices. In this chapter we shall consider at an introductory level two multijunction structures, the bipolar junction transistor and *p-n-p-n* four-layer devices. The transistor has the basic features of an amplifier and can be employed as a switch or as a logic element. It is the semiconductor device that revolutionized electronics in the 1950s, transforming the industry from a vacuum tube technology to the solid-state technology. There are other types of transistors such as the field-effect transistors that we shall study in Chapter 11. Despite this fact, the word *transistor* without descriptive adjectives usually refers to the bipolar junction transistor. A field-effect transistor is usually referred to as an *FET*.

Four-layer devices are used as switches.

10.1 BIPOLAR JUNCTION TRANSISTOR

Two coupled *p-n* junctions in the same host crystal constitute a device known as the *bipolar junction transistor*. It may have either of the forms *n-p-n* or *p-n-p*. The three extrinsic regions are designated the *emitter*, the *base*, or the *collector* depending on the functional behavior of the carrier densities and currents under normal biasing conditions. The base region is common to both junctions. Under normal operating conditions the junction between the emitter and the base is forward-biased. Thus excess minority carriers are injected into both the emitter and the base at the edges of the emitter-base space-charge layer. The base-collector junction is reverse-biased, and therefore its electric field extracts minority carriers from the base and the collector. The critical difference between two uncoupled junctions in isolated regions of the same crystal and a transistor lies in the width of the common region, the base. If the base is thin compared with the minority carrier diffusion length in it, then most of its excess minority carriers injected by the forward-biased emitter junction will be extracted from the base region by the reverse-biased collector junction. In effect, because the emitter junction voltage enhances the minority carrier concentration in the base, it also enhances the reverse bias current across the collector junction. That is, the emitter junction voltage controls the collector junction current. The term *bipolar* refers to the fact that both majority and minority carriers contribute to the current, as we shall see.

For a specific example of this transistor action, consider the energy band diagrams of Fig. 10-1 for an *n-p-n* transistor. The band diagram for an unbiased transistor is sketched in Fig. 10-1(b). The equilibrium contact potentials ϕ_{0BE} and ϕ_{0BC} have been assumed to be the same. In practice, however, the transistor regions may be asymmetrically doped, causing the contact potentials to differ. Figure 10-1(c) shows the energy diagram for normal biasing conditions with the instantaneous voltage applied to the emitter junction $v_{BEj} > 0$ and the instantaneous voltage applied to the collector junction $v_{BCj} < 0$. By convention the junction voltages refer to the potential on the *p*-side relative to that on the *n*-side. Lowering the potential barrier in the emitter junction injects electrons from the *n*-type emitter into the *p*-type base and holes from the base into the emitter. The junction current that flows in response to the applied voltage v_{BEj} is denoted i_{BEj}. If the emitter is much more heavily doped than the base, the electron current into the base is much greater than the hole current into the emitter. Let us assume for this discussion that the hole current into the emitter is negligibly small. We shall return to this problem later when we introduce the concept of emitter efficiency. The excess electrons in the base diffuse away from the edge of the emitter junction space-charge region. If recombination in the base can be neglected, these excess electrons will all be collected by the reverse-biased collector junction. The base width is made small compared with a diffusion length to minimize recombination. The extent to which minority carriers do not survive the transport across the base and its effect on the transistor characteristics will be discussed shortly. The presence of excess electrons in the base dramatically affects the current at the collector since all electrons that diffuse to the edge of this space-charge layer are extracted from the base independent of the magnitude of the reverse bias voltage, provided it is several times the thermal voltage $\kappa T/e$.

Sec. 10-1 Bipolar Junction Transistor   313

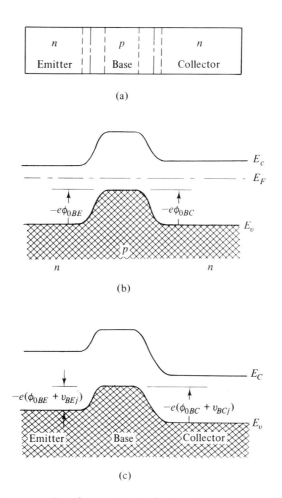

Fig. 10-1 n-p-n Transistor: (a) physical configuration, and energy diagrams for (b) unbiased and (c) normally biased conditions.

The Two-Diode Model

Let us now develop a mathematical model to describe the physical concepts discussed above. For these purposes we shall assume planar junctions with minority carrier flow normal to these planes. That is, we shall consider a one-dimensional model for the transistor. Such a model conforms to the physical geometries of many practical devices, although it neglects some two-dimensional effects that may become important in high-frequency or high-power applications. Furthermore, we shall concentrate on the p-n-p configuration because the flow of minority holes in the n-type base is in the direction of electrical current flow. We could just as easily choose to study *n-p-n* transistors for learning purposes but the sign relationship between the direction of electron motion and electrical current flow adds a needless complication.

To apply bias to the two junctions, leads are attached to each of the three extrinsic regions. In general currents flow in each of these leads. The conventional directions of these currents and the bias supplies required for both *n-p-n* and *p-n-p* configurations are indicated in Fig. 10-2. Also shown in

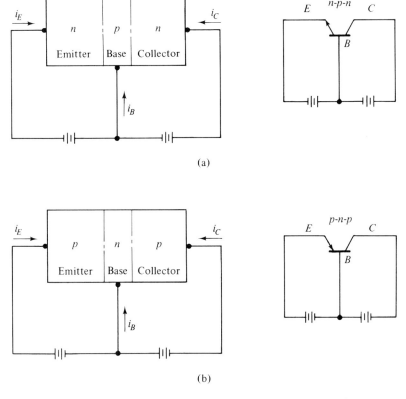

Fig. 10-2 Normal biasing conventions and circuit symbols of transistors: (a) *n-p-n*, (b) *p-n-p*.

the figure are the circuit symbols for the two types of junction transistor. The emitter is distinguished from the collector by the arrowhead, which points in the direction of current flow under normal bias conditions, thus serving to differentiate between the two types as well. Figure 10-3 shows the positive sign conventions for the junction voltages and currents in a *p-n-p* transistor. The directions of the total instantaneous junction voltages and currents conform to the conventions of Chapter 9.

Sec. 10-1 Bipolar Junction Transistor

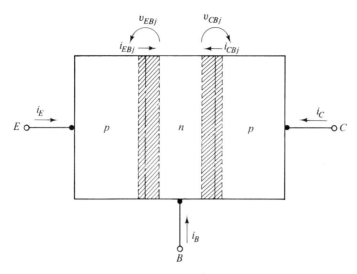

Fig. 10-3 Sign conventions for a *p-n-p* transistor.

From our physical description we know that the collector terminal current consists of two parts, the ordinary current of the reverse-biased base-collector junction and the additional current due to the excess minority holes that are injected into the base by the emitter-base junction voltage and that reach the collector. The latter current is proportional to the emitter-base junction current so we can write the collector current as

$$i_C = i_{CBj} - \alpha_F i_{EBj} \qquad (10\text{-}1)$$

where α_F is the proportionality coefficient coupling the two diodes. The signs in Eq. (10-1) result from the conventions of Fig. 10-3. The factor α_F is an important transistor parameter. If very little recombination occurs in the base, then *alpha* approaches unity. We shall shortly examine the dependence of α_F on the transistor materials and construction. By the symmetries evident in Fig. 10-3 the emitter current is also the sum of two currents, that of the emitter-base junction and the base-collector junction current that reaches the emitter junction. We may write

$$i_E = i_{EBj} - \alpha_R i_{CBj} \qquad (10\text{-}2)$$

where α_R is the fraction of the collector junction current that passes through the emitter junction. Finally by Kirchhoff's current law we may write

$$i_B = -(i_E + i_C) \qquad (10\text{-}3)$$

An equivalent circuit composed of ideal circuit elements that is also represented by Eqs. (10-1)–(10-3) is shown in Fig. 10-4. This circuit is an idealized two-diode model for a *p-n-p* transistor commonly known as the Ebers-Moll model. Coupled with the appropriate equations relating the junction

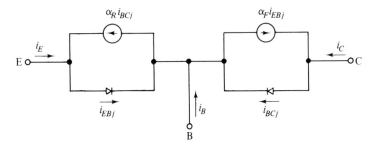

Fig. 10-4 Idealized two-diode model of a *p-n-p* transistor.

currents to their respective junction voltages, this circuit describes low-frequency large-amplitude behavior of transistors with considerable accuracy. For example, the junction relationships may be written using Eq. (9-65):

$$i_{EBj} = I_{sE}[\epsilon^{ev_{EBj}/\kappa T} - 1] \tag{10-4}$$

$$i_{CBj} = I_{sC}[\epsilon^{ev_{CBj}/\kappa T} - 1] \tag{10-5}$$

Equations (10-4) and (10-5) are the characteristics of the respective junctions with no voltage applied to the other junction. The currents I_{sE} and I_{sC} are the saturation currents of the appropriate junction with the opposite junction shorted. Under normal operating conditions

$$\frac{ev_{EBj}}{\kappa T} \gg 1, \quad \frac{e|v_{CBj}|}{\kappa T} \gg 1, \quad v_{CBj} < 0 \tag{10-6}$$

so that the junction currents controlled by their respective voltages are

$$i_{EBj} \simeq I_{sE}\epsilon^{ev_{EBj}/\kappa T} \tag{10-7}$$

and

$$i_{CBj} \simeq -I_{sC} \tag{10-8}$$

Using Eqs. (10-7) and (10-8) we can obtain expressions for the emitter and collector terminal currents from Eqs. (10-1) and (10-2). Furthermore, by eliminating the emitter junction current in Eq. (10-1) using Eq. (10-2) we can obtain an expression relating the terminal currents and v_{EBj}. The results of these manipulations are

$$\boxed{-i_C = \alpha_F i_E + (1 - \alpha_F \alpha_R)I_{sC}} \tag{10-9}$$

and

$$\boxed{i_E = I_{sE}\epsilon^{ev_{EBj}/\kappa T} + \alpha_R I_{sC}} \tag{10-10}$$

Equations (10-9) and (10-10) indicate that under normal operating conditions the emitter terminal current is essentially the forward bias emitter junction current and the collector terminal current is dominated by the emit-

Sec. 10-1 Bipolar Junction Transistor

ter current, which in turn depends exponentially on the emitter-base junction voltage.

With the emitter junction forward-biased and the collector junction reverse-biased, as in Eq. (10-6), the transistor is said to be operating in the *normal* region. The two-diode model is also applicable under other biasing conditions. For example, *cutoff* is the condition when both junctions are reverse-biased. With both junctions forward-biased, the transistor is biased in the *saturation* region. In *inverse* operation the emitter junction is reverse-biased and the collector junction is forward-biased, thus essentially reversing the roles of the emitter and collector from those of the normal operating condition.

The Transistor Alpha

The critical parameter in transistor design is α_F, the transistor alpha. The reverse alpha α_R is of little importance because I_{sC} in Eqs. (10-9) and (10-10) is small. From Eq. (10-9) we can write

$$\alpha \equiv \alpha_F = -\frac{\partial i_C}{\partial i_E}\bigg|_{v_{CB}=\text{constant}} = \frac{-i_C}{i_E} \quad (10\text{-}11)$$

The reason we must restrain the collector junction voltage from changing in Eq. (10-11) is that v_{CB} affects the width of the collector junction space-charge layer and hence I_{sC}. Furthermore, α depends on v_{BC} indirectly because the effective width of the base region is reduced by an increased collector junction space-charge layer. In words, alpha is the incremental change in collector current for a change in emitter current. It is called the *common-base forward current transfer ratio*. The negative sign in Eq. (10-11) results from the sign conventions for the terminal currents.

Physically the transistor α depends on three factors, the efficiency of the injection process at the emitter junction, the efficiency of the transport of minority carriers across the base, and the efficiency of the extraction of minority carriers from the base at the collector junction. We shall now investigate these processes to determine the dependence of α on the transistor parameters.

Emitter junction current in a *p-n-p* transistor in normal operating conditions is composed of two parts, the injection of holes into the base and of electrons into the emitter. If we designate the electric current due to the former i_{EBp} and to the latter i_{EBn}, we have

$$i_E = i_{EBp} + i_{EBn} \quad (10\text{-}12)$$

An indication of the effectiveness of the emitter for injecting minority holes into the base is given by

$$\gamma \equiv \frac{\partial i_{EBp}}{\partial i_E}\bigg|_{v_{CB}=\text{constant}} \quad (10\text{-}13)$$

To make γ large, the electron current i_{EBn} from base to emitter should be made small. We shall see later that this is accomplished by doping the emitter heavily and the base lightly and by making the base thin. To calculate γ we need to know the profiles of the minority carrier concentrations on either side of the emitter junction since in the neutral regions the currents are driven by diffusion. If the emitter is many diffusion lengths wide, the profile of excess electron density in the emitter is exponential, just as in Section 9.3. At the edge of the space-charge layer i_{EBn} is given by

$$i_{EBn} = \frac{eAD_{eE}n_{E0}}{L_{eE}}[\epsilon^{ev_{EBj}/\kappa T} - 1]$$

$$\simeq \frac{eAD_{eE}n_{E0}}{L_{eE}}\epsilon^{ev_{EBj}/\kappa T} \qquad (10\text{-}14)$$

which is obtained from Eq.(9-59). In Eq. (10-14) n_{0E} is the equilibrium concentration of minority carrier electrons in the p-type emitter with D_{eE} and L_{eE} also being emitter parameters. The current injected into the base i_{EBp} depends on the hole gradient at the base edge of the emitter junction space-charge layer. Because the base region is narrow, the previous exponential solutions of Chapter 9 are not valid in the base. We shall postpone our calculation of the spatial dependence of the minority density in the base until we have completed our discussion of the factors that determine alpha.

The efficiency of the base in transporting the holes injected at the emitter junction across the base to the collector junction is defined as

$$B \equiv \frac{\partial i_{CBp}}{\partial i_{EBp}}\bigg|_{v_{CB}=\text{constant}} \qquad (10\text{-}15)$$

The *base-transport factor* B is the change in the collector junction hole current per unit change in the emitter junction hole current. To estimate B we shall again need the minority density profile in the base.

The collector junction current, like the emitter junction current, is the sum of two currents,

$$i_C = i_{CBp} + i_{CBn} \qquad (10\text{-}16)$$

where i_{CBn} is the current due to the extraction of minority electrons from the p-type collector region. The collector efficiency δ is defined as

$$\delta \equiv \frac{\partial i_C}{\partial i_{CBp}}\bigg|_{v_{CB}=\text{constant}} \qquad (10\text{-}17)$$

The electron current i_{CBn} can be reduced so that δ approaches unity by doping the collector region more heavily than the base. For large reverse voltages across the base-collector junction, avalanche breakdown can occur in this space-charge layer, multiplying the available charge. In these instances $\delta > 1$.

Sec. 10-1 Bipolar Junction Transistor

Using Eqs. (10-13), (10-15), and (10-17), it is easy to see that

$$\alpha = \gamma B \delta \qquad (10\text{-}18)$$

For a good transistor $\delta \simeq 1$ so that $\alpha \simeq \gamma B$.

To estimate the dependence of α on transistor parameters we now turn to the problem of minority carrier diffusion in the base.

Minority Carriers in the Base

The bias voltages on the emitter and collector junctions determine the minority carrier densities at the edges of the respective space-charge layers. A forward bias at the emitter junction creates excess minority concentrations on either side of that junction, while a reverse bias at the collector junction creates a deficiency in the minority carrier densities at that junction. The spatial variations of the minority concentrations throughout the bulk of the emitter, base, and collector are determined by a continuity relationship dominated by diffusion. The appropriate relationship is the minority carrier diffusion equation. If the emitter and collector are both wide compared with their minority diffusion lengths, the minority densities in each will vary exponentially with distance away from the edges of the space-charge layer. These variations are indicated in Fig. 10-5. The base, however, is usually

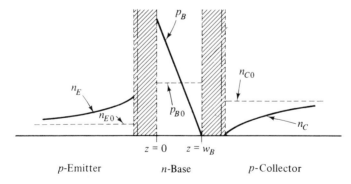

Fig. 10-5 Minority carrier concentrations in a p-n-p transistor with forward-biased emitter junction and reverse-biased collector junction.

narrow. Therefore we shall have to solve the minority carrier diffusion equation specifically in the base region to find the charge density and diffusive current.

The steady-state diffusion equation for the hole density in an n-type base,

p_B, is

$$D_{hB}\frac{\partial^2 p_B}{\partial z^2} = \frac{p_B - p_{B0}}{\tau_{hB}} \tag{10-19}$$

where D_{hB} is the diffusion coefficient for holes, p_{B0} is the equilibrium hole density, and τ_{hB} is the hole lifetime, all evaluated in the base. The left-hand side of this equation is the divergence of the hole current; the right-hand side is the net recombination rate. The solution to Eq. (10-19) is of the familiar form

$$\delta p_B(z) = K_1 \epsilon^{-z/L_{hB}} + K_2 \epsilon^{z/L_{hB}} \tag{10-20}$$

where $\delta p_B(z) \equiv [p_B(z) - p_{B0}]$ is the excess hole density in the base and

$$L_{hB} = (D_{hB}\tau_{hB})^{1/2}$$

The constants of integration K_1 and K_2 are determined by the boundary conditions, the value of p_B at either end of the base. If we denote the edge of the emitter space-charge layer as $z = 0$ and the edge of the collector space-charge layer as $z = w_B$, the minority concentrations p_B at $z = 0$ and $z = w_B$ are determined by the respective junction bias voltages. That is,

$$\delta p_B(0) = p_{B0}[\epsilon^{ev_{EBJ}/\kappa T} - 1] \tag{10-21}$$

and

$$\delta p_B(w_B) = p_{B0}[\epsilon^{ev_{CBJ}/\kappa T} - 1] \tag{10-22}$$

After some algebraic manipulation to solve for K_1 and K_2 we obtain

$$\delta p_B(z) = \frac{\delta p_B(w_B)\sinh(z/L_{hB}) - \delta p_B(0)\sinh[(z - w_B)/L_{hB}]}{\sinh(w_B/L_{hB})} \tag{10-23}$$

as the spatial variation of the hole density in the base, $0 \leq z \leq w_B$. For a small base, $w_B \ll L_{hB}$, we can expand Eq. (10-23) using the approximation

$$\sinh x \simeq x + \cdots, \qquad x \ll 1$$

obtaining

$$p_B(z) = [p_B(w_B) - p_B(0)]\frac{z}{w_B} + p_B(0) \tag{10-24}$$

That is, for a narrow base the minority carrier density varies linearly from emitter to collector. Therefore the gradient of this density and hence the diffusion current through the base is constant. Since the hole diffusion current is constant if Eq. (10-24) is valid, there is essentially no recombination in the base under these conditions.

To find the actual diffusive hole current in the n-type base we shall use the more general expression, Eq. (10-23), in

$$i_p = -eAD_{hB}\frac{\partial p_B}{\partial z}, \qquad 0 \leq z \leq w_B$$

obtaining

$$i_p(z) = \frac{-eAD_{hB}}{L_{hB}\sinh(w_B/L_{hB})}\left[\delta p_B(w_B)\cosh\left(\frac{z}{L_{hB}}\right) - \delta p_B(0)\cosh\left(\frac{z - w_B}{L_{hB}}\right)\right]$$

$$\tag{10-25}$$

Sec. 10-1 Bipolar Junction Transistor

If the collector junction is reverse-biased with a potential several times the thermal voltage, Eq. (10-22) indicates that $p_B(w_B)$ is very small. Therefore we shall neglect the first term on the right-hand side of the equation above. If $w_B \ll L_{hB}$, we can expand the hyperbolic sine function as before and use Eq. (10-21) to obtain

$$i_p(z) \simeq \left(\frac{eAD_{hB}p_{B0}}{w_B}\right) \cosh\left(\frac{z - w_B}{L_{hB}}\right) \epsilon^{ev_{EBj}/\kappa T} \qquad (10\text{-}26)$$

We shall now use this result to estimate the emitter efficiency γ and the base-transport factor B.

The base-transport factor as defined by Eq. (10-15) is the derivative of the hole current at the base-collector junction with respect to that at the emitter-base junction for fixed value of v_{CB}. Because the currents are linearly proportional to the same function, the derivative is just their ratio. In the present notation

$$B = \frac{i_p(z = w_B)}{i_p(z = 0)}\bigg|_{v_{CB}=\text{constant}} \qquad (10\text{-}27)$$

is equivalent to Eq. (10-15). Using the expansion

$$\cosh x \simeq 1 + \frac{x^2}{2} + \cdots, \qquad x \ll 1$$

and Eq. (10-26), we find that for $w_B \ll L_{hB}$

$$i_p(w_B) \simeq \left(\frac{eAD_{hB}p_{B0}}{w_B}\right)\epsilon^{ev_{EBj}/\kappa T} \qquad (10\text{-}28)$$

and

$$i_p(0) \simeq \left(\frac{eAD_{hB}p_{B0}}{w_B}\right)\left(1 + \frac{w_B^2}{2L_{hB}^2}\right)\epsilon^{ev_{EBj}/\kappa T} \qquad (10\text{-}29)$$

Therefore

$$\boxed{B \simeq 1 - \frac{w_B^2}{2L_{hB}^2}} \qquad (10\text{-}30)$$

Thus to improve the base-transport factor the width of the base should be as small as possible compared with the minority diffusion length. Typical transistors have values for B in the range 0.90 to 0.99. The base cannot be made too thin, however, or *punch-through* might occur. Punch-through is the term applied to conditions that exist when the space-charge layer of the collector junction becomes so wide under reverse bias conditions that it extends through to the emitter depletion region, resulting in direct conduction between the emitter and collector.

The emitter efficiency γ as defined by Eq. (10-13) is the derivative of the emitter junction hole current with respect to the total emitter current with v_{CB} fixed. Because we have assumed that no recombination occurs in the space-charge layers, the junction hole current i_{EBp} is just the base hole current evaluated at the edge of the emitter space-charge layer $i_p(z = 0)$. Again each of these currents is proportional to the same function so that the derivative is just their ratio. Therefore, using Eq. (10-11),

$$\gamma = \frac{i_p(0)}{i_p(0) + i_{EBn}}\bigg|_{v_{CB}=\text{constant}}$$

Using Eq. (10-14) for i_{EBn} and Eq. (10-28) for $i_p(0)$, we find after some manipulation

$$\gamma = \frac{1}{1 + (D_{eE}n_{E0}w_B/D_{hB}p_{B0}L_{eE})} \quad (10\text{-}31)$$

So that γ will be as large as possible the design of a transistor should have $w_B \ll L_{eE}$ and $n_{E0} \ll p_{B0}$ for the p-n-p configuration. Because the equilibrium minority carrier densities are inversely proportional to the majority concentrations through

$$np = n_i^2$$

the minority ratio can be written in terms of the majority densities as

$$\frac{n_{E0}}{p_{B0}} = \frac{n_{B0}}{p_{E0}} = \frac{N_{DB}}{N_{AE}} \quad (10\text{-}32)$$

Thus to make this ratio small the p-type emitter should be more heavily doped than the n-type base. The ratio of the diffusion coefficients may be written in terms of the mobilities using Einstein's relation

$$\frac{D_{eE}}{D_{hB}} = \frac{\mu_{eE}}{\mu_{hB}} \quad (10\text{-}33)$$

Furthermore, if the doping levels are not so high as to affect the mobilities of the carriers, the mobility of a given carrier will be the same in both the base and the emitter. Recognizing that the factor differing from unity in Eq. (10-31) is small we can rewrite this expression using Eqs. (10-32) and (10-33), obtaining

$$\gamma \simeq 1 - \left(\frac{n_{B0}\mu_e w_B}{p_{E0}\mu_h L_{eE}}\right) \quad (10\text{-}34)$$

Equation (10-34) may be written in terms of the conductivities of the base and emitter regions σ_B and σ_E by multiplying both the numerator and the denominator of the ratio on the right-hand side by q, giving

$$\boxed{\gamma \simeq 1 - \left(\frac{\sigma_B w_B}{\sigma_E L_{eE}}\right)} \quad (10\text{-}35)$$

Now that we have explored the physics of the transistor and determined the parameters that effect the transistor alpha we shall turn to its representation in circuit configurations that are frequently encountered.

The Common-Base Configuration

Viewed as a circuit element the transistor is a three-terminal device. Therefore it has two network ports, an input and an output, with one terminal common to both. A transistor that is connected to external circuitry with the base lead in both the input and the output ports is said to be in the *common-base configuration*. Figure 10-6 shows such a configuration.

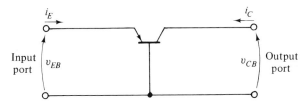

Fig. 10-6 Common-base configuration for the *p-n-p* transistor.

The relationships between the input parameters, i_E and v_{EB}, and the output parameters, i_C and v_{CB}, have been established on physical principles earlier in this section assuming that the terminal voltages appear across the respective junctions. Now we shall indicate the connection between our physical model and the circuit representation of the transistor in the common-base configuration.

As a two-port network, two of the four transistor variables i_E, v_{EB}, i_C, and v_{CB} may be treated as dependent variables, while the other two are independent. The so-called *hybrid* representation treats the input current i_E and the output voltage v_{CB} as independent variables so that the input voltage v_{BE} and the output current i_C may be expressed functionally as

$$v_{EB} = f_1(i_E, v_{CB}) \qquad (10\text{-}36)$$

$$i_C = f_2(i_E, v_{CB}) \qquad (10\text{-}37)$$

The specific functional forms implied in Eqs. (10-36) and (10-37) can be determined by manipulating Eqs. (10-9) and (10-10) of the two-diode model if the transistor is biased in the normal mode. The dependence on v_{CB} is not explicit in the latter relations but appears through the effect of base width modulation on the alphas and on I_{sC}. On the other hand, the functional relationships of Eqs. (10-36) and (10-37) can be measured directly. The results of a typical measurement, sketched in Fig. 10-7, are the static common-base terminal characteristics of a *p-n-p* transistor. The only feature of these characteristics not predicted by the equations of the two-diode model is the nonzero slopes of the i_C curves with v_{CB}. Base width modulation accounts for this feature.

To develop a low-frequency small-signal circuit model in the hybrid rep-

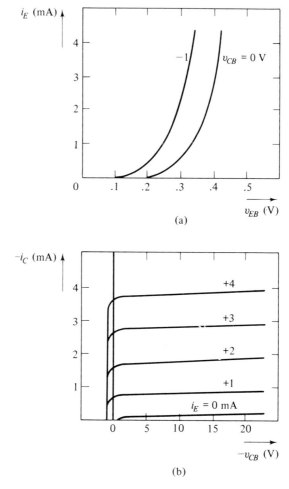

Fig. 10-7 Typical common-base p-n-p transistor characteristics.

resentation we shall consider the relationships among small changes in the input and output parameters about a d.c. operating point. For example, a small change in the total emitter-base voltage δv_{EB} about its average value V_{EB} determined by the d.c. biases is the same as its small-amplitude a.c. signal denoted v_{eb} with lowercase subscripts. From Eqs. (10-36) and (10-37) we can write

$$v_{eb} = h_{ib}i_e + h_{rb}v_{cb} \qquad (10\text{-}38)$$

and

$$i_c = h_{fb}i_e + h_{ob}v_{cb} \qquad (10\text{-}39)$$

where h_{ib}, h_{rb}, h_{fb}, and h_{ob} are called the *common-base hybrid parameters*. Obviously,

$$h_{ib} = \left.\frac{\partial v_{EB}}{\partial i_E}\right|_{v_{CB}=\text{constant}} \qquad (10\text{-}40)$$

The parameter h_{ib} relates input variables and has dimensions of an impedance; it is therefore called the *common-base input impedance*. Similarly,

$$h_{rb} = \left.\frac{\partial v_{EB}}{\partial v_{CB}}\right|_{i_E=\text{constant}} \qquad (10\text{-}41)$$

relates the input voltage to output voltage; it is the *reverse voltage transfer function*. Also,

$$h_{fb} = \left.\frac{\partial i_C}{\partial i_E}\right|_{v_{CB}=\text{constant}} \qquad (10\text{-}42)$$

and is called the *common-base forward current transfer ratio*. Finally,

$$h_{ob} = \left.\frac{\partial i_C}{\partial v_{CB}}\right|_{i_E=\text{constant}} \qquad (10\text{-}43)$$

This parameter is an *output admittance*. Notice that Eqs. (10-38) and (10-39) in conjunction with the definitions of the hybrid parameters are mathematically equivalent to Taylor expansions of the dependent variables about their steady-state values and that only the linear terms of the expansions are retained. Therefore the parameters are evaluated at the d.c. operating point. By using the hybrid parameters we therefore are restricting the signals to small excursions about the d.c. operating point where the static terminal characteristics can reasonably be represented by straight lines.

A linear incremental equivalent circuit based on the hybrid relationships of Eqs. (10-38) and (10-39) is shown in Fig. 10-8. Notice the similarity between the form of this circuit and that of the idealized two-diode mode illustrated in Fig. 10-4. First, as is evident upon comparison of Eq. (10-22) with

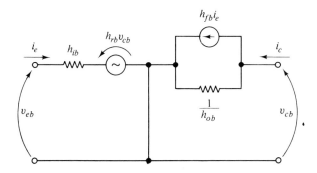

Fig. 10-8 Common-base h-parameter representation of the transistor of Fig. 10-6.

Eq. (10-11),

$$h_{fb} = -\alpha \quad (10\text{-}44)$$

Furthermore, assuming that the terminal voltage v_{EB} appears across the emitter junction, inspection of Eq. (10-40) reveals that h_{ib} is the incremental resistance of the emitter junction. Similarly, h_{ob} is the incremental conductance of the collector junction. Finally, the reverse voltage transfer ratio h_{rb} is related to α_R, the common-base reverse current transfer ratio. The series combination of h_{ib} and $h_{rb}v_{cb}$ in the emitter segment of the circuit of Fig. 10-8 is equivalent to a parallel combination of h_{ib} and a current generator because of the correspondence between the Thevenin and Norton equivalent circuits. The equivalent current source is just $\alpha_R i_c$. In the hybrid representation, however, i_c is considered a dependent variable so the voltage generator appears in the circuit of Fig. 10-8. Therefore the hybrid representation of the common-base configuration is equivalent to a two-diode model with the incremental junction resistances taken into account. It is useful only for small-amplitude variations about the operating point where the h-parameters are evaluated. And it is valid only at low frequencies since dynamic effects have not been included.

An equivalent circuit for the common-base transistor configuration that accounts for dynamic effects at moderate frequencies is shown in Fig. 10-9.

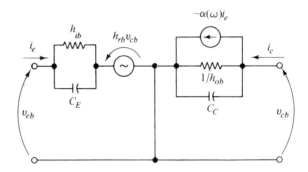

Fig. 10-9 Common-base equivalent circuit of a transistor at moderate frequencies.

The incremental resistance of the forward-biased emitter junction is shunted by a diffusion capacitance. In addition the frequency dependence of the forward current transfer ratio $h_{fb} = -\alpha$ is also indicated. That the transistor α should be a function of frequency can be understood on the basis of the physical picture we have already developed. The alpha indicates the portion of emitter terminal current that reaches the collector. At high frequencies some of the a.c. component of the terminal current i_e charges the emitter junction capacitance C_E. This component does not result in collector current

Sec. 10-1 Bipolar Junction Transistor

since the charge associated with it is stored near the emitter junction space-charge layer. Instead the collector current is proportional to the current i_r that flows through h_{ib}, which is

$$\frac{i_r}{i_e} = \frac{1/j\omega C_E}{h_{ib} + (1/j\omega C_E)}$$

$$\frac{i_r}{i_e} = \frac{1}{1 + j\omega h_{ib} C_E}$$

Thus

$$\boxed{\alpha(\omega) = \frac{-h_{fb}}{1 + j\omega h_{ib} C_E}} \qquad (10\text{-}45)$$

The frequency

$$\omega_{\alpha b} = \frac{1}{h_{ib} C_E} \qquad (10\text{-}46)$$

is called the *common-base alpha cutoff* frequency. At $\omega_{\alpha b}$ the transistor alpha is 0.707 times its low-frequency value. A transistor designed to operate at moderate frequencies may have an alpha cutoff on the order of tens of megahertz. It can be shown that $\omega_{\alpha b}$ is proportional to the minority carrier diffusion coefficient in the base and inversely proportional to the square of the base width w_B. Thus a high-frequency transistor must have a thin base.

The Common-Emitter Configuration

The common-emitter configuration of a *p-n-p* transistor is shown in Fig. 10-10. The input variables are i_B and v_{BE}, while the output variables are i_C

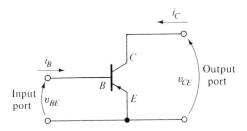

Fig. 10-10 Common-emitter configuration for a *p-n-p* transistor.

and v_{CE}. In the hybrid representation of common-emitter configuration it is customary to consider the input voltage v_{BE} and output current i_C as the dependent variables, while the input current i_B and output voltage v_{CE} are treated as independent. That is,

$$v_{BE} = f_3(i_B, v_{CE}) \qquad (10\text{-}47)$$

$$i_C = f_4(i_B, v_{CE}) \qquad (10\text{-}48)$$

The functional relationships expressed can be determined from the two-diode equations, as was the case for the common-base configuration. However, in this instance the process is complicated by the introduction of v_{CE}, the potential between the collector and emitter terminals, rather than the collector junction voltage. Plots of typical terminal characteristics displayed in Fig. 10-11 indicate the functional dependences.

The common-emitter hybrid parameters are defined by the following equations for the a.c. components of the currents and voltages:

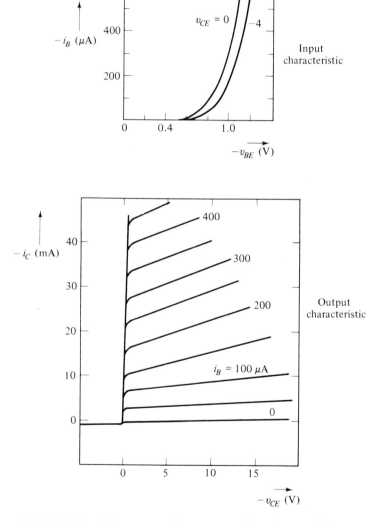

Fig. 10-11 Typical common-emitter p-n-p transistor characteristics.

Sec. 10-1 Bipolar Junction Transistor

$$\boxed{v_{be} = h_{ie}i_b + h_{re}v_{ce}} \tag{10-49}$$

and

$$\boxed{i_c = h_{fe}i_b + h_{oe}v_{ce}} \tag{10-50}$$

The *common-emitter input impedance* is

$$h_{ie} = \left.\frac{\partial v_{BE}}{\partial i_B}\right|_{v_{CE}=\text{constant}} \tag{10-51}$$

while

$$h_{re} = \left.\frac{\partial v_{BE}}{\partial v_{CE}}\right|_{i_B=\text{constant}} \tag{10-52}$$

is the *common-emitter reverse voltage transfer ratio*. The *common-emitter forward current transfer ratio* is

$$h_{fe} = \left.\frac{\partial i_C}{\partial i_B}\right|_{v_{CE}=\text{constant}} \tag{10-53}$$

and the *output conductance* is

$$h_{oe} = \left.\frac{\partial i_C}{\partial v_{CE}}\right|_{i_B=\text{constant}} \tag{10-54}$$

The linear incremental equivalent circuit based on the hybrid relationships of Eqs. (10-49) and (10-50) is shown in Fig. 10-12.

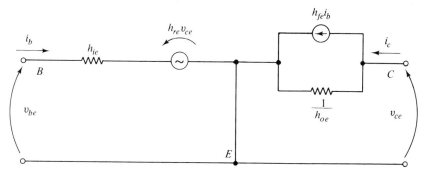

Fig. 10-12 Common-emitter *h*-parameter representation of the transistor of Fig. 10-10.

The most important of the common-emitter hybrid parameters is h_{fe}, which represents the current gain from input to output. It is easy to show using Eqs. (10-3) and (10-9) and the definition of h_{fe} that

$$\boxed{h_{fe} = \frac{\alpha}{1-\alpha}} \tag{10-55}$$

Because the value of α is usually very close to unity, h_{fe} is a very sensitive function of the transistor parameters. The other common-emitter hybrid parameters can also be expressed in terms of the common-base parameters with the results

$$h_{ie} \simeq \frac{h_{ib}}{1-\alpha} \tag{10-56}$$

$$h_{oe} \simeq \frac{h_{ob}}{1-\alpha} \tag{10-57}$$

$$h_{re} \simeq \frac{h_{ib}h_{ob}}{1-\alpha} - h_{rb} \tag{10-58}$$

The equivalent circuit of Fig. 10-12 is a low-frequency small-signal model. Figure 10-13 shows an equivalent circuit useful at higher frequencies. The

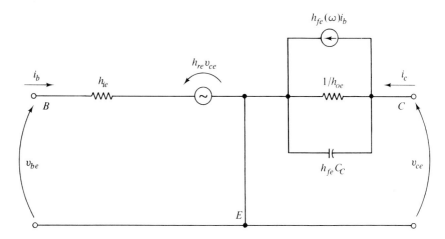

Fig. 10-13 Common-emitter equivalent circuit of a transistor at moderate frequencies.

two most important frequency-dependent elements are the collector capacitance $h_{fe}C_C$ and the forward current-transfer ratio $h_{fe}(\omega)$. Using Eqs. (10-45) and (10-55), we find

$$h_{fe}(\omega) = \frac{-h_{fb}}{1 + h_{fb} + j(\omega/\omega_{ab})} \tag{10-59}$$

At the common-emitter cutoff frequency ω_{ae} the parameter $h_{fe}(\omega)$ has its value reduced from that of the low-frequency limit by a factor of 0.707. That is,

$$\frac{\omega_{ae}}{\omega_{ab}} = 1 + h_{fb}$$

Sec. 10-1 Bipolar Junction Transistor

or
$$\omega_{\alpha e} = \omega_{\alpha b}(1 - \alpha) \qquad (10\text{-}60)$$

In other words, the cutoff frequency in the common-emitter configuration is much lower than in the common-base configuration since α is almost unity.

Yet another high-frequency model of a transistor is the *hybrid-π* equivalent circuit drawn in Fig. 10-14. This model takes into account a potential

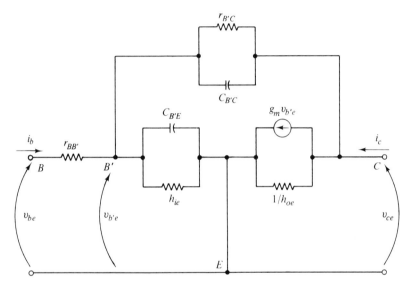

Fig. 10-14 Hybrid-π equivalent circuit of a transistor.

drop in the low-conductivity base that causes the emitter-base junction voltage to differ from the emitter-to-base terminal voltage. The junction node in the base is denoted B' so that the actual a.c. junction voltage is $v_{b'e}$. Furthermore, the equivalent current generator is expressed in terms of the actual base-emitter junction voltage via a transconductance g_m. The incremental resistance of the base-collector junction and space-charge layer capacitance are the primary components in $r_{B'C}$ and $C_{B'C}$, respectively.

The hybrid-π and other models of the bipolar transistor are studied in detail in most books on electronic circuits and in the more advanced texts on physical electronics.

Noise

The important noise sources in a transistor are similar to those in a diode. The motions of the carriers across the potential barriers of the two junctions in a transistor, being comprised of independent random events, constitute currents with full shot noise. As discussed in Section 9.6, the spectral density of this noise is

$$\frac{\overline{I_{ns}^2}}{\Delta f} = 2eI \qquad (10\text{-}61)$$

an expression that applies to each component of the current. Recall that noise sources always add in a mean square sense, independent of the signs of the components of the currents. That is, noise described by Eq. (10-61) is evident for the current supported by each distinct motion of the carriers. For example, the hole current in a p-n-p transistor consists of four components. The largest component is the result of holes injected from the emitter into the base and extracted by the collector. Another group of holes is comprised of those injected from the emitter into the base that recombine in the base. Of secondary importance are holes thermally generated in the base region and subsequently extracted either by the emitter or by the collector. The current associated with each of these motions has associated with it full shot noise with spectral density given by Eq. (10-61). Under various conditions of operation some of these noise sources may either be ignored if small or combined to account for correlations between sources.

An additional noise source is sometimes required to account for the thermal noise of the base resistance $r_{BB'}$. The description of this noise source is complicated by the necessity of representing the base resistance as an equivalent distributed element rather than as a lumped one because of the geometries of practical devices.

10.2 FOUR-LAYER DEVICES

In addition to the bipolar transistor, another class of multijunction device that is of significant importance in electronics is the four-layer device, consisting of four regions of alternating p- and n-type materials forming three p-n junctions. These devices take a variety of forms, all operating as switches. One, the silicon controlled rectifier (SCR), has features that make it useful in switching circuits capable of handling relatively high powers.

In this section we shall consider the fundamental principles of operation of three types of four-layer devices. First we shall study a two-terminal device called the Shockley or p-n-p-n diode, which is switched on or triggered by the voltage across it. Next we shall investigate the problem of triggering by other means, principally by an electrical signal applied to a third terminal. This leads us to the SCR. Finally, we shall introduce a symmetrical or bilateral switch.

The Shockley or p-n-p-n Diode

The physical configuration, circuit symbol, and I-V characteristic of a two-terminal, three-junction, four-layer p-n-p-n diode are illustrated in Fig.

10-15. For the convenience of discussion the four layers are labeled with numbers in circles starting at the anode end of the diode. That is, region $p_{①}$ is the anode and region n_4 the cathode. The interior regions, $n_{②}$ and $p_{③}$, are sometimes called base regions or gates. The junctions are also labeled from the anode end, the $p_{①}$-$n_{②}$ junction being called j_1, the middle junction, j_2, and the $p_{③}$-$n_{④}$ junction, j_3. The circuit symbol resembles both a diode symbol and the number 4.

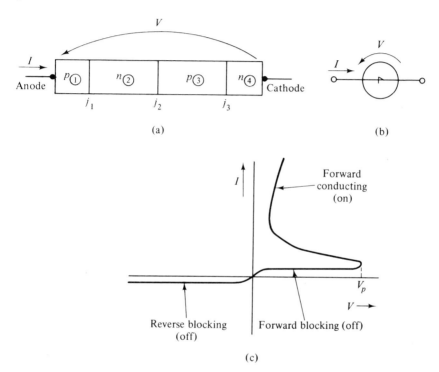

Fig. 10-15 *p-n-p-n* Diode: (a) physical configuration, (b) circuit symbol, (c) *I-V* characteristic.

The *I-V* characteristic of Fig. 10-15(c) is divided into three regions in which the diode operates in different states, the reverse blocking, the forward blocking, and the forward conducting states. In either of the blocking states the diode switch is off, passing negligibly small current for a wide range of voltages. In the forward conducting state the switch is on, capable of passing very large currents with only a small voltage drop across the device. The state of the diode when forward-biased, on or off, depends on its past history. As the positive voltage across the diode is increased from zero the diode assumes the forward blocking state until the switching voltage V_p is reached, at which time it switches rapidly to the forward conducting state.

It is this switching mechanism that we shall be most concerned with in our description of the mechanisms that underlie the operation of the diode.

In the *reverse blocking state*, the potential difference imposed across the diode causes the two outer junctions, j_1 and j_3, to be reverse-biased while the middle junction, j_2, is forward-biased. While j_2 is forward-biased, the current through it is low because the carrier flows into regions $n_{②}$ and $p_{③}$ are determined by the reverse-biased junctions, j_1 and j_3. The reverse-biased junctions supply majority carriers to regions $n_{②}$ and $p_{③}$, which in turn are injected across j_2, constituting the forward current at that junction. Thus the current through the diode is limited by the saturation currents of j_1 and j_2, which are determined by thermal generation of hole-electron pairs near the respective junctions. Reverse-bias breakdown can occur in j_1 and j_3 if a large negative bias is applied, in which case the current through the diode is limited only by the external circuitry. Practical high-voltage devices have been made to withstand reverse voltages in excess of 1000 V.

In the *forward blocking state*, the middle junction is reverse-biased while j_1 and j_3 are forward-biased. Most of the applied voltage appears across j_2. The current across j_2 is the saturation current of that junction. The charges that carry this current are the minority carriers in regions $n_{②}$ and $p_{③}$. These are supplied by thermal generation near j_2 and by injection from regions $p_{①}$ and $n_{④}$ across the forward-biased junctions j_1 and j_3, respectively. However, if the interior regions $n_{②}$ and $p_{③}$ are wide relative to the diffusion lengths of their respective minority carriers and if the forward biases on j_1 and j_3 are low so that the relative number of minority carriers injected into regions $n_{②}$ and $p_{③}$ are low, then the contributions of these injected carriers to the current across j_2 are negligible. Under these circumstances, the current across j_2 is due to thermal generation of carrier pairs near j_2. Figure 10-16(a) shows the profiles of the minority carrier concentrations in the *p-n-p-n* device in the forward blocking state with low applied voltage. The figure indicates that the injection levels across j_1 and j_3 are very low, that the base regions are wide compared with the minority carrier diffusion lengths, and that j_2 is reverse-biased. The minority carrier densities in the respective regions are denoted with numerical subscripts without circles. For example, the minority carrier electron density in region $p_{①}$ is designated n_1 while in Fig. 10-16(c) its equilibrium value is n_{10}.

In the *forward conducting state*, all three junctions are forward biased and a large current flows. Because the polarity of the forward voltage across j_2 is opposite to that of the forward voltages across j_1 and j_3, the net voltage difference across the diode is small, usually less than 1 V.

We shall now consider the mechanisms by which the diode switches from the forward blocking to the forward conducting state, which results when j_2 switches from being reverse-biased to being forward-biased.

With the diode in the forward blocking state, minority carriers are in-

Sec. 10-2 Four-Layer Devices

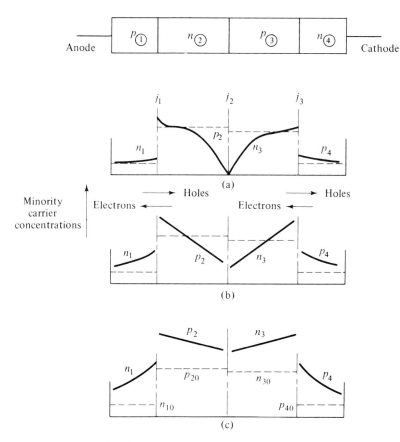

Fig. 10-16 Minority carrier concentrations in a *p-n-p-n* diode: (a) forward blocking, low voltage; (b) forward blocking, $V \leq V_p$; (c) forward conducting.

jected into the $n_{\text{②}}$ and $p_{\text{③}}$ regions by the forward bias voltages on junctions j_1 and j_3. As the forward voltage across the device is increased, the forward bias voltages across j_1 and j_3 also increase, leading to more injection of minority carriers into regions $n_{\text{②}}$ and $p_{\text{③}}$. This injection process is enhanced if the anode and cathode regions, $p_{\text{①}}$ and $n_{\text{④}}$, respectively, are heavily doped. When the forward biases across j_1 and j_3 are large enough for the injection levels to be high, the recombination processes in regions $n_{\text{②}}$ and $p_{\text{③}}$ become saturated and the injected carriers diffuse to the middle junction j_2. At j_2 they are collected by the field of the reverse-biased junction and contribute to the current across that junction. The minority carrier concentrations under these circumstances are sketched in Fig. 10-16(b).

Switching occurs in the following way. Forward bias on junctions j_1 and

j_3 injects minority carriers into regions $n_{②}$ and $p_{③}$. The minority holes in region $n_{②}$ are swept up by the field of the reverse-biased junction j_2 and injected into region $p_{③}$ where they are majority carriers. Similarly, the minority electrons in $p_{③}$ are extracted from that region and inserted into region $n_{②}$ where they are majority carriers. The presence of these excess majority carriers in regions $n_{②}$ and $p_{③}$ increases the forward biases across j_1 and j_3, leading to increased injection of minority carriers across j_1 and j_3. With increased forward injection across j_1 and j_3 the process cycles again, leading to even further increases in the minority carrier densities at j_2. When the minority carrier densities on either side of j_2 exceed their equilibrium values, j_2 is forward-biased and the diode is on. The condition that must exist before switching can occur is that the forward biases across j_1 and j_3 inject sufficient numbers of minority carriers into regions $n_{②}$ and $p_{③}$ to saturate the recombination processes. This saturation depends on the width of these base regions and the concentrations of impurities.

In the switching process the p-n-p-n structure may be considered as two transistors with one common junction, j_2. The regions $p_{①}$-$n_{②}$-$p_{③}$ form one transistor with $p_{①}$ the emitter and $p_{③}$ the collector. The other transistor is composed of $n_{④}$-$p_{③}$-$n_{②}$ with $n_{④}$ the emitter and $n_{②}$ the collector. The configuration of this two-transistor model is illustrated in Fig. 10-17. The collector junction of each transistor is j_2. The descriptions of the physical processes involved in switching that were presented in the foregoing paragraphs can be viewed in terms of the behavior of these two equivalent transistors. In the forward blocking state both transistors operate in normal biasing con-

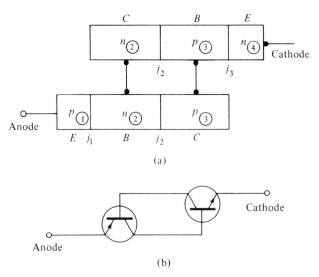

Fig. 10-17 Two-transistor model of a p-n-p-n diode: (a) physical analog, (b) equivalent circuit.

ditions. However, the forward current transfer ratios, the transistor α's, are low because the base regions are wide compared with their respective diffusion lengths. When the injection of minority carriers into the base regions is sufficient to saturate the recombination processes, the transistor α's are increased, leading to coupling between their respective collector and base currents. In the forward conducting state both transistors are saturated, the emitter and collector junctions of each being forward-biased.

There are several ways in which the process of switching from the forward blocking to the forward conducting state may be initiated or *triggered*. The most common method for triggering the two-terminal device is to raise the diode voltage to the threshold value, V_p. With the diode bias at V_p the reverse bias across j_2 initiates an avalanche breakdown that injects majority carriers into the base regions, initiating the regenerative process that ends when the device is in the forward conducting state as described above. Other methods of producing hole-electron pairs at j_2 to initiate switching are heating and optical absorption. A three-terminal *p-n-p-n* device can be triggered by an electrical signal to the third trigger or *gate* lead. The most important of these three-terminal electrically triggered devices is the SCR.

Silicon Controlled Rectifier

The SCR is a *p-n-p-n* device with a gate lead connected to the *p*-type base region, as illustrated in Fig. 10-18. This figure also shows the circuit symbol and *I-V* characteristic of the device. The *I-V* characteristic depends on the gate current, the switching voltage being lowered for positive gate current. Conversely, with the SCR biased in the forward blocking state, application of a positive pulse of gate current can switch the device to the forward conducting state. The amount of gate current required for turn-on is small, usually of the order of milliamperes, while the anode current may be hundreds of amperes. Thus a small current can be used to control a large current. The three-terminal device is therefore a controlled switch or rectifier. It is commonly fabricated from silicon, and hence SCR.

Gate triggering occurs with the application of positive current through the gate lead attached to base region $p_{(3)}$. The holes injected into region $p_{(3)}$ by the gate current increase the forward bias across j_3, thus initiating the transistor action of the $n_{(4)}$-$p_{(3)}$-$n_{(2)}$ transistor. Electrons are injected into the $p_{(3)}$ base across j_3 and diffuse to the collector junction j_2. At j_2 these electrons are swept into region $n_{(2)}$ where they are majority carriers and where they contribute to increased forward bias of the emitter junction, j_1, of the $p_{(1)}$-$n_{(2)}$-$p_{(3)}$ transistor. Holes are then injected across j_1 into the base region $n_{(2)}$. These holes diffuse to j_2 where they are swept into region $p_{(3)}$ and the regenerative cycle begins again.

The time required for switching an SCR from the forward blocking to the

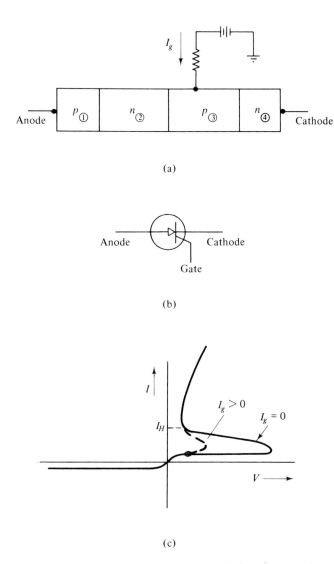

Fig. 10-18 Silicon-controlled rectifier: (a) physical configuration, (b) circuit symbol, (c) I-V characteristic.

forward conducting state is of the order of the transit times of the minority carriers across the base regions $n_{②}$ and $p_{③}$. This turn-on delay time in typical SCRs ranges from 0.1 to 1.0 μsec. Once the SCR is switched to the forward blocking state, the gate current loses control of the anode current. The device is turned off only by reducing the anode current below a critical value called the holding current I_H. When the current is driven below I_H, the device switches to the forward blocking condition with a turn-off time in the range of 1 to 10 μsec. Because the switching times of SCRs are long compared with those of fast switching transistors, they are useful only for relatively low-frequency switching applications. On the other hand, they can handle high power levels and are therefore useful in power switching circuits.

A Bilateral Switch

The p-n-p-n diode and the SCR have I-V characteristic that are not symmetrical with respect to the zero volts-zero amperes condition. In some applications in electronics it is useful to have a device that will switch symmetrically with either reverse or forward bias. One example of such a bilateral switch based on the concepts of the p-n-p-n diode is shown in Fig. 10-19. There are two significant differences between this device and the ordinary p-n-p-n diode. First the $n_{④}$ region extends over only half of the width of the cathode contact. Second, a fifth region, $n_{⑤}$, is diffused into half of the anode region. The whole structure may be considered as two parallel p-n-p-n devices, one consisting of the regions $p_{①}$-$n_{②}$-$p_{③}$-$n_{④}$ and the other of $p_{③}$-$n_{②}$-$p_{①}$-$n_{⑤}$, as illustrated in Fig. 10-19(c). When a positive potential is applied to the anode, the former diode is forward-biased while the latter is reverse-biased. The I-V characteristic that results for $V > 0$ is due primarily to the $p_{①}$-$n_{②}$-$p_{③}$-$n_{④}$ diode. When a negative voltage is applied to the anode the roles of the two p-n-p-n structures are reversed and the I-V characteristic for $V < 0$ is due primarily to the forward-biased $p_{③}$-$n_{②}$-$p_{①}$-$n_{⑤}$ diode. The I-V characteristic sketched in Fig. 10-19(d) is symmetrical about the origin.

BIBLIOGRAPHY

Further readings on the semiconductor devices described in this chapter as well as other devices may be found in the bibliographic listings in Chapters 8 and 9. Additional material on specialized topics may be found in

GENTRY, F. E., F. W. GUTZWILLER, N. HOLONYAK, JR., and E. E. VON ZASTRO, *Semiconductor Controlled Rectifiers: Principles and Application of p-n-p-n Devices.* Englewood Cliffs, N.J.: Prentice-Hall, Inc., 1964.

PROBLEMS

10-1. Derive Eqs. (10-9) and (10-10).

10-2. Obtain Eq. (10-14) from the results of Chapter 9.

10-3. Obtain Eq. (10-23) for the spatial variation of the minority hole density in the base of a p-n-p transistor.

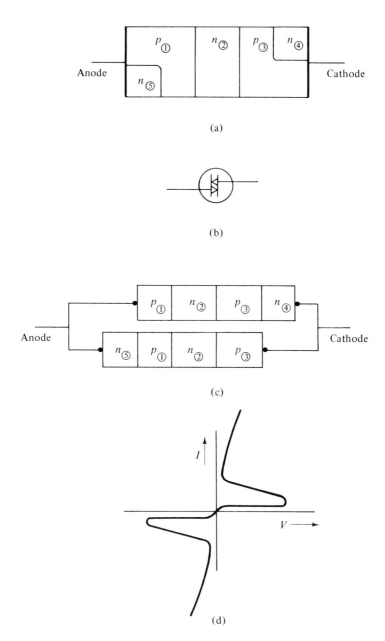

Fig. 10-19 Bilateral switch: (a) physical configuration, (b) circuit symbol, (c) equivalent model of two p-n-p-n diodes, (d) I-V characteristic.

10-4. Find the minority hole diffusion length L_{hB} in the base of a *p-n-p* transistor if $B = 0.98$ and the width of the base is 10^{-3} cm. Estimate the transistor α if the conductivity of the emitter is 100 times greater than that of the base.

10-5. Estimate the common-base hybrid parameters for the transistor whose terminal characteristics are those of Fig. 10-7.

10-6. Derive Eq. (10-55) for the common-emitter forward current transfer ratio h_{fe}.

10-7. Estimate h_{fe} for the transistor of Problem 10-4.

10-8. Estimate the common-emitter hybrid parameters for the transistor whose terminal characteristics are those of Fig. 10-11.

11

Field-Effect Devices

A field-effect transistor (abbreviated FET) is a device in which the bulk conductance of an extrinsic semiconductor is controlled by an electric field applied externally. Two separate mechanisms are used to achieve control of the conductance in practical FET devices. One of these employs the depletion layer of a reverse-biased p-n junction to restrict the physical dimension of a high-conductivity channel within the bulk material. Such a device is called a junction field-effect transistor (JFET). The second mechanism involves the modulation of the conductivity of the medium by the induction or depletion of charge densities in the device through application of a voltage to a capacitive structure. A device utilizing this technique is called either an insulated-gate field-effect transistor (IGFET) or a metal-oxide-semiconductor field-effect transistor (MOSFET). The former term is a generic one, while the latter specifies the physical configuration of the materials employed. In this chapter we shall study the JFET, a metal-insulator-semiconductor capacitive structure, and its application to devices such as the MOSFET.

The fundamental principles of operation of both types of FET differ from those of the ordinary junction transistor. Most importantly the modulation of the conductance in the JFET is a majority carrier effect. The minority carriers effectively do not participate in the transistor action. Carrier flow depends on the majority carrier mobility. In contrast the n-p-n and p-n-p junction transistors are minority carrier devices dominated by minority carrier diffusion. In ordinary transistors both minority and majority carriers participate in the current and must be taken into account in the analysis. Because two types of charge carrier are involved, such a device is said to be *bipolar*. Field-effect devices, on the other hand, are *unipolar* since only the majority carriers need be considered.

11.1 THE JUNCTION FIELD-EFFECT TRANSISTOR

One possible configuration of a discrete JFET is sketched in Fig. 11-1. The device shown consists of an *n*-type semiconductor with ohmic contacts at either end called, respectively, the *source* and the *drain*. On the sides of the bar,

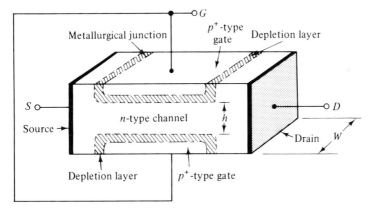

Fig. 11-1 Basic structure of an *n*-channel JFET with $v_{DS} = 0$ and $i_D = 0$.

heavily doped p^+-regions called *gates* are formed. The ohmic contacts to the gates are ordinarily connected together through external circuitry so that bias applied at the gate terminal appears across both of the gate-bulk depletion layers. The region of the *n*-type material between the two depletion layers is called the conduction *channel*. The channel has length L in the direction of current flow between source and drain, width W, and height h between the two depletion layers. The device illustrated in the figure is called an *n*-channel JFET. A junction FET can also be constructed with a *p*-channel and n^+-gates.

The *n*-channel JFET operates in the following way. The primary current flows through the drain contact, the channel, and the source. The terms source and drain are descriptive of majority carrier (electron) flow from source to drain. Field effects occur when bias is applied both to the gate and to the drain. Bias applied to the gate regions appears across the gate-channel depletion layers, affecting their width. For a reverse-biased abrupt junction the width of these space-charge layers varies as the square root of the gate-channel junction voltage. Because the gates are much more heavily doped than the channel, the space-charge layers of width w_n lie entirely on the channel sides of the metallurgical junctions. Furthermore, depletion of mobile charge carriers in the space-charge layers restricts the drain current to flow in the channel between them. Since the resistance of the semiconductor between the source and drain contacts depends on the effective cross-sectional area through which current may flow, this resistance can be modulated by application of bias to the gate. Increasing the magnitude of the reverse bias junction voltage increases the width of the space-charge layers, decreases the channel height, and increases the source-to-drain resistance.

Bias applied between the source and the drain also affects the channel width so that in general the source-drain terminal characteristics are not ohmic. When the potential of the drain is made positive with respect to that of the source, current flows and a spatially dependent potential appears in the channel. The channel near the drain is more positive than the channel near the source. The positive channel potential contributes to reverse-biasing the gate-to-channel junctions. Therefore, since the potential in the gates is spatially uniform, the magnitude of the potential difference between the channel and the gates varies with position in the device. At the drain end of the channel the gate-channel junctions have greater reverse bias voltage across them and hence the depletion layers are wider, as indicated in Fig. 11-2. Therefore increasing the drain potential decreases the height of the conduction channel near the drain and increases the channel resistance.

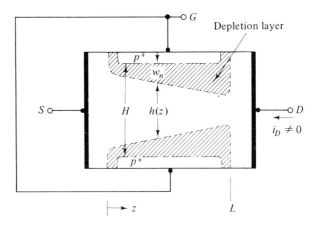

Fig. 11-2 n-Channel JFET with $v_{DS} > 0$ showing variation of the channel height with position.

Common-Source Terminal Characteristics

The JFET is most frequently encountered in circuits in the common-source configuration illustrated in Fig. 11-3. The circuit symbol includes an

Sec. 11-1 The Junction Field-Effect Transistor 345

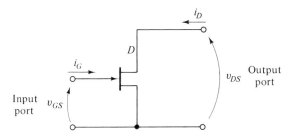

Fig. 11-3 Common-source configuration of an *n*-channel JFET.

arrow/line combination that indicates a *p-n* gate-to-source junction. For a *p*-channel JFET the direction of the arrow is reversed. If, unlike the one shown in Fig. 11-1, the practical device is constructed with asymmetries between the source and drain, the drain lead will be indicated by a letter *D*. The symbol of Fig. 11-3 represents therefore an *n*-channel asymmetrical JFET.

The input gate current of an FET under ordinary operating conditions is very small. The input impedance of a typical silicon JFET at room temperature may be 10^9 to 10^{10} Ω. The high gate impedance occurs because the gate junctions are always reverse-biased and current flows readily between the source and the drain. Under dynamic conditions, the a.c. gate current charges the junction space-charge layer capacitance.

The d.c. drain characteristic of a typical symmetrical *n*-channel JFET, shown in Fig. 11-4, is a plot of the drain current i_D vs. the voltage of the drain relative to the source v_{DS} for several values of the voltage of the gate relative to the source v_{GS}. The curves are shown only in the first quadrant but also

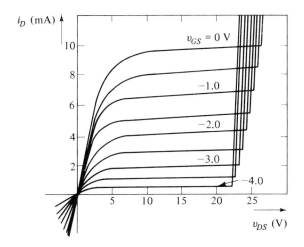

Fig. 11-4 Typical output characteristics of a common-source *n*-channel JFET.

appear in the third quadrant. The characteristic of a symmetrical device is symmetrical about the origin.

From the characteristic it is seen that for low values of drain voltage v_{DS}, the source-to-drain resistance, given by the inverse of the slope of a characteristic curve, depends strongly on the gate bias v_{GS}. Operated in this part of the characteristic the FET is a voltage-controlled resistor, the control being v_{GS}. For fixed v_{GS} and increasing values of v_{DS} the drain resistance increases until the current saturates. Saturation occurs when the channel height reaches its smallest value at the drain end of the channel. It turns out that the minimum channel height is small but not zero. Finite current flows through a very narrow channel. *Pinch-off* is the term used to describe conditions under which the current channel is so restricted. The pinch-off voltage is defined as the lowest value of v_{DS} at which pinch-off occurs with $v_{GS} = 0$. That is, at the pinch-off voltage the channel is constricted to its minimum height at the drain end. For higher values of drain voltage, the drain current remains almost constant, but the length of that part of the channel over which pinch-off occurs increases. Figure 11-5 shows the configuration of a

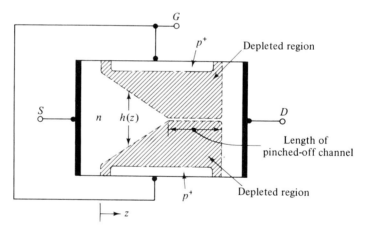

Fig. 11-5 n-Channel JFET under conditions of drain pinch-off.

JFET under pinch-off conditions. The drain current is relatively insensitive to high drain voltages in pinch-off because the current of the carriers injected from channel through the pinched-off region is space-charge-limited. At even higher drain voltages avalanche breakdown occurs in the space-charge layers, giving rise to rapid increases in drain current. Notice how the values of v_{DS} at which pinch-off and breakdown begin depend on the gate bias v_{GS}. As an amplifier the FET is biased to operate in the pinch-off region where small changes in the gate voltage produce large changes in the drain current and voltage, depending on the load line.

Pinch-off may also occur if sufficient reverse bias is applied to the gates, reducing the channel height to its minimum value. During gate pinch-off, however, the channel is uniformly constricted. The drain current flowing under gate pinch-off conditions may be as little as 10^{-9} A.

Quantitative Analysis of the JFET

An analysis of the characteristic of the JFET can be based on the dependence of the resistance of the drain circuit on the channel height. Referring to Fig. 11-2 we find that the channel height h is

$$h(z) = H - 2w_n(z) \tag{11-1}$$

where H is the separation of the gate metallurgical junctions and $w_n(z)$ is the width of each depletion layer. For an abrupt planar junction $w_n(z)$ is given by Eq. (9-34),

$$w_n = \left\{ \frac{-2\varepsilon[\phi_0 + v_{GC}(z)]}{eN_d[1 + (N_d/N_a)]} \right\}^{1/2} \tag{11-2}$$

where $v_{GC}(z)$ is the potential of the p-type gate with respect to the n-type channel. The potential $v_{GC}(z)$ depends not only on the gate voltage v_{GS} but also on the drain voltage v_{DS}, which establishes a potential gradient in the n-channel. If the potential at a position z in the channel relative to the source is designated $v_{CS}(z)$, then from the diagram of Fig. 11-6 it is clear that

$$v_{GC}(z) = v_{GS} - v_{CS}(z) \tag{11-3}$$

For simplicity we shall assume that the potential difference v_{DS} appears only across the region between the space-charge layers and that there are no potential drops in the channel near the source and drain contacts. In mathematical

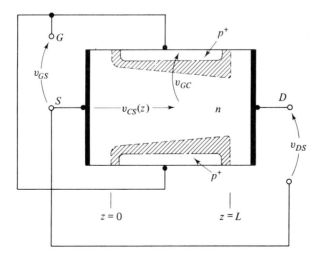

Fig. 11-6 Potentials in an n-channel JFET.

terms

$$v_{CS}(z=0) = 0 \atop v_{CS}(z=L) = v_{DS}.\rbrace \quad (11\text{-}4)$$

Applying Eqs. (11-2) and (11-3) to Eq. (11-1) we find for heavily doped gates, $N_a \gg N_d$,

$$h(z) = H - 2\left\{\frac{-2\varepsilon[\phi_0 + v_{GS} - v_{CS}(z)]}{eN_d}\right\}^{1/2} \quad (11\text{-}5)$$

This result is only valid provided it does not predict a negative value for h since its minimum value cannot be less than zero at pinch-off.

Gate pinch-off occurs when the channel height is reduced effectively to zero due to the application of gate bias with $v_{DS} = 0$. If the drain is short-circuited to the source, there will be no potential gradient along the channel. The gate voltage at which gate pinch-off occurs is denoted by V_{GP} and can be found by setting $h = 0$ and $v_{CS} = 0$ in Eq. (11-5). That is,

$$V_{GP} = -\phi_0 - \frac{eN_d H^2}{8\varepsilon} \quad (11\text{-}6)$$

We shall define the combination of constants in the second term of the right as V_P

$$V_P = \frac{eN_d H^2}{8\varepsilon} \quad (11\text{-}7)$$

and note that ordinarily $V_P \gg \phi_0$ so that

$$V_{GP} \simeq -V_P \quad (11\text{-}8)$$

The definition Eq. (11-7) may also be used to rewrite Eq. (11-4) for $h(z)$, resulting in

$$h(z) = H\left\{1 - \left[\frac{v_{CS}(z) - \phi_0 - v_{GS}}{V_P}\right]^{1/2}\right\} \quad (11\text{-}9)$$

Again we should re-emphasize that Eq. (11-9) is valid only when $h \geq 0$. Since the maximum potential in the channel is v_{DS}, the restriction on the validity of this expression is

$$v_{DS} - \phi_0 - v_{GS} \leq V_P \quad (11\text{-}10)$$

Recall that $\phi_0 < 0$ and that $v_{GS} \leq 0$ for an n-channel JFET.

Near the origin of the drain characteristics of a JFET the curves of Fig. 11-4 are approximately straight lines. Because of this linear relationship between i_D and v_{DS}, this is called the ohmic region of operation. The drain resistance, inversely proportional to the slope of the characteristic, depends on the gate voltage. An analysis of the ohmic region proceeds as follows.

We shall assume that v_{DS} is much smaller than the magnitude of v_{GS} or equivalently that the potential in the channel v_{CS} can be neglected. As a consequence the channel height h is assumed to be unaffected by v_{DS} and hence

Sec. 11-1 The Junction Field-Effect Transistor

spatially uniform. The resistance of the conducting channel of length L, width W, and height h is

$$R = \frac{L}{\sigma W h} \quad (11\text{-}11)$$

where σ is the n-channel conductivity due to the majority electrons of density n and mobility μ. Using Eq. (11-9) for h with $v_{CS} = 0$ and Eq. (11-11) for R, Ohm's law for the channel predicts

$$i_D = \frac{\sigma W h}{L}\left[1 - \left(\frac{-\phi_0 - v_{GS}}{V_P}\right)^{1/2}\right]v_{DS} \quad (11\text{-}12)$$

provided $v_{DS} \ll |v_{GS}|$. It is in this region of the characteristic that a JFET can be used as a voltage-controlled resistance.

As the drain voltage increases for fixed gate bias the channel resistance increases due to the constriction of the current channel by the potential drop between the drain and the source. Eventually the channel resistance becomes very large when the channel height is reduced to its minimum value near the drain and the current saturates. To analyze the characteristic of the JFET between the ohmic region and saturation we shall divide the channel into thin slices of differential length dz, width W, and height $h(z)$. The resistance of each slice is

$$dR = \frac{dz}{\sigma W h(z)} \quad (11\text{-}13)$$

The drain current flows through each of these segments in series so that the differential voltage drop in the channel is

$$dv_{CS} = i_D dR \quad (11\text{-}14)$$

Solving for i_D, we find

$$i_D = \sigma W h(z)\frac{dv_{CS}}{dz} \quad (11\text{-}15)$$

The dependence of i_D on v_{DS} and v_{GS} may be determined by integrating Eq. (11-15) along the active length of the channel:

$$\int_0^L i_D\, dz = \int_0^L \sigma W h(z)\frac{dv_{CS}}{dz} dz \quad (11\text{-}16)$$

Because i_D is independent of position, the left-hand side of Eq. (11-16) is $i_D L$. The right-hand side is equivalent to an integral over the channel voltage,

$$\int_{v_{cs}(z=0)}^{v_{cs}(z=L)} \sigma W h(z)\, dv_{CS}$$

where the limits of integration are given by the assumption of Eq. (11-4). Using Eq. (11-9) for h, Eq. (11-16) is

$$i_D L = \sigma W H \int_0^{v_{DS}}\left\{1 - \left[\frac{v_{CS}(z) - \phi_0 - v_{GS}}{V_P}\right]^{1/2}\right\}dv_{CS} \quad (11\text{-}17)$$

Performing the required integration and substituting for H from Eq. (11-9) yields

$$i_D = \frac{\sigma W}{L}\left(\frac{8\varepsilon}{eN_d}\right)^{1/2}\left[V_P^{1/2}\, v_{DS} - \frac{2}{3}(v_{DS} - \phi_0 - v_{GS})^{3/2}\right.$$
$$\left. + \frac{2}{3}(-\phi_0 - v_{GS})^{3/2}\right] \qquad (11\text{-}18)$$

which is a rather complicated expression. Equation (11-18) indicates that i_D increases with v_{DS} to a maximum value and then decreases with a further increase in v_{DS}, whereas the actual experimental characteristic is a monotonically increasing function of drain voltage. Fortunately it is easy to show that the maximum drain current predicted by Eq. (11-18) occurs for

$$v_{DS}\big|_{max} = V_P + v_{GS} + \phi_0 \qquad (11\text{-}19)$$

Comparison of Eq. (11-19) with Eq. (11-10) reveals that this value of v_{DS} is the maximum one permitted in this analysis. When v_{DS} satisfies Eq. (11-19) pinch-off occurs at the drain end of the channel. Further increase in v_{DS} causes the pinch-off region to extend toward the source without substantial change in drain current. Therefore Eq. (11-18) is valid only for values of v_{DS} equal to or less than the maximum value given by Eq. (11-19).

The saturated drain current is determined from Eq. (11-18) with the value of v_{DS} given by Eq. (11-19). After some manipulation we find

$$i_{D,\text{sat.}} = \frac{\sigma W H}{L}\left\{\frac{1}{3}V_P + (\phi_0 + v_{GS})\left[1 + \frac{2}{3}\left(\frac{-\phi_0 - v_{GS}}{V_P}\right)^{1/2}\right]\right\} \qquad (11\text{-}20)$$

which depends on the gate bias v_{GS}.

The drain voltage at which saturation occurs for $v_{GS} = 0$ is called the drain pinch-off voltage V_{DP}. From Eq. (11-19) we see that

$$V_{DP} \simeq V_P \qquad (11\text{-}21)$$

since V_P is ordinarily much greater than the contact potential. Notice that the magnitudes of the gate and drain pinch-off voltages are theoretically the same. In practice it is difficult to determine the drain pinch-off voltage from the terminal characteristics.

Circuit Representations

In most applications JFETs are biased in the region of saturated drain current and operated as a two-port network in the common-source configuration of Fig. 11-3. If we consider the gate and drain currents to be functions of the gate and drain voltages, we can write

$$i_D = f_1(v_{DS}, v_{GS}) \qquad (11\text{-}22)$$

and

$$i_G = f_2(v_{DS}, v_{GS})$$

Sec. 11-1 The Junction Field-Effect Transistor

In practice the input impedance to an FET is so high that the gate current can be ignored. Furthermore, we should be cautioned that if the device is biased in the saturation region of the characteristic, the relationship of Eq. (11-22) is not that given by Eq. (11-18) because the latter is valid only for voltages below pinch-off.

For small-amplitude changes in current and voltage about the d.c. operating point determined by the biasing circuit we can perform a Taylor expansion of the functional form of Eq. (11-22) to write

$$i_d = \frac{1}{r_d} v_{ds} + g_m v_{gs} \qquad (11\text{-}23)$$

where the lowercase subscripts indicate a.c. components of current and voltage. The parameter r_d is the *drain resistance* defined by

$$\frac{1}{r_d} \equiv \left. \frac{\partial i_D}{\partial v_{DS}} \right|_{v_{GS}=\text{constant}} \qquad (11\text{-}24)$$

and g_m is the *transconductance* between the input voltage and the output current:

$$g_m = \left. \frac{\partial i_D}{\partial v_{GS}} \right|_{v_{DS}=\text{constant}} \qquad (11\text{-}25)$$

Both r_d and g_m are evaluated at the d.c. operating point of the device. Two incremental equivalent circuits for small-amplitude low-frequency a.c. signals based on Eq. (11-23) are shown in Fig. 11-7. These two circuits are the Thevenin and Norton equivalent circuits of one another. As is evident from the

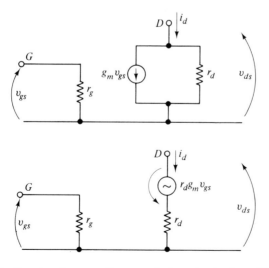

Fig. 11-7 Two low-frequency linear equivalent circuits for a JFET.

second of these circuits the product $r_d g_m$ acts like an open drain voltage amplification factor. An input gate impedance r_g is shown in each circuit, but it is so high in practical cases that the input may be considered an open circuit.

At higher frequencies capacitive effects must be taken into account in constructing a circuit model of the device. The space-charge layer capacitances of the reverse-biased gate-channel junctions are distributed along the length of the junction interfaces. In addition, stray capacitances exist between the leads and contacts. All of these capacitances may be taken into account by adding three capacitances between the three terminals as indicated in Fig. 11-8. The depletion layer capacitances are included in C_{gd} and C_{gs}.

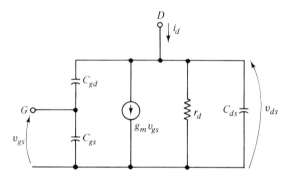

Fig. 11-8 High-frequency incremental equivalent circuit for a JFET.

Noise

The JFET produces less noise than the junction transistor. The primary noise source in an FET is the thermal noise of the channel resistance. Shot noise does not occur in a JFET because no current flows across the gate-channel junctions. An important source of the thermal noise in the JFET is the resistance of the channel near the source since this noise is amplified by the transistor action. The noise can be reduced by offsetting the gates toward the source, thus decreasing the length of the resistive channel and the resistance producing the noise that is amplified.

Junction FETs have better thermal stability than ordinary transistors because the former are unipolar and the latter bipolar devices. Minority carrier transistors are sensitive to thermal generation of intrinsic hole-electron pairs, but the majority carrier densities in unipolar devices are less sensitive to temperature, the impurities being fully ionized. What sensitivity that does occur is due to the inverse dependence of the majority carrier mobility on the three-halves power of the temperature.

11.2 MOS CAPACITOR

In addition to those of the *p-n* junction, the electrical properties of the metal-insulator-semiconductor structure are of technological and economic importance to practical device applications in electronics. The structure consists of a thin layer of insulating material between a semiconductor and a metal plate in a planar geometry. Because the insulator is in many cases an oxide of the semiconductor, the configuration is more commonly referred to as a metal-oxide-semiconductor (MOS) structure. Silicon nitride, Si_3N_4, is also widely used as an insulating layer.

An MOS device is constructed in the following manner. An extrinsic semiconductor is prepared and its surface oxidized by heating the semiconductor in an oxygen-enriched atmosphere. Silicon is the most commonly used semiconductor because its properties are appropriate and silicon technology is highly developed. Silicon dioxide is an excellent insulator with an energy band gap of about 8 V, a relative dielectric constant of 3.9, and a dielectric breakdown field of 6×10^6 V/cm. The metal, ordinarily aluminum, is deposited on top of the oxide layer by evaporation in an evacuated chamber. Additional details on the fabrication of MOS devices will be described in the next chapter.

In this section we shall study the electrical properties of the MOS configuration and in the next section those of the two types of field-effect transistors utilizing the MOS structure to provide insulation between the gate lead and a conducting channel. These devices are variously referred to as the *insulated-gate field-effect transistor* (IGFET) or the *metal-oxide-semiconductor field-effect transistor* (MOSFET).

The MOS configuration illustrated in Fig. 11-9 is essentially that of two

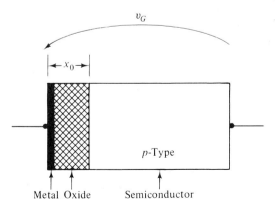

Fig. 11-9 Geometry of an MOS capacitor with a *p*-type semiconductor.

regions of high conductivity separated by a dielectric insulator. A metal plate and an extrinsic semiconductor comprise the regions of high conductivity. The figure indicates a p-type semiconductor but an n-type would do as well. The capacitance of such a structure depends on the voltage v_G applied to the metal plate relative to the semiconductor.

An electron energy band diagram for an unbiased MOS capacitor is sketched in Fig. 11-10. This diagram is representative of the practical case

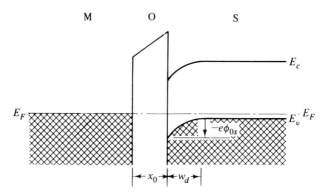

Fig. 11-10 Energy band diagram of an unbiased MOS capacitor with a p-type semiconductor.

utilizing aluminum and p-type silicon with silicon dioxide as the insulator. In equilibrium a Fermi energy E_F is defined and is a constant throughout the structure. The energies at the top of the valence band and the bottom of the conduction band in the semiconductor are denoted E_v and E_c, respectively. Differences in the work functions of the metal, the oxide, and the semiconductor create contact potential fields in the oxide and near the surface of the semiconductor. The contact fields cause the electronic energy bands to bend. In the oxide the field is uniform because the conductivity is very low; thus the slope of the energy bands is constant. In the semiconductor, however, a redistribution of the mobile charge carriers screens the interior of the semiconductor from this electric field so that it appears only near the surface.

For the configuration illustrated in Fig. 11-10 the contact field creates a depletion layer of width w_d at the surface of the p-type semiconductor. This depletion layer is deficient in mobile holes and thus has a net electric space charge residing on the "ionized" acceptor impurity atoms. Assuming the complete depletion of majority carriers in the region, the width of the space-charge layer w_d turns out to be the same as that on the p-side of an abrupt p-n^+ junction. If we denote the equilibrium contact potential in the semiconductor as ϕ_{os} and adapt Eq. (9-33) to the present purposes, we can write the width of the depletion layer in equilibrium as

$$w_d = \left[\frac{2\varepsilon_s \phi_{0s}}{eN_a}\right]^{1/2}, \quad v_G = 0 \quad (11\text{-}26)$$

Sec. 11-2 MOS Capacitor

where ε_s is the dielectric constant of the semiconductor. Because of the voltage convention of Fig. 11-9 it is convenient to measure ϕ_{0s} as the potential of the metal relative to that of the semiconductor in equilibrium. Therefore for a p-type semiconductor ϕ_{0s} is positive.

The capacitance of the structure of Fig. 11-10 is that of two conductors separated by two nonconducting layers in series, one the thickness w_o of the insulating oxide and the other the thickness w_d of the carrier-depleted space-charge layer. That is, the effective capacitance between the metal and the bulk of the semiconductor is given by

$$\frac{1}{C} = \frac{1}{C_{\text{oxide}}} + \frac{1}{C_d} \tag{11-27}$$

where the capacitance of the oxide layer is

$$C_{\text{oxide}} = \frac{\varepsilon_{\text{oxide}} A}{w_o} \tag{11-28}$$

and the capacitance of the depletion region is

$$C_d = \frac{\varepsilon_s A}{w_d} \tag{11-29}$$

In Eq. (11-28) the dielectric constant of the oxide is denoted $\varepsilon_{\text{oxide}}$ while its width is w_o.

When a bias voltage v_G is applied to the metal plate some of this potential appears across the space-charge layer of the semiconductor and some across the oxide. The portion of v_G that appears in the semiconductor, v_s, alters the width of the depletion layer according to

$$w_d = \left[\frac{2\varepsilon_s(\phi_{0s} + v_s)}{eN_a}\right]^{1/2} \tag{11-30}$$

provided w_d remains positive. Thus the capacitance of the MOS structure, determined by Eq. (11-27) with Eqs. (11-28) – (11-30), depends on the voltage applied across it.

The effect of a positive voltage applied to the metal contact of an MOS capacitor utilizing a p-type semiconductor is illustrated by the energy band diagrams of Fig. 11-11. The widening of the depletion layer is indicated in both sketches. Figure 11-11(a) is drawn for small positive bias voltage v_G. This condition is called the *depletion mode* of operation. As the magnitude of the positive bias is increased a new phenomenon occurs known as *inversion*. As the potential drop in the semiconductor is increased the bending of the bands increases. The minority carrier electrons find conduction band states at lower energy near the surface of the semiconductor where the bands are strongly bent. Not only is it natural for these electrons to occupy these states because they are lower in energy than conduction band states within the bulk of the semiconductor, but also the electric field in the space-charge layer is of the polarity to attract electrons to the surface of the semiconductor.

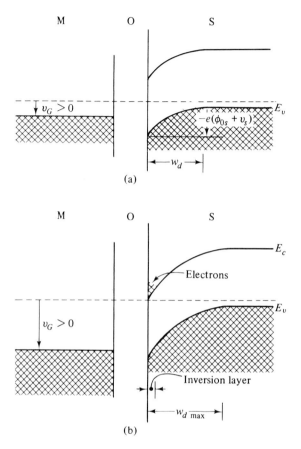

Fig. 11-11 Energy band diagrams for the MOS capacitor of Fig. 11-9 biased in the (a) depletion and (b) inversion modes.

When the density of electrons at the surface exceeds the density of acceptor impurities, the surface layer of the semiconductor just under the oxide has the properties of an n-type semiconductor since the density of mobile electrons is much greater than the concentration of holes there. Thus in the steady state the applied voltage has induced an n-type layer in the p-type semiconductor, giving rise to the term inversion. The distribution of charge in the MOS capacitor of Fig. 11-9 in the steady-state inversion mode is indicated in Fig. 11-12.

Inversion occurs when the density of the mobile electrons is of the order of the density of the immobile negative charge on the ionized acceptor atoms. Because the density of the ionized acceptors is the same as that of the majority carrier holes in the bulk of the semiconductor, the density of mobile electrons in the inversion layer is approximately equal to or in excess of that

Sec. 11-2 MOS Capacitor

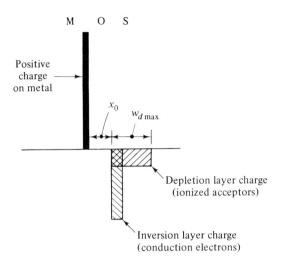

Fig. 11-12 Charge distribution in the MOS capacitor of Fig. 11-9 under inversion conditions.

of the holes in the bulk. Therefore the conductivity of the inversion layer is high.

The applied bias at which inversion first occurs is called the *turn-on voltage*, $v_G = V_T$. The potential drop in the semiconductor when $v_G = V_T$ is approximately equal to the energy gap measured in volts. Actually the potential across the depletion layer when inversion occurs is given by

$$e(\phi_{0s} + v_{si}) = E_g - 2E_F \quad \text{inversion} \quad (11\text{-}31)$$

where E_F is the Fermi energy of the equilibrium capacitor measured relative to the top of the valence band in the bulk. If the Fermi energy were a valid concept under these nonequilibrium conditions, we would say that the Fermi level at the edge of the inversion layer is as far below the bottom of the conduction band as the Fermi level is above the top of the valence band far from the surface. If the bulk material is *p*-type, the inversion layer is *n*-type.

As the bias voltage is increased beyond the turn-on voltage the bending of the bands increases, increasing the number of electrons in the inversion layer and hence increasing its conductivity. Because the additional positive charge placed on the metal by the increased bias can easily be matched by an increase in electrons in the inversion layer, the width of the depletion layer is unaffected by v_G in the inversion mode. That is, the maximum value of w_d is given by Eq. (11-30) with Eq. (11-31):

$$w_{d_{max}} = \left[\frac{2\varepsilon_s(E_g - 2E_F)}{e^2 N_a}\right]^{1/2} \quad (11\text{-}32)$$

The capacitance of the ideal MOS configuration under inversion operation is therefore independent of the bias voltage.

The foregoing description of the capacitance of an MOS structure under inversion is valid provided the time scale of the application of the potential is shorter than that required for thermal generation of significant quantities of carrier pairs. That is, for frequencies higher than the generation frequency, the capacitance of the MOS structure is that of two insulators in series, the oxide and the depletion layer of width as given by Eq. (11-32). At lower frequencies the presence of the thermally-generated carriers alters the form of the space-charge layer, making it much thinner than in the absence of these carriers and thus causing the capacitance of the structure to be essentially that of the oxide alone.

If the bias voltage of the metal relative to the semiconductor for the capacitor of Fig. 11-9 is made negative, the width of the space-charge layer decreases until it vanishes. When v_G is increased even further beyond this flatband condition the energy band diagram resembles that of Fig. 11-13. The

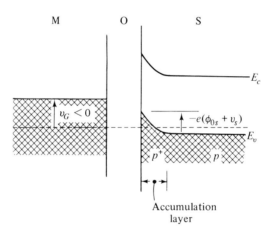

Fig. 11-13 Energy band diagram for the MOS capacitor of Fig. 11-9 biased in the accumulation mode.

electric field applied to the semiconductor across the oxide is of the opposite polarity to that of the depletion or inversion modes. This field causes valence band states near the oxide to rise above their energy levels in the bulk. The majority holes tend to accumulate in these states of lower hole energy. Furthermore, the electric field near the surface of the semiconductor is of the polarity to attract holes to the surface. The conductivity of the semiconductor in the region where the holes accumulate is increased over that of the bulk due to the increased concentration of mobile charge carriers in the accumulation layer. The accumulation layer has the properties of a p^+-type semiconductor. This condition of operation, as illustrated in Fig. 11-13, is called the *accumulation mode* since the majority carriers accumulate near the surface, enhancing the conductivity of this thin layer.

The capacitance of the MOS structure operated in the accumulation mode is due to the oxide layer separating the metal and the accumulation region. Since the depletion layer does not exist, the total capacitance is C_{oxide}. The bulk of the semiconductor acts as a resistive material between the accumulation layer and its ohmic contact to the external biasing circuit.

The variation of an MOS capacitance with bias voltage is shown in Fig. 11-14. This plot is for a capacitor structure with a p-type semiconductor. A

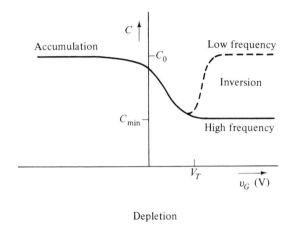

Fig. 11-14 MOS capacitance as a function of gate voltage.

similar figure is obtained for n-type MOS capacitors with the polarity of v_G reversed. That is, for an MOS capacitor utilizing an n-type semiconductor, inversion occurs for negative bias, $v_G < 0$, when the magnitude of v_G exceeds a turn-on voltage. Similarly, accumulation occurs for positive bias.

The MOS structure has two technologically important uses other than as a capacitor. One of these is as a charge-control element. We shall return to the subject of charge-control devices in Section 11.4. The second is the MOSFET, which utilizes the ability of the biasing voltage to modulate the conductance of the semiconductor by affecting the concentration of mobile charges near its surface under the oxide. Because the MOSFET is the more important of these two applications, we shall study it next.

11.3 METAL-OXIDE-SEMICONDUCTOR FIELD-EFFECT TRANSISTORS

There are two types of MOSFET. One is designed to utilize depletion and accumulation modes of operation; it is called the depletion-type MOSFET. The second is designed to operate in an inversion mode; this type is known as the enhancement-type MOSFET. We shall consider the operating principles and characteristics of each in turn.

Depletion-Type MOSFET

A possible physical configuration of an n-channel depletion-type MOSFET is illustrated in Fig. 11-15, which also shows the circuit symbol for this device in a common-source configuration. The source and drain are

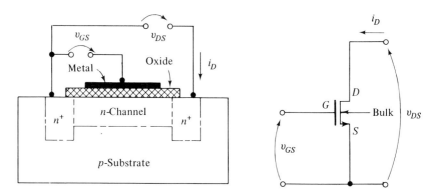

Fig. 11-15 Physical configuration and circuit symbol of an n-channel depletion-type MOSFET.

heavily doped n^+-regions diffused into a p-type substrate. An n-type channel connects the source and drain. The channel may be formed by diffusion or by direct implantation of energetic impurity ions. The latter process is called ion implantation and will be described in Chapter 12 in connection with the fabrication of integrated circuits. An MOS capacitive structure is formed using the n-type channel. The metal electrode of the capacitor is the gate. It is also possible to fabricate a depletion-type MOSFET with a p-type channel on an n-type substrate with p^+ source and drain. The circuit symbol of this device distinguishes between the n-channel and p-channel versions by employing an arrowhead to indicate the direction from p to n at the channel-substrate junction. For the purposes of illustrating the operating principles of the device we shall concentrate on the n-channel configuration of Fig. 11-15. Similar effects occur in p-channel devices with appropriate changes in polarity.

With the gate voltage v_{GS} equal to zero, drain current flows between the n^+ drain and source regions through the conductive n-channel in response to an applied drain voltage v_{DS}. Current does not flow through the substrate because of the p-n junction separating the source, drain, and channel from it. The resistance of the channel, and hence the drain current, depends on the

Sec. 11-3 Metal-Oxide-Semiconductor Field-Effect Transistors 361

drain voltage in a manner that we shall now describe. Because of the differences in the work functions of the metal, oxide, and semiconductor, a depletion layer, illustrated in Fig. 11-16(a), generally exists in the semiconductor

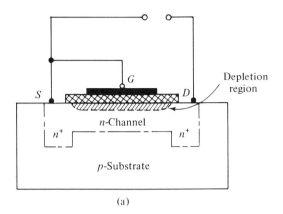

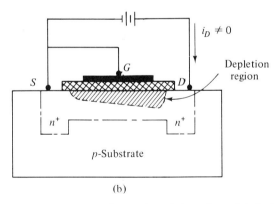

Fig. 11-16 Depletion mode of operation of the n-channel depletion-type MOSFET of Fig. 11-15: (a) without and (b) with drain current.

just under the oxide even when the applied voltage is zero. The width of this depletion layer determines the height of the conducting channel between the source and the drain and hence determines the channel resistance. As positive drain voltage is applied the potential drop in the channel due to the drain current causes the gate-to-channel voltage to become a function of position. If positive bias is applied to the drain with the gate shorted to the source, the potential of the gate relative to the channel will be negative and of greater magnitude at the drain end. Thus the width of the depletion layer will be a function of position along the channel, as pictured in Fig. 11-16(b). As the drain end of the channel is constricted with increasing v_{DS} the resistance of the channel increases in much the same manner as occurs in the JFET. Eventually the MOS depletion layer comes close to the space-charge layer of the channel-substrate junction and the current saturates. These comments also apply for nonzero gate voltage v_{GS} provided this bias is applied so that the MOS structure is operating in the depletion mode. For an n-channel device the depletion mode of operation essentially requires v_{GS} less than or equal to zero. Actually the depletion layer may still be maintained for slightly positive v_{GS}.

On the other hand, if v_{GS} is made sufficiently positive that the MOS structure operates in an accumulation mode, the physics of the conduction mechanism changes. In the accumulation mode a thin layer of enhanced electron density is formed just under the oxide surface. If these electrons are highly mobile, the accumulation layer will have n^+-properties and therefore will offer a lower resistance to current flow between the source and drain than the remainder of the channel. The degree to which the accumulated carriers are

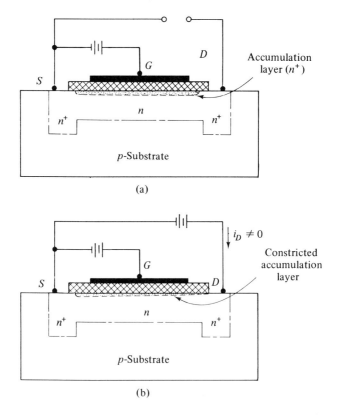

Fig. 11-17 Accumulation mode of operation of the n-channel depletion-type MOSFET of Fig. 11-15: (a) without and (b) with drain current.

mobile depends on the presence of imperfection states that act as traps near the surfaces of both the semiconductor and the oxide. An objective of the technological art is the fabrication of clean surfaces with few traps so that most of the accumulated charge can contribute to enhanced conductivity.

Figure 11-17(a) shows the accumulation layer in a n-channel MOSFET with the drain circuit open. When drain voltage v_{DS} is applied, current flows and a potential drop appears along the channel. If v_{DS} is positive, the channel potential causes the gate to be at a smaller positive potential with respect to the channel at the drain end than at the source end. That is, the accumulation layer at the drain end becomes constricted and the channel resistance increases. At a particular drain voltage that depends on the gate bias the accumulation layer at the drain end of the channel vanishes and the current saturates. Further increases in v_{DS} result in little change in i_D. The current is determined by the space-charge-limited flow of carriers in the accumulation layer injected through the constricted region to the drain.

The drain characteristics of a typical n-channel depletion-type MOSFET are shown in Fig. 11-18. The regions of the characteristic in which operation is determined by accumulation or depletion in the channel are indicated in the figure.

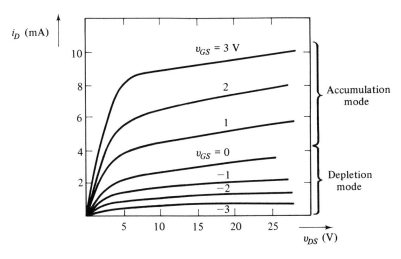

Fig. 11-18 Typical drain characteristic of an n-channel depletion-type MOSFET.

Enhancement-Type MOSFET

The physical configuration of an enhancement-type MOSFET is illustrated in Fig. 11-19. This type of MOSFET differs from the depletion type of Fig. 11-15 by the absence of the diffused or implanted n-channel between

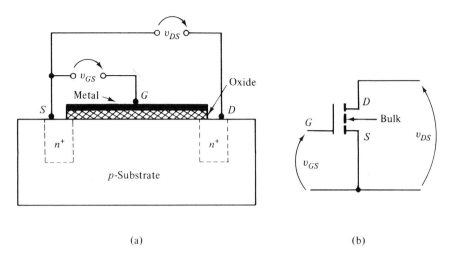

Fig. 11-19 Enhancement-type MOSFET: (a) physical configuration, (b) circuit symbol.

the source and the drain. A p-type enhancement MOSFET can be constructed by diffusing a p^+ source and drain into an n-type substrate. We shall, however, concentrate on the device illustrated in Fig. 11-19.

The MOS capacitive gate structure of the enhancement-type MOSFET is designed to operate in the inversion mode. That is, gate bias v_{GS} is applied to create an n-type inversion layer under the oxide, thus forming a conductive channel between the source and drain. Without the induced inversion layer very little drain current flows because a postive drain voltage v_{DS} results in reverse bias across the drain-to-substrate junction, especially if the substrate is electrically connected to the source. To induce an inversion layer in the p-type substrate a positive voltage exceeding the turn-on voltage must be applied to the gate. An inversion layer in the MOSFET of Fig. 11-19 is illustrated in Fig. 11-20(a) for $i_D = 0$. If drain voltage v_{DS} is applied, drain current will flow through the inversion layer provided it extends from source to drain. For this reason the metal gate must be made to align perfectly with or to overlap the n^+ source and drain regions. The proximity of the gate to the highly conducting source and drain results in harmful capacitances. In depletion-type MOSFETs the gate need not overlap the drain or source so the degrading effects of these capacitances are avoided.

Modulation of the gate voltage v_{GS} affects the resistivity of the inversion layer not only through the width of the conduction path but also through changes in the electron density. If the device is carefully fabricated, most of the carriers in the inversion layer contribute to the current. Drain current also affects the resistance of the inversion channel because of the potential drop between source and drain. As in the accumulation mode of a depletion

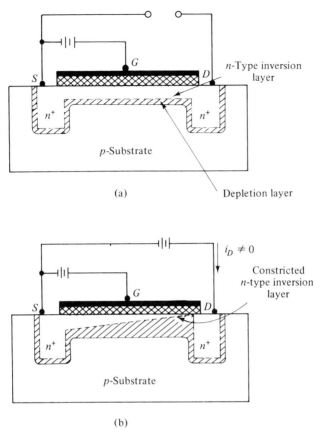

Fig. 11-20 Inversion mode of operation of the enhancement-type MOSFET of Fig. 11-19: (a) without and (b) with drain current.

MOSFET, drain current causes the gate-to-substrate voltage to decrease toward the drain end of the MOS capacitor. An inversion layer for $v_{DS} > 0$ is shown in Fig. 11-20(b). As v_{DS} increases the inversion layer is constricted, and the channel resistance increases, until at saturation the inversion layer disappears at the drain end. For higher values of drain voltage the current is relatively unaffected. Under saturated conditions the current is due to carriers in the inversion layer near the source that are injected through the depleted region into the drain.

The drain characteristics of a typical enhancement-type MOSFET with a p-type substrate are shown in Fig. 11-21. Notice that drain current flows only for positive gate voltages large enough to create an inversion layer. The terminal characteristics of the two types of MOSFET, illustrated in Figs. 11-18 and 11-21, are very similar in structure. Although the physical mech-

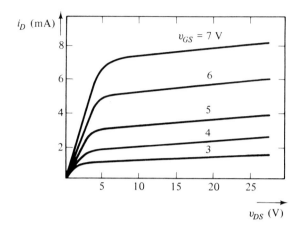

Fig. 11-21 Typical drain characteristic of an enhancement-type MOSFET with a p-type substrate.

anisms of depletion, accumulation, and inversion differ, the variations of the source-to-drain resistances by the field effect each have their source in the spatial dependence of the channel potential due to the flow of drain current.

Circuit Representations of MOSFETs

The circuit representations of either type of MOSFET biased in the saturation region are similar to those of the JFET since like the JFET the drain current in a MOSFET is considered to be a function of the gate voltage and the drain voltage. The low-frequency gate current is negligibly small.

The linear incremental equivalent circuit for small-amplitude a.c. signals about the d.c. operating point of a MOSFET has the same structure as that for a JFET. Two equivalent circuits are shown in Fig. 11-7. It is interesting to compare the range of parameters of MOSFETs and JFETs. The drain resistances r_d of typical MOSFETs are lower than those of JFETs as is evident upon comparison of the I-V characteristics of Figs. 11-4, 11-18, and 11-21. In MOSFETs r_d is on the order of 10^3 to 10^5 Ω while in JFETs it is 10^5 to 10^6 Ω. The transconductances g_m of the two types of FET are about the same, lying in the range of 10^3 to 10^4 μmhos. Some MOSFETs have slightly larger values. The input gate resistances of MOSFETs are much higher than those of JFETs because of the oxide layer in the MOS structure. The input resistance of MOSFETs lie in the range from 10^{12} to 10^{15} Ω while those of JFETs are 10^9 to 10^{10} Ω.

An equivalent circuit for MOSFETs valid at high frequencies is shown in Fig. 11-8. Again this is structurally the same circuit that is valid for JFETs. However, the physical mechanisms that are represented by the capacitors

Sec. 11-4 Charge-Coupled Devices

are different in the two cases. In MOSFETs the capacitance between the gate and the source, C_{gs}, and that between the gate and the drain, C_{gd}, are due to the MOS structure as well as stray lead capacitance. These capacitances are less troublesome in the depletion MOSFET than in the enhancement MOSFET because the metal gate must overlap the source and drain regions in the latter device. The gate-to-drain capacitance, C_{gd}, is particularly undesirable since it provides a feedback path from the output to the input of the device.

11.4 CHARGE-COUPLED DEVICES

A charge-coupled device (CCD) is composed of an array of MOS capacitive structures sharing a common semiconductor substrate. Each of the isolated metal gates is accessible through an electrical contact. Bias voltages are applied to the gates so that the capacitive structures operate in a depletion or inversion mode. In the inversion mode, the electric field in the contact region of the semiconductor creates a potential well that attracts minority carriers to the inversion layer. By suitable pulsing of the bias voltages on the gates the charge stored in the inversion layer of the capacitor can be transferred to an adjacent capacitor. If the presence of charge under a gate represents information, then some type of processing of this information can be performed by applying gate bias pulses. Charge-coupled devices have wide applicability in circuits that perform logical functions on signals, as delay lines, in image-sensing schemes, and as memories.

Principles of Operation

The capacitive structures in CCDs are biased in the depletion and inversion modes of operation. Under these circumstances the minority carriers are attracted to the surface of the semiconductor by the electric field in the depleted region under the oxide. The formation of an inverted charge layer in an MOS structure takes a certain amount of time depending, for example, on the rate of generation of hole-electron pairs, the recombination rate, the drift of minority carriers toward the inversion layer, and the diffusion of these carriers toward the bulk of the semiconductor. Proper operation of charge-coupled devices depends on a technology that minimizes the effects that shorten the time required to form an inversion layer thermally. It is desirable to produce a relatively strong electric field and hence a deep potential well and yet keep the surface layer depleted. Under these circumstances minority carriers that somehow are injected into a specific structure can accumulate in the potential well of that capacitor. These injected charges can then be controlled by the application of pulsed biases to the various gates. However, the processing must proceed on a time scale that is short compared with the time required for a true inversion to form from the minority carriers generated within the semiconductor itself. By the reduction of surface states and bulk states that act as centers for generation of charge pairs it is possible in the present state of the art to create MOS structures that will remain depleted for times of the order of seconds after application of an inverting pulse. Charge-coupled devices then are used to store and to manipulate externally injected charge carriers in the potential wells of depleted regions for times less than that required for the accumulation of inversion charge from thermal processes.

There are several aspects of the operation of charge-control devices. Among these are the injection of minority carriers, the detection of these carriers, and the process of transferring the charge from one location to another. First it is necessary to supply the minority carriers that are to be stored and processed by some external means. Among the possibilities are initiation of an avalanche breakdown in the depletion layer of the semiconductor by application of a large voltage pulse to a gate. Another method is to inject carriers across a *p-n* junction close to the input MOS capacitor. Yet another method is to generate hole-electron pairs optically at or near the depletion region. The latter injection technique makes CCDs useful as image-sensing devices.

Detection of the presence of the injected charges can be accomplished by measuring the incremental capacitance of the MOS structure or its surface potential, both of which depend on the presence or absence of charges in the well. Another method is to use a reverse-biased junction near the output capacitor to collect the minority carriers.

The process of charge transfer is illustrated in Fig. 11-22. This schematic representation of a CCD shows four gates over a common insulator and a *p*-type silicon substrate. Let us assume that electrons have been injected and are temporarily stored under the second gate in the potential well formed by application of a potential to that gate. The potential of the second gate must be larger than that applied to the other gates to localize the well and the charge as indicated in Fig. 11-22(a). In Fig. 11-22(b) a pulse has been applied to the third gate, creating a wider depletion layer, a stronger electric field, and a deeper potential well under that electrode. The electrons are transferred by this field to the semiconductor under the third gate. Figure 11-22(c) shows the gate biases reduced to values that will hold the electrons under the third gate.

There are other types of charge-coupled devices, including a form in which charge may be stored for long periods of time between two dielectric layers.

The development of charge-coupled devices is in its earliest stages. Because of the simplicity of the basic structure, they are easy and hence inexpensive to fabricate. Furthermore, they can be made smaller and can operate

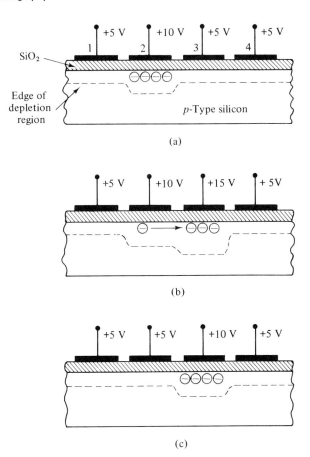

Fig. 11-22 Charge transfer between adjacent MOS structures in a charge-coupled device.

at higher frequencies than more conventional transistor circuits that perform similar functions.

BIBLIOGRAPHY

Field-effect devices are discussed in several of the texts referenced in Chapter 9, especially those by Gibbons, Sze, and Chirlian. Also see

WALLMARK, J. T., and H. JOHNSON, eds., *Field-Effect Transistors: Physics, Technology, and Applications*. Englewood Cliffs, N.J.: Prentice-Hall, Inc., 1966.

GOSLING, W., *Field-Effect Transistor Applications*. New York: John Wiley & Sons, Inc., 1965.

GOSLING, W., W. G. TOWNSEND, and J. WATSON, *Field Effect Electronics*. London: Butterworth & Co. (Publishers) Ltd., 1971.

FITCHEN, F. C., *Electronic Integrated Circuits and Systems*. New York: Van Nostrand Reinhold Company, 1970.

PROBLEMS

11-1. Explain why the values of drain voltage v_{DS} at which pinch-off and breakdown occur in a JFET depend on the gate voltage V_{GS}.

11-2. For the JFET whose drain characteristics are shown in Fig. 11-4, estimate the gate pinch-off voltage V_{GP} and the drain pinch-off voltage V_{DP}. Compare these values.

11-3. The drain characteristics of an n-channel JFET shown in Fig. 11-4 indicate that the device can be operated with positive gate voltage. Usually, however, the gate voltage is kept negative. Explain why positive values of v_{GS} are ordinarily avoided.

11-4. Estimate the transconductance of the n-channel JFET with the drain characteristic of Fig. 11-4. Try to estimate the drain resistance.

11-5. Sketch energy band diagrams for an MOS capacitor with an n-type semiconductor showing the depletion, inversion, and accumulation modes of bias.

11-6. Estimate the drain resistance and the transconductance of the n-channel depletion-type MOSFET with the drain characteristic of Fig. 11-18.

11-7. Estimate the drain resistance and the transconductance of the enhancement-type MOSFET with the drain characteristics of Fig. 11-25. Compare the results obtained with those of Problems 11-14 and 11-16.

11-8. Suppose that you wish to design a linear amplifier and anticipate that the a.c. gate voltage amplitude may be as much as 1 V. Compare the typical characteristics of the JFET and the two types of MOSFET to determine which of these devices you might use to obtain performance that is as close to being linear as possible.

12

Integrated Circuits

An *integrated circuit* (*IC*) is a network of interconnected passive and active circuit elements manufactured as a unit on a supporting material. The supporting material, called the *substrate*, may be either an insulator or an extrinsic semiconductor, either one of which may provide mechanical support for and electrical isolation of the circuit elements. The circuit elements fabricated on the substrate are resistors, capacitors, distributed RC networks, diodes, and transistors. All of the discrete solid-state devices we have studied in this book have their counterparts in integrated form. In addition there are devices in common use that are practical only in an integrated form.

The attraction of the electronics industry to integrated circuits is due not only to their small size but also to the greater flexibility in the design of sophisticated circuits, better reliability, and lower cost. The elements of an integrated circuit may be very small, thus permitting large numbers of devices to be fabricated on a single substrate. It is not uncommon to produce several hundred transistors, diodes, resistors, and capacitors in a semiconductor with an active surface area of 0.01 in^2. Practical limits to the density of elements on an IC are determined by the size of individual elements such as resistors and capacitors and by the power-handling capabilities required of the circuit. Because of the availability of such large numbers of elements, the circuit designer can create highly sophisticated networks at low cost. Furthermore, large numbers of identical integrated circuits can be manufactured at the same time using the well-controlled processes of the highly developed silicon and thin and thick-film technologies. This batch processing is ordinarily highly automated and produces low cost ICs with negligible deviation in performance from piece to piece. Adding to the reliability of integrated circuits relative to that of discrete circuits is the incorporation of the interconnections between the individual elements in the design and manufacture of the unit.

A wide variety of integrated circuits is available commercially. Most, however, fall into one of three classes of IC. One is the *monolithic IC* whose elements are formed on a semiconducting substrate. The word monolithic means single stone, referring to the semiconductor wafer or chip into which the circuit is built. Monolithic silicon ICs utilize many of the semiconducting devices we have studied in this book and their characteristics are the subject of Section 12.2. Another type is the *film IC* whose elements are formed on top of an insulating substrate. The most prevalent of this type is the thin-film IC utilizing resistors, capacitors, and thin-film transistors, a type of insulated-gate FET. We shall consider thin-film ICs in Section 12.3. The third class of IC is the *hybrid IC*, composed of combinations of integrated circuits or of an IC with discrete components using thin-film or thick-film technology for some of the passive components. Features of the latter class of IC will be taken up in Section 12.4.

Problems that significantly affect the design, fabrication, and performance of integrated circuits in addition to the availability of large numbers of circuit elements are the physical layout and electrical isolation of the elements of the IC. These problems ordinarily are not important ones in discrete circuits. Isolation of neighboring elements is a particularly important problem in monolithic silicon ICs because of the conductive properties of the substrate. The physical layout is of concern to the designer because of the necessity of building the interconnecting leads into the IC. The various techniques used to provide isolation and interconnections can introduce parasitic resistance and capacitance into the network and affect the performance of the IC. We shall discuss the standard techniques of isolation and intraconnection as well as methods of fabrication in Section 12.1.

The material in this chapter is intended merely to be descriptive of integrated circuit technology. In some instances the technology is well founded in scientific understanding, while in other instances it is essentially empirical. In all cases, however, it is a rapidly developing technology. Therefore we shall discuss only the most fundamental principles of the fabrication and design of the primary types of ICs. The continuing development of the design, manufacture, and utilization of integrated circuits is best followed in the periodical literature referenced in the Bibliography at the end of this chapter.

12.1 TECHNOLOGY OF INTEGRATED CIRCUITS

The manufacture of integrated circuits is accomplished by a variety of processes utilizing the electrical, mechanical, thermal, chemical, and optical properties of materials. In some instances these properties are understood on the basis of various theories of materials science. Such is the case with the electrical properties of materials studied in this book. On the other hand, some properties of materials used in the fabrication of ICs are not yet understood scientifically but are known from empirical observation. All types of materials are used, semiconductors, insulators, and conductors.

By the judicious selection of materials and manufacturing processes, a variety of physical structures can be fabricated, each with its own peculiar characteristics. Among the processes that are commonly employed are diffusion of impurity atoms to create extrinsic semiconductors, epitaxial growth of extrinsic semiconductors, implantation of impurity ions, deposition of thin layers of materials by evaporation or sputtering, thermal oxidization, photolithography, and chemical etching.

In this section we shall describe these techniques and the common solutions to the problems of isolation and interconnection of individual elements. Our emphasis will be on those processes utilized in the manufacture of monolithic silicon and thin-film integrated circuits.

Fabrication

The fabrication of monolithic silicon integrated circuits depends on the technology of selectively doping predetermined regions of the silicon wafer. For example, a junction transistor consists of alternating regions of an n-type and a p-type semiconductor. An n-channel depletion-type MOSFET consists of an n-type channel between the n^+-type source and drain. Selective variation of the impurity concentrations that provide the appropriate extrinsic behavior may be accomplished either by diffusion, ion implantation, or epitaxial growth.

The process of *thermal diffusion* was introduced in Chapter 3, and its application to the fabrication of semiconductor devices was described in Chapter 9. In practice impurity atoms are introduced at the surface of a portion of a semiconductor crystal. The impurities diffuse into the bulk of the solid with a characteristic diffusion coefficient that depends on the host lattice, the impurity, and the temperature. For diffusions to occur in times that are practical it is necessary to heat the wafer to temperatures that typically lie in the range from 600° to 1200°C.

The spatial variation of the density of the dopant atoms in the crystal depends on how the impurities are introduced at the surface and the length of time during which the system is heated. There are two general profiles that can be obtained. One results when a limited quantity of the impurity material is placed on the surface. As time increases the dopant profile extends farther into the bulk, but because the net number of impurity atoms is fixed, the concentration at the surface decreases with time. The process is governed by a time-dependent diffusion equation, the solution of which satisfies an initial condition at the surface and the requirement that the total number of impurity atoms remains independent of time. Density profiles resulting from this *limited* or *fixed source* diffusion are sketched in Fig. 12-1(a).

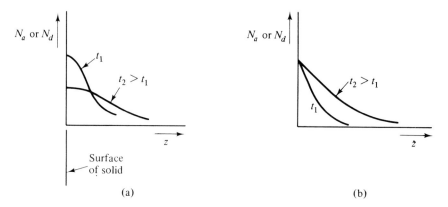

Fig. 12-1 Spatial variation of the concentration of impurity atoms diffusing into a semiconductor: (a) fixed source, (b) constant source.

An alternative class of profiles may be obtained using the so-called *constant source* diffusion process in which the density of impurity atoms at the surface of the crystal is maintained at a constant value. As time increases the dopant profile extends farther into the bulk and the net number of impurities introduced into the solid increases. The concentration of impurities at the surface is determined by the equilibrium condition established between the solid and, for example, a vapor of the dopant material. The upper limit on the surface concentration is determined by the solubility of the material in the solid. Figure 12-1(b) shows the density profiles that result for two diffusion times with a constant impurity source.

While impurity diffusion is a highly refined technology, its use in the manufacture of semiconductor devices gives rise to two problems. First the density of impurities diffused into the bulk of a semiconductor depends on position, as illustrated in Fig. 12-1. The diffusion is carried out under conditions of elevated temperature with a gradient in the density of the diffusing impurity. When the temperature of the material is reduced to room temperature, the diffusion process essentially stops and the density gradient is

Sec. 12-1 Technology of Integrated Circuits

effectively frozen in the bulk. Thus the impurity concentrations in diffused devices are nonuniform. As a result of the spatial variation of the density, practical p-n junctions formed by diffusion are not abrupt stepjunctions as idealized in Chapter 9. Neither are they linearily graded junctions. Of course, the fact that the density profiles of diffused impurities are nonuniform need not be a disadvantage provided the physical design of the elements takes this problem into account. It can be an advantage in some instances.

The second problem arising from impurity diffusion is related to the necessity of forming alternating regions with extrinsic properties of different types. Suppose, for example, that we wish to diffuse an n-type region into the surface of a p-type region to form a p-n junction. The concentration of the donor impurities diffused into the n-region must be higher than that required to compensate for the acceptor impurities already present. If the original p-region was also formed by diffusion, the greatest concentration of acceptor impurities lies near the surface just where compensation must be achieved to form the n-layer. Thus the density of the diffused donor atoms must be rather high. This type of compensation problem can be alleviated somewhat if the original p-region is of the limited-source type. The limit to the number of successive diffusions that can be performed to make alternating extrinsic regions is determined by the effect of the impurity concentrations on the mobilities and lifetimes of the charge carriers. In practical cases, degradation of the carrier properties occurs for more than three successive diffusions.

To overcome problems associated with impurity compensation in diffused semiconductors the lowest-lying electrically active layer, which may also be the substrate, is made extrinsic at the start. The impurity concentration in the substrate can be made spatially uniform by pulling the crystal from a melt seeded with impurity atoms or by a process known as *epitaxial growth*. Epitaxy is the process of formation of a crystal by deposition of atoms from a vapor onto a heated crystalline surface. As the individual atoms strike the surface some may adhere to it. Because the crystal is heated, these atoms may move along the surface to a position where the binding forces with the crystalline atoms establish an equilibrium configuration. The deposited atoms form a crystalline array that reflects the periodicity of the original crystal. If the atoms of the vapor are the same as those of the original substrate, a single crystal can be grown that is a continuation of the original crystal. If the atoms of the vapor differ from those of the substrate, a new crystal may still be grown on top of the original one provided the lattice of the new crystal has properties that are the same or very similar to those of the substrate.

An extrinsic crystal can be grown epitaxially by introducing the appropriate impurities into the vapor. The resulting crystal can be uniformly doped if appropriate precautions are taken. The extrinsic character of the crystal can be changed as it is grown by substitution of different impurities into the atmosphere. This can be done without compensation by removing the original impurity from the vapor before injecting the new impurity. Thus uniformly doped p-n junctions can be formed in epitaxially grown crystals. In practice, however, it is usually desirable that p-n junctions be highly localized in space so that the junction devices will be small. It is difficult to grow crystals epitaxially on only selected portions of a crystalline substrate. Therefore junctions are often formed by diffusion of a compensating impurity into an epitaxially grown extrinsic crystal. That is, both epitaxy and diffusion are used in the production of a single integrated circuit.

An alternative method of doping a semiconductor is by the direct implantation of impurity atoms in the bulk. Energetic atoms or ions of the impurity impinging on the surface of a semiconductor penetrate the solid and become embedded within it. The profile of the dopants and the depth of penetration are determined by the materials and the energy of the impurity particles. In practice high-energy beams of ions are directed at the crystal surface, which is at the ambient temperature. It is often necessary to accelerate the ions through a potential difference on the order of 100 kV to produce a beam of the appropriate energy. After implantation the crystal is heated to reduce the structural defects created by the penetrating ions and to cause the impurities to assume appropriate positions in the lattice. Temperatures are typically below those required for diffusions. The charge of the ions is neutralized by connecting the substrate to an electrical circuit that supplies electrons. The technique, called *ion implantation*, has certain advantages in principle over diffusion. First, implantation is not sensitive to temperature, as is diffusion. Second, by measuring the current of the compensating electrons it is possible to monitor the implantation process as it occurs and therefore to provide greater control over the actual number of impurities introduced. Third, ion implantation has the advantage that the location of the implanted ions is well determined. The peak density of the implanted impurities can be below the surface, forming a buried layer. Fourth, it is possible to implant the impurities through a thin insulating layer on the surface of the wafer. Implantation has disadvantages too. These disadvantages arise mainly from technological problems associated with reliable production of high-energy beams of sufficient density so that exposure times are reasonable. At the present time the great majority of ICs are fabricated using diffusion processes.

Another process used in fabricating ICs is the vapor deposition of thin layers of materials on top of a crystal. *Thin-film deposition* may or may not form single crystals of the deposited materials. Metals, semiconductors, and insulators may be vapor-deposited in films as thin as 100 Å. If the surface onto which the film is to be deposited is heated, the deposited atoms will wander on the surface and aggregate at preferred locations. To form a continuous, uniformly thin film the surface is cooled. The vapor atoms do not wander along the cooled surface but remain where they strike it. It is clear that the deposited atoms may not form a single crystal. The exposure time for known deposition rate determines the thickness of a deposited film. The electrical properties of thin films may differ from those of the same material in bulk because of both the noncrystalline form and surface effects of the film.

Vapor deposition of thin films is an important step in the manufacture of both thin-film and monolithic integrated circuits. In thin-film ICs, insulators, semiconducting materials, and metals are deposited in thin layers to form conductors, resistors, capacitors, and thin-film transistors. In monolithic ICs, metals are deposited to provide conducting paths between the individual components, and dielectric materials are deposited to provide insulation between conducting paths. The gaseous state from which the atoms are deposited can be formed by evaporating the material in a heated crucible somewhere in the vacuum system, although this process frequently introduces impurity materials from the crucible itself. Metallic vapors are frequently formed by *sputtering* from the cathode of a low-pressure glow discharge in an inert gas. In sputtering, metallic atoms are dislodged from the cathode by the impact of ions of the inert gas.

The formation of insulating layers is a very important step in the fabrication of integrated circuits. In monolothic silicon ICs, advantage can be made of the dielectric properties of silicon dioxide. There are several methods by which a layer of SiO_2 can be formed on top of a silicon wafer. One is simply by heating silicon in an oxygen-enriched atmosphere, i.e., thermal oxidization. Another is to expose the surface of the silicon to an oxygen plasma. Yet another is the vapor deposition of silicon dioxide from an oxygen-enriched silicon atmosphere introduced into the vacuum chamber. The latter process is related to epitaxy, but the resulting layer of SiO_2 is ordinarily amorphous.

To create large numbers of small devices in a single integrated circuit, diffusions and depositions must be performed in small selected regions of the silicon wafer of a monolithic IC. To do this the processes of photolithography and selective chemical etching have been developed and employed. These techniques are also used in the fabrication of thin-film elements.

It is known, for example, that the diffusion coefficients of most common doping impurities in silicon dioxide are much smaller than they are in silicon itself. Therefore to exclude impurity diffusion into regions of the silicon chip where it is undesirable a layer of oxide is formed on the surface. To expose the regions of the semiconductor where impurity diffusion is desired, the following processes, illustrated in Fig. 12-2, are used. A thin film of a light-sensitive material called *photoresist* is coated on an SiO_2 layer. The photoresist becomes polymerized when exposed to ultraviolet radiation. After exposure it is highly resistant to a substance called an *etchant* that dissolves the unexposed material. To expose the desired surface of the wafer a mask is placed over it, and the unshielded photoresist is subjected to ultraviolet light. As an example of the procedure, the exposure of the masked photoresist is

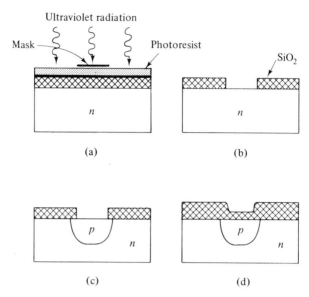

Fig. 12-2 Steps in the process of selective diffusion using photolithography and chemical etching.

illustrated in Fig. 12-2(a). After illumination in the ultraviolet light, the mask is removed and a chemical developer is applied to dissolve the unexposed photoresist. Then an etchant is used to dissolve the silicon dioxide in the uncoated areas. Next another chemical agent is applied to remove the exposed photoresist without affecting the oxide layer under it. Figure 12-2(b) shows the wafer at this stage of the process after the SiO_2 layer has been selectively etched. The wafer is now ready for the vapor diffusion. Figure 12-2(c) illustrates the configuration after the diffusion of p-type impurities into the unmasked region of the n-type wafer. After the diffusion is completed a layer of oxide is again applied to the surface as shown in Fig. 12-2(d), and the chip is ready for the next stage in the fabrication of the IC.

A similar method is used to deposit metal electrodes and interconnecting paths. After the regions where metal is to contact the semiconductor are exposed, a thin film of conductor, usually aluminum, it deposited over the entire wafer. To remove the undesired deposits another photoresist is applied, masked, and then selectively etched, leaving only the desired deposits.

Isolation

Electrical isolation is a critical problem in monolithic silicon integrated circuits. After all, the substrate of such an IC is doped silicon, a moderately good conductor. Therefore if individual elements are to be formed in neighboring regions of the substrate, some means of providing electric isolation between these regions is necessary. Otherwise conducting paths through the substrate will undermine the performance of the circuit.

Three methods of isolation are commonly encountered in practical monolithic integrated circuits. One uses a reverse-biased *p-n* junction to achieve isolation. Another uses a layer of a dielectric material as insulation between active regions. The third technique actually removes the semiconducting substrate material between active regions and therefore requires a different means of mechanical support. Each method has its own advantages and disadvantages and each has its own peculiar effect on the performance of the circuit. We shall discuss each in turn.

The simplest of these schemes to implement is the one employing *reverse-biased p-n junction isolation*. In junction-isolated ICs, each region of the semiconductor wafer containing devices is of the same type of extrinsic material. Each region, called an *island*, is surrounded by extrinsic material of the opposite kind. The resulting *p-n* junction is reverse-biased so that only the very small reverse saturation currents flow between adjacent islands. Figure 12-3

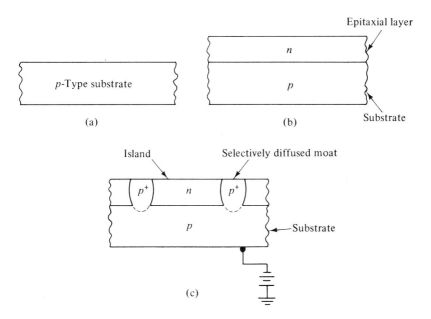

Fig. 12-3 Steps in fabricating junction-isolated islands.

illustrates the fabrication of junction-isolated *n*-type islands on a *p*-type substrate. First a uniformly doped *n*-type layer is grown epitaxially on a *p*-type substrate. Then, by the methods described earlier in this section, relatively narrow but heavily doped p^+-regions are selectively diffused through the epitaxial layer into the substrate. These p^+-regions form *moats* around the n-type islands. Each island is surrounded by extrinsic material of the opposite kind, in the substrate and the moats. To effect the isolation a negative potential is applied to the substrate and moats to ensure that the junctions between them and the islands are always reverse-biased. In the case of n-type islands, the substrate is biased at the most negative potential available, short of one causing breakdown.

The effectiveness of junction isolation is limited by the characteristics of reverse-biased p-n junctions. The leakage current is the reverse bias saturation current. At higher frequencies the depletion layer capacitance reduces the effectiveness of the isolation by shunting the high incremental resistance of the junction. An equivalent circuit representing the high-frequency behavior of the junction isolation of neighboring islands is shown in Fig. 12-4. The

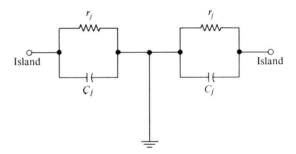

Fig. 12-4 Equivalent circuit for junction isolation.

resistances of the moat and substrate regions are ordinarily so much smaller than the incremental resistances of the reverse-biased junctions that they can be neglected.

The primary disadvantages of junction isolation are the limitation of its use to lower-frequency applications because of the junction capacitance and the necessity of supplying a bias voltage to the substrate. The *dielectric isolation* scheme does not suffer from these disadvantages. It generally has a much higher leakage resistance and better frequency performance than junction isolation. And it requires no external bias. However, the manufacturing process for achieving dielectric isolation is more complicated.

In dielectric isolation schemes, each island is surrounded by an insulating oxide layer. The very low conductivity of silicon dioxide provides very high resistance isolation. This resistance is, of course, shunted by a capacitance resulting from the separation of the two conducting islands by the oxide dielectric. However, this capacitance can be made to be much smaller than the corresponding space-charge layer capacitances of a junction-isolated IC. Therefore dielectric isolation is more effective than junction isolation at higher as well as at lower frequencies.

Sec. 12-1 Technology of Integrated Circuits

The steps required in a typical process for accomplishing dielectric isolation are illustrated in Fig. 12-5. First the *n*-type silicon in which the circuit elements ultimately will be placed is made. Then grooves are etched into it,

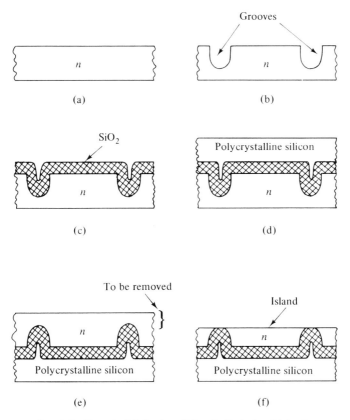

Fig. 12-5 Steps in fabricating dielectrically isolated islands.

as illustrated in Fig. 12-5(b). Next the surface of the material is oxidized, forming the dielectric that will serve as the insulating medium. Polycrystalline silicon is then deposited on top of the oxide. The polycrystalline silicon has no electrical function and serves only as a mechanical support. The structure is then turned upside down as indicated in Fig. 12-5(e), and a portion of the *n*-type silicon is removed mechanically by a processes called lapping. The resulting product, shown in Fig. 12-5(f), has *n*-type islands entirely surrounded by the dielectric SiO_2.

The third isolation scheme has even better performance characteristics at all frequencies than does dielectric isolation. Fabrication of beam-lead air-isolated ICs is, however, even more complex. Isolation is achieved after the various circuit elements have been formed in their respective locations on the semiconducting chip without regard for providing isolation. The individual elements are interconnected with structurally heavy leads. The chip is then turned over and masked. Finally, the portions of the substrate that separate the circuit elements are selectively etched away, leaving the islands containing the elements isolated from one another by air. The stray capacitance between adjacent islands is very small, and the leakage resistance through the air is very large. Thus electrical isolation is excellent even at very high frequencies. Unfortunately, removal of the portion of the substrate that does not participate in the electrical activity of the IC also reduces the mechanical support of the system. Frequently the top of the IC is supported by a glass or ceramic substrate before the bottom is etched away.

Interconnection

Interconnection of individual circuit elements of an IC is achieved by the vapor deposition of a metallic conductor and the removal of unwanted deposits by appropriate masking and etching techniques. Before metallization the chip is covered with an insulating layer that is then selectively etched to expose the regions of the IC where the metal contacts are to be applied. The most important problem in the design of integrated circuits related to interconnection is to provide a physical layout of the elements that minimizes the number of places leads must cross. When leads cross, insulation in the form of a dielectric layer must be provided between them if they are not to be electrically connected.

There are two ways in which crossovers can be fabricated. One method requires deposition of alternate layers of conductor, insulator, and conductor, as illustrated in Fig. 12-6(a). This *multilayer metalization* is technologically difficult to perform, the primary problem being the deposition of the oxide layer after the first metalization. To obtain a thin stable layer of dielectric it is usually necessary to perform the deposition at high temperatures. But after the first metalization, the circuit cannot be heated above the eutectic temperature of the metal-semiconductor system. Generally multilayer metalization is used only in the most complex circuits and then only if no other choice is available.

The second technique of providing insulation between crossing leads is illustrated in Fig. 12-6(b). At the location of the crossing a resistor is diffused into the chip and covered with an oxide layer. One conducting path crosses under the other via this resistor. Such a *diffused crossunder* may utilize a resistor already designed into the circuit or it may require the placement of a low-value resistor especially for this purpose provided this resistance does not materially alter the characteristics of the circuit.

Two leads separated by a dielectric are coupled by a capacitance that can

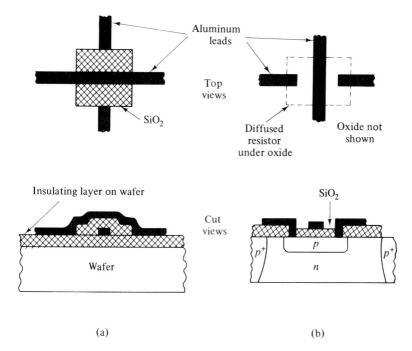

Fig. 12-6 Crossover of leads: (a) multilayer metalization, (b) diffused crossunder.

affect the high-frequency performance of the circuit. This capacitance can be reduced either by making the oxide layer thick or by reducing the area of the crossover. The latter of these possibilities increases the resistance of the lead. The capacitive coupling between crossed leads may be put to good advantage if the layout is properly effected. For example, capacitive coupling between two specific leads may be desirable. In this case the capacitance might be designed into the crossover by adjusting the thickness of the dielectric and the area of the conductors.

12.2 MONOLITHIC SILICON INTEGRATED CIRCUITS

A monolithic silicon integrated circuit consists of interconnected circuit elements fabricated in electrically isolated islands on a single wafer of silicon. Either junction, dielectric, or air and beam-lead isolation may be employed in a monolithic IC. To fabricate the individual elements the processes of epitaxial growth, diffusion, ion implantation, oxidation, vapor deposition, photolithography, and chemical etching are utilized. The circuit elements usually found in monolithic ICs are diffused resistors, capacitors, distributed

RC elements, diodes, and bipolar or unipolar transistors or both. In this section we shall consider the physical form of these elements and some of the problems associated with their design and manufacture.

Transistors

One of the many practical forms of an integrated bipolar transistor is illustrated in Fig. 12-7. This example utilizes junction isolation, but the form

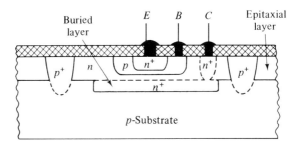

Fig. 12-7 Integrated n-p-n transistor with junction isolation.

of the transistor could be the same with dielectric or beam-lead isolation. The processing steps required to fabricate the n-p-n transistor illustrated in the figure are as follows. A heavily doped n^+-layer is selectively diffused into a p-type substrate. Then an epitaxial n-layer is grown on top of the substrate, burying the n^+-layer. The purpose of this *buried layer* is to increase the conductivity of the volume of the chip that serves as the collector. The remainder of the epitaxial layer is grown as n-type so that the subsequent impurity diffusions forming the base and emitter regions do not encounter excessive problems with impurity compensation. A limited-source diffusion of acceptor material into the epitaxial layer is performed to fabricate the base, and then a constant-source diffusion of donor atoms forms the heavily doped emitter region. Also diffused into the chip are the p^+-isolating moats and an n^+-region that serves as a low-resistance path between the buried layer of the collector and the collector contact. After a final oxide layer is grown on the surface and selectively etched, aluminum contacts are attached to the emitter, base, and collector regions.

It is evident that the geometry of the transistor depicted in Fig. 12-7 differs from that of Fig. 10-1. The latter is a planar transistor with one-dimensional carrier motion perpendicular to the planes of the transistor junctions. Current flow in the integrated form of the junction transistor as illustrated in Fig. 12-7 is two-dimensional. Analysis of the performance of such a device is difficult, although it follows the same physical reasoning as the analysis of the planar transistor studied in Section 10.1. Of particular importance is the effect of the geometry of the base region on performance.

It is interesting to note the volume of a monolithic IC occupied by a bipolar transistor like the one of Fig. 12-7. Typically the substrate is 200 microns or 0.2 mm thick and the epitaxial layer on the order of 10 microns or 0.01 mm. The base region is of the order of 1 micron or 10^{-3} mm thick. The surface area of the island containing such a device is on the order of 0.02 (mm)^2.

Physical configurations other than that shown in Fig. 12-7 are also used as monolithic bipolar transistors. Of the various geometries that are possible there are two types. In one type, active carrier flow is primarily perpendicular to the surface of the IC. The example of Fig. 12-7, called the standard buried collector configuration, is one of these. In the second type, carrier flow is parallel to the surface of the IC. Transistors of the latter class are called lateral transistors.

One form of a unipolar JFET used in monolithic ICs is illustrated schematically in Fig. 12-8 in both top and cut views. This device is formed by two

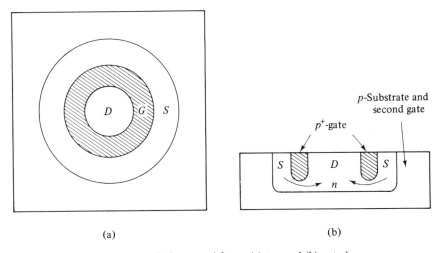

Fig. 12-8 JFET in integrated form: (a) top and (b) cut views.

diffusions into a p-type substrate and has an n-type channel between the source and drain contacts. It is therefore referred to as a double-diffused, n-channel monolithic JFET. The arrows in Fig. 12-8(b) indicate the directions in which majority electrons flow from source to drain under the gate. For the illustrated device the p-type substrate acts as the second gate region. If the substrate is biased separately from the upper gate, the circuit can be designed so that the channel conductivity is relatively insensitive to the substrate potential. The substrate junction itself then serves as an isolating junction, and additional isolation is not required. In other forms of the monolithic JFET, however, a separate isolation scheme may be required.

An essential requirement of the design of any monolithic JFET is to ensure that the gate bias voltage has control of the source-to-drain current. That is, there must be no leakage path for current that avoids the gate. Therefore the diffused gate region must completely surround the surface of either the source or the drain regions, as indicated in Fig. 12-8(a).

Possible configurations of integrated depletion and enhancement MOSFETs are illustrated in Figs. 12-9 and 12-10, respectively. The channel under the gate of the depletion-type MOSFET may be formed by either diffusion or

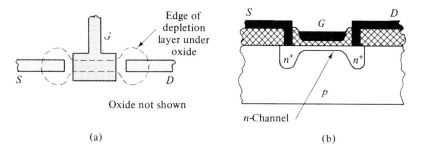

Fig. 12-9 n-Channel depletion-type MOSFET in integrated form: (a) top and (b) cut views.

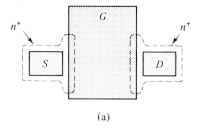

(a)

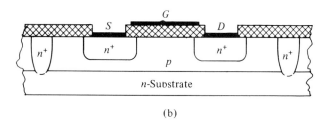

(b)

Fig. 12-10 Enhancement-type MOSFET in integrated form: (a) top and (b) cut views.

ion implantation. The reverse-biased junction with the substrate restricts the current to the channel. In the enhancement-type MOSFET a conducting path exists between the source and drain only when the gate is biased in the inversion mode. Notice in Fig. 12-10 that the metal gate overlaps the heavily doped source and drain regions and that there is no conducting path between the source and the drain that is not under the influence of the gate voltage.

It is possible and sometimes desirable to construct monolithic ICs with bipolar transistors, JFETs and MOSFETs. However, MOS integrated circuits have certain advantages. Among these are simpler fabrication, because MOSFETs usually require fewer processing steps, and lower cost. In addition a MOSFET usually occupies less surface area than a bipolar transistor and so their packing density can be higher.

Diodes

Monolithic junction diodes can be fabricated in ICs with dielectric isolation by diffusion, as illustrated in Fig. 12-11. However, the simple structure of Fig. 12-11 cannot be adapted to junction-isolated ICs because, for example, a p-type substrate and moat must be biased negatively with respect to the

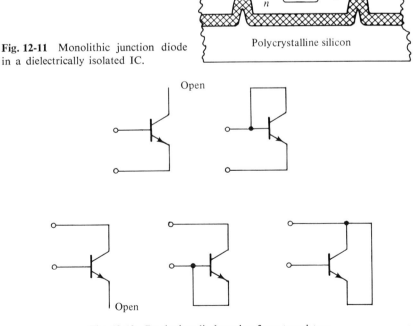

Fig. 12-11 Monolithic junction diode in a dielectrically isolated IC.

Fig. 12-12 Producing diode action from transistors.

n-region of the diode to maintain isolation. However, the potential of the n-region of the diode as required by the circuit may not be compatible with this requirement. In addition transistor action might occur between a forward-biased diode and the reverse-biased substrate junction.

In junction-isolated ICs, diodes may be produced by fabricating a junction transistor and utilizing the emitter-base junction or the base-collector junction as a diode. Five possible diodes formed from transistors are illustrated in Fig. 12-12. Each of these diode connections has its own characteristics, although generally it is the breakdown voltage and the charge storage time that are most dependent on the electrode arrangement.

Schottky barrier diodes are also employed in ICs. As majority carrier devices these diodes have very fast switching times. Furthermore, their fabrication is simple, requiring only the metallization of a semiconductor to form the metal-semiconductor contact.

Resistors

Resistors are formed in monolithic ICs by controlling the resistivity of bulk semiconductor material. Such a resistor is called a *diffused resistor* because the impurity concentrations diffused into the bulk determine the resistivity of the material. The resistance of bulk material of cross-sectional area A and length L is given by

$$R = \frac{L}{\sigma A}, \tag{12-1}$$

where σ is an average bulk conductivity. Recall that the conductivity is the inverse of the resistivity ρ. In IC applications the resistive material forms a relatively thin layer of thickness T at the surface of the silicon wafer. Figure 12-13 indicates such a resistor in a junction-isolated IC. If the effective cross-sectional area for the current flow has thickness T and width W so that $A = WT$, Eq. (12-1) may be written as

$$R = R_s \frac{L}{W} \tag{12-2}$$

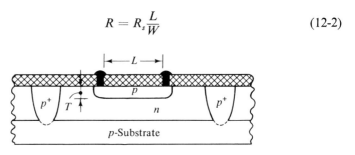

Fig. 12-13 Diffused silicon resistor.

where

$$R_s \equiv \frac{\rho}{T} \qquad (12\text{-}3)$$

is called the *sheet resistance*. The sheet resistance is a useful concept because it is a function only of the thickness and conductivity of the resistive layer. Once the sheet resistance of a particular diffused layer is determined, the values of resistance are determined according to Eq. (12-2) by the surface geometry of the resistor. Furthermore, the resistance of a sheet whose width W is equal to its length L is just R_s. That is, the resistance of a square sheet, regardless of its dimension, is determined by the conductivity and thickness of the sheet. Thus the sheet resistance R_s while having the dimensions of ohms is more commonly expressed in *ohms per square*. Sheet resistances that are commonly available lie in the range from a few to several hundred ohms per square.

Two examples of the surface layout of diffused resistors are illustrated in Fig. 12-14. The width of the diffusion pattern cannot readily be made less

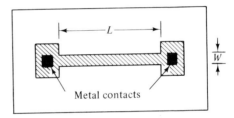

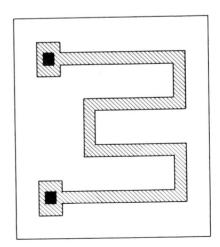

Fig. 12-14 Surface layouts of diffused resistors.

than about a thousandth of an inch, and hence large-valued resistors require long circuitous patterns and occupy large areas of the chip. The range of values for practical diffused resistors is about 50 Ω to 50 kΩ.

The absolute tolerance in manufacturing a specified value of resistance depends on control of the masking and diffusion processes and is on the order of 20%. However, relative tolerances for resistors within the same IC are much better so that circuit designs usually take advantage of resistance ratios rather than absolute values. Additional disadvantages of junction-isolated diffused resistors are the parasitic junction capacitances and the possibility of transistor action in the *p-n-p* structure of Fig. 12-13.

An alternative method of obtaining a resistive function in a monolithic IC is to use active elements to simulate a resistance. The most common of these circuits uses an enhancement-type MOSFET with the drain shorted to the gate. It is called an *MOS resistor*. The circuit diagram of an MOS resistor is shown in Fig. 12-15. The width of the conducting channel is adjusted by

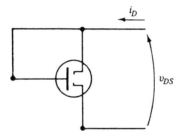

Fig. 12-15 Circuit schematic diagram of an enhancement-type MOS resistor.

design to obtain the desired resistance value between the drain and the source contacts. An MOS resistor requires no additional steps in fabrication, provides much higher resistance, and occupies much less space on the chip than does a diffused resistor. Equivalent sheet resistances of an MOS resistor are typically of the order of 20 kΩ/square, or at least two orders of magnitude greater than that of diffused resistors occupying the same surface area. Thus for resistors of comparable value the MOS resistor requires one one-hundredth of the surface area required by a diffused resistor. Furthermore, resistances greater than 50 kΩ are not practical except in MOS form. However, the resistor is not linear over a wide range.

Capacitors

Either a reverse-biased *p-n* junction or an MOS structure can be used as a capacitor in monolithic silicon integrated circuits. The two types with junction isolation are illustrated in Fig. 12-16. The capacitor using the *p-n* junction has several disadvantages. First it must be reverse-biased. Second, its backward resistance, although high, still permits a leakage current to flow. Third, the capacitance is nonlinear. The MOS type, however, has a very high

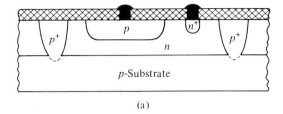

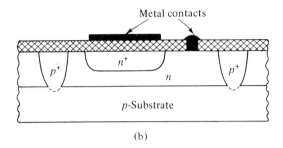

Fig. 12-16 Monolithic capacitors: (a) reverse-biased p-n junction, (b) MOS structure.

leakage resistance and does not require a bias. An oxide capacitor can also be constructed so as to be fairly insensitive to the voltage applied across it; that is, it may be a linear capacitor. Furthermore, the value of this capacitance is easily adjusted by controlling the geometry and the thickness of the oxide layer. The unique charge storage capabilities of an MOS capacitive structure make it particularly useful in integrated digital circuits.

Inductors

Practical inductors cannot be integrated into monolithic silicon ICs if values greater than a few microhenries are required. For this reason either discrete inductors are externally connected to the chip forming a hybrid IC or active networks are employed to simulate inductive characteristics over the range of frequencies desired. These circuits can be quite complicated but active inductorless filters are practical in integrated form because of the availability of very large numbers of active and passive elements at low cost on an IC.

Distributed Networks

A distributed RC network may also be incorporated in a monolithic IC. Composed of a capacitive structure with one of the conductive electrodes replaced by a resistive material, a distributed RC element has the equivalent circuit of Fig. 12-17(b). The resistors and capacitors of the figure represent the resistance and capacitance per unit length along the structure shown in Fig. 12-17(a) from contact ① to contact ③. Such a device can be analyzed using transmission line technques. It is found to have filtering characteristics.

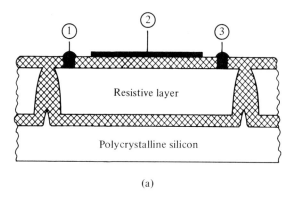

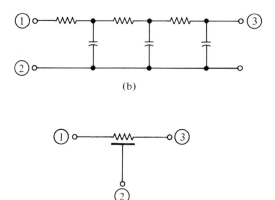

Fig. 12-17 Distributed RC element: (a) structure, (b) equivalent circuit, (c) circuit symbol.

The cross-sectional area of the distributed RC element can be either uniform along the distance between the ends of the resistive layer or non-uniform. In the latter case the RC line is said to be *tapered*. A tapered line can be used either for impedance matching or for its unique filtering characteristics.

12.3 THIN-FILM INTEGRATED CIRCUITS

The second of the two types of integrated circuits is the film IC. Film ICs are fabricated by the placement of layers of conductors, semiconductors, and insulators on a ceramic substrate to form a network of interconnected circuit elements. There are several processes by which the films are produced. One process consists of the preparation of conductive, resistive, or dielectric powders that are applied to the surface with an ink. Layers produced in this way have thicknesses on the order of 25 microns or more. Integrated circuits fabricated in this way are called *thick-film* ICs. While thick-film ICs are easy to manufacture, the processes are difficult to control so that tolerances on the values of circuit element parameters are not good. An alternative process for producing film ICs utilizes vacuum deposition of evaporated or sputtered material. These processes can produce films on the order of 1 micron in thickness or less. Integrated circuits fabricated in this way are called *thin-film* ICs. Thin-film techniques are more costly than thick-film processing, but the resulting circuits are more stable and give better performance.

In this section we shall describe the physical configurations of circuit elements of importance in thin-film ICs. Thin-film elements are produced under vacuum by the deposition and selective etching of thin layers of metals, semiconductors, and insulators. Some vapors can be obtained from an evaporated source, while others must be sputtered from the cathode of a glow discharge of an inert gas. The thickness of the films can be controlled by adjusting the exposure time for known deposition rates. Films can be removed from undesirable locations by chemical etchants using photolithographic techniques. Electrical isolation is provided by removal of the film material between adjacent elements, leaving only the nonconducting glass substrate for mechanical support.

Thin-Film Transistors

The thin-film transistor is an insulated-gate field-effect transistor utilizing a thin layer of an extrinsic semiconducting material rather than a bulk diffused semiconductor. Figure 12-18 shows two possible physical configurations of such a device. The most frequently used semiconductor is cadmium sulfide (CdS), while silicon monoxide (SiO) is used to insulate the gate from the semiconductor. The source, gate, and drain contacts are shown as being gold. Aluminum and copper are also used for contacts and interconnections. The thicknesses of each of the films of Fig. 12-18 are 0.1 to 0.2 microns. The linear dimensions of these devices on the surface of the substrate are 5 to 10 microns.

The thin-film transistor can be designed for either depletion-type or enhancement-type operation. For depletion-type operation, a high-conductivity

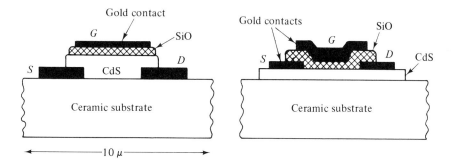

Fig. 12-18 Thin-film transistors.

semiconductor is used. Application of gate bias either constricts the conducting channel between source and drain by increasing the width of the depletion layer under the surface of the oxide or creates an accumulation layer of increased conductivity through which the current flows. For enhancement-type operation, a low-conductivity extrinsic semiconductor is used. A conducting channel appears when the gate is biased to create an inversion layer under the surface of the oxide. Terminal characteristics of these thin-film transistors are similar to those of the depletion and enhancement MOSFETs of Section 11.3. Thin-film transistors have the disadvantage of poor heat dissipation and therefore are susceptible to internal heating. Furthermore, the mobility is generally poor in deposited semiconductor films. Because of the difficulties encountered in manufacturing satisfactory thin-film transistors, they are usually not employed in practice. Instead, hybrid combinations of thin-film passive elements and monolithic transistors are used to take advantage of the superior characteristics of monolithic transistors and thin-film resistors and capacitors.

Thin-Film Resistors

Resistors on thin-film integrated circuits are fabricated by the deposition of thin layers of resistive materials on the surface of the substrate. The most commonly used materials are tantalum and nickel-chromium alloys. These materials have sheet resistances of from 5 to 400 Ω/square depending on the thickness of the film and the fabrication process. Other materials have sheet resistances up to 10,000 Ω/square. An important requirement of the materials used is stability of their properties with time and temperature. Because of the control that can be exercised during deposition, thin-film resistors can be manufactured with very small-value tolerances. Their resistance values are also less sensitive to temperature than those of diffused semiconductor resistors used in monolithic ICs. Examples of some geometrical patterns used for thin-film resistors are illustrated in Fig. 12-19.

Sec. 12-3 Thin-Film Integrated Circuits   395

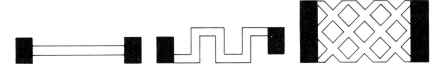

Fig. 12-19   Surface patterns of thin-film resistors.

The values of tantalum resistors can be adjusted after fabrication by thermal oxidation or by anodizing the tantalum. These processes increase the resistance so that ICs are designed with low resistances if these trimming processes are anticipated. Tantalum resistors can be trimmed to less than 0.1% of desired values.

Thin-Film Capacitors

Thin-film capacitors are fabricated by the successive deposition of conductive, dielectric, and conductive layers on the glass substrate of the IC. The most commonly used dielectrics are silicon monoxide (SiO), tantalum oxide (Ta_2O_5), and certain organic polymers. Figure 12-20 shows two examples of thin-film capacitors in cross section.

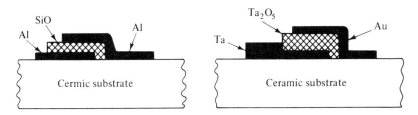

Fig. 12-20   Thin-film capacitors.

Both SiO and Ta_2O_5 can be prepared in uniform, reproducible, well-controlled thicknesses to produce high-tolerance capacitors. In the fabrication of thin-film capacitors using Ta_2O_5 as the dielectric, one electrode of the capacitor is made of tantalum, a reasonably good conductor. The surface of the tantalum is electrically anodized to form the Ta_2O_5 layer, and the second capacitor plate is deposited on top. Dielectric thicknesses range from 0.5 to 3.0 microns.

Distributed Networks

High-precision distributed RC elements can be fabricated in thin-film form by stacking resistive, dielectric, and conducting layers. An RC element can also be made with two resistive layers separated by the dielectric. A distributed LC delay line consists of two conducting layers separated by a dielectric. The distributed LC element is called a strip line.

12.4 HYBRID INTEGRATED CIRCUITS

A hybrid integrated circuit is the combination of one or more monolithic chip ICs with thin- (or thick-) film and/or discrete elements. The purpose of building and using hybrid circuits is to take advantage of the best characteristics of each component of the hybrid. For example, some circuit elements cannot be miniaturized or integrated. Inductors and large-value capacitors are not practical in integrated form. Elements that dissipate large quantities of power are also difficult to integrate because of temperature stability and heat sink problems. These discrete elements in combination with a monolithic or film IC form a hybrid integrated circuit. Other kinds of devices can also be combined with integrated circuits to form useful hybrids. Among these are electro-optic devices such as light-emitting diodes and radiation detectors, electromechanical devices such as piezoelectric crystals, and surface wave devices.

In addition to the combination of discrete elements with integrated devices, the field of hybrids includes combinations of monolithic and thin-film ICs and of thin- and thick-film ICs. The active devices of monolithic silicon ICs are superior to thin-film transistors but thin-film resistors are more precise, have better temperature stability, and can be made with higher values than their monolithic counterparts. A hybrid combination can combine the best attributes of both kinds of integrated device. In some cases thin-film and diffused elements can be fabricated within the same hybrid integrated circuit.

Hybridization between thin- and thick-film technologies may also be advantageous. While thick-film elements may have poorer tolerances, they are mechanically more rugged and cheaper to fabricate than thin-film elements.

As the technologies of monolithic and thin- and thick-film ICs advance it is reasonable to expect that the areas of advantage and disadvantage of each one relative to the others will shift. Yet it is also reasonable to expect that, as long as no one type of integrated circuit can do it all, the interfacing of separate technologies into hybrid circuits will be an important engineering problem. Furthermore, as integration and miniaturization of electronics proceeds, the engineer will encounter many more problems whose solutions require hybridization of electronics with devices that interface with the nonelectrical world such as electro-optical, electromechanical, and electrochemical devices.

BIBLIOGRAPHY

Some references for additional reading about the fundamentals of integrated circuit technology are

GROVE, A. S., *Physics and Technology of Semiconductor Devices.* New York: John Wiley & Sons, Inc., 1967.

WARNER, R. M., JR., and J. N. Fordenwalt, eds., *Integrated Circuits—Design Principles and Fabrication.* New York: McGraw-Hill Book Company, 1965.

BURGER, R. M., and R. P. DONOVAN, *Fundamentals of Silicon Integrated Circuit Technology,* Vols. I and II. Englewood Cliffs, N.J.: Prentice-Hall, Inc., 1967 and 1968.

SORKIN, R. B., *Integrated Electronics.* New York: McGraw-Hill Book Company, 1970.

FITCHEN, F. C., *Electronic Integrated Circuits and Systems.* New York: Van Nostrand Reinhold Company, 1970.

GHANDHI, S. K., *The Theory and Practice of Microelectronics.* New York: John Wiley & Sons, Inc., 1968.

BERRY, R. W., P. M. HALL, and M. T. HARRIS, *Thin Film Technology.* New York: Van Nostrand Reinhold Company, 1968.

To follow the state of the art of integrated circuits, see IEEE Spectrum, Electronics, and the appropriate IEEE transactions and journals.

PROBLEMS

12-1. The thickness of the silicon dioxide layer employed as dielectric insulation between adjacent islands in a monolithic silicon IC is typically 1 micron. Find the capacitance per square centimeter if the relative dielectric constant of SiO_2 is 3.9.

12-2. What should be the sheet resistance of a diffused semiconductor layer to make a resistor of 4000 Ω if the surface of the diffusion pattern is 0.020 in. long by 0.001 in. wide?

12-3. Calculate the capacitance per square centimeter of a thin-film capacitor using a Ta_2O_5 dielectric 0.2 microns thick. The relative dielectric constant of Ta_2O_5 may be taken to be 21. What is the capacitance if the effective cross-sectional area of a capacitor is a square with sides of 0.001 in.?

12-4. Sketch an equivalent circuit for the tapered distributed RC element illustrated in Fig. 12P-4. Notice that this element has two resistive layers separated by the dielectric rather than one resistive and one conducting layer. Indicate how the distributed elements of the equivalent circuit vary between the input and output of this four-terminal device.

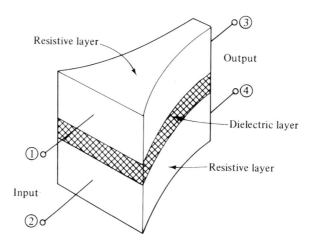

Fig. 12P-4 Tapered thin-film distributed RC element with two resistive layers.

A

Constants

Base of natural logarithms,
$$\epsilon = 2.718\ldots$$
Boltzmann's constant,
$$\kappa = 1.3805 \times 10^{-23} \text{J}/°\text{K}$$
Electronic charge,
$$e = 1.6021 \times 10^{-19} \text{ C}$$
Free electron mass,
$$m_e = 9.1091 \times 10^{-31} \text{ kg}$$
Permittivity of free space,
$$\varepsilon_0 = 8.854 \times 10^{-12} \simeq (\tfrac{1}{36}\pi) \times 10^{-9} \text{ F/m}$$
Planck's constant,
$$h = 6.6256 \times 10^{-34} \text{ J-sec}$$
Planck's constant reduced,
$$\hbar = \frac{h}{2\pi} = 1.0545 \times 10^{-34} \text{ J-sec}$$
Proton mass,
$$m_p = 1.6725 \times 10^{-27} \text{ kg}$$
Speed of light in free space,
$$c = 2.998 \times 10^8 \text{ m/sec}$$

B

Properties of Silicon and Germanium at Room Temperature

	Germanium	Silicon
Relative dielectric constant, $\varepsilon/\varepsilon_0$	16	11.8
Energy band gap, E_g (eV)	0.67	1.11
Intrinsic carrier density, n_i (cm^{-3})	2.4×10^{13}	1.5×10^{10}
Electron mobility, μ_e (cm^2/V-sec)	3900	1500
Hole mobility, μ_h (cm^2/V-sec)	1900	600
Effective density of states in valence band, N_v (cm^{-3})	5.5×10^{18}	1.1×10^{19}
Effective density of states in conduction band, N_c (cm^{-3})	9.9×10^{18}	2.8×10^{19}
Electron effective mass, m_e^*/m_e	0.55	1.10
Hole effective mass, m_h^*/m_e	0.37	0.59

C

Answers to Selected Problems

Chapter 1

1-4. current gain = 50
1-5. $V = E_b - IR$

Chapter 2

2-1. $\overline{(\text{K.E.})} = \tfrac{3}{2}\kappa T$
$v_{\text{rms}} = \left(\dfrac{3\kappa T}{m}\right)^{1/2}$
v_{rms} (electrons, room T) $= 1.15 \times 10^7$ cm/sec

2-4. a) $f_{MB}(v) = nV\left(\dfrac{2}{\pi}\right)^{1/2}\left(\dfrac{m}{\kappa T}\right)^{3/2} v^2 \epsilon^{-mv^2/2\kappa T}$
b) $v_{mp} = \left(\dfrac{2\kappa T}{m}\right)^{1/2}$

2-6. $E_{mp} = \tfrac{1}{2}\kappa T$
2-7. $L = 2.5 \times 10^{-6}$ cm
2-8. 0.3174
0.0454
0.0026

Chapter 3

3-2. 9.6 mA
3-3. $v_{rms} = 1.15 \times 10^7$ cm/sec
$v_D = 7.35 \times 10^{-3}$ cm/sec
$I = 1.56 \times 10^9$ A!
3-4. $\rho = 6.25 \times 10^9$ C/m^3
$n = 3.9 \times 10^{28}$ electrons/m^3
3-5. $\sigma = 9.6 \times 10^{-7}$ mhos/cm
$J = 9.6 \times 10^{-2}$ A/cm^2
3-6. $J = 640$ A/cm^2
3-8. $\lambda = 6.17 \times 10^{-6}$ cm
$\tau = 1.36 \times 10^{-10}$ sec
3-11. $x_{rms} = (2Dt)^{1/2}$
$\tau = 5 \times 10^{-4}$ sec
3-14. $D_e = 97.5$ cm^2/sec
3-15. $V_n = 4$ μV

Chapter 4

4-1. at 100 eV: $v = 5.94 \times 10^6$ m/sec
at 10^4 eV: $v = 5.94 \times 10^7$ m/sec
at 290°K: $v_{rms} = 1.15 \times 10^5$ m/sec
at 2400°K: $v_{rms} = 3.31 \times 10^5$ m/sec
4-2. $V_{max} = 37.9$ V
4-4. $V(x) = V_0 \left(\dfrac{x}{d}\right)^{4/3}$
4-5. $n(x = d) = 7.86 \times 10^{12}$/m^3
$I = 0.844$ mA
4-6. $\lambda_e = 690$ m
4-7. $\lambda_e = 4.04 \times 10^{-10}$ cm
4-8. $\lambda_e = 4.24 \times 10^{-6}$ cm
4-10. $\omega_{pe} = 5.64 \times 10^7$ rad/sec
$\omega \simeq \omega_{pe}$
4-11. $\omega_{pe} = 1.64 \times 10^{16}$ rad/sec
4-12. $\omega_{pe} = 1.323 \times 10^{13}$ rad/sec
4-13. $n_e = 1.264 \times 10^{11}$/cm^3

Chapter 5

5-1. $v_{min} = 1.09 \times 10^{15}$ Hz
$\lambda_{max} = 276$ nm
5-2. $\lambda = 6.45 \times 10^{-33}$ cm

Answers to Selected Problems   403

5-3. $\lambda = 0.788$ Å
5-5. $\Psi_{\text{packet}}(z, t = 0) = \dfrac{2\delta \sin(\delta z)}{(\delta z)} \epsilon^{jk_0 z}$
      $\delta \Delta z = 2.8$
5-7. 1 state has $n = 1$
      4 states have $n = 2$
      9 states have $n = 3$
      16 states have $n = 4$
5-8. $\lambda_{\min} = 367$ nm
5-10. $E_F = \left(\dfrac{\hbar^2}{2m}\right)\left(\dfrac{3\pi^2 N}{V}\right)^{2/3}$
5-12. $E = 4.4\,\kappa T$
5-13. for $E - E_F = 3\,\kappa T$, error is 4.8%

Chapter 6

6-1. $\mathbf{J} = +e\mathbf{v}_s$
6-4. $m^* = m_e$
6-6. $\omega_c \tau > 2\pi$
6-7. $m_t^* = 0.913 \times 10^{-31}$ kg

Chapter 7

7-1. $\tau = 2.54 \times 10^{-14}$ sec
7-2. $\nu_{\min}$ about 9.9 to 10.4×10^{14} Hz
7-6. $\mathcal{E}_z = 2.04 \times 10^7$ V/m

Chapter 8

8-3. $\Delta E = -0.0117$ eV
8-4. for silicon, $N_v = 1.09 \times 10^{19}/\text{cm}^3$
              $N_c = 2.78 \times 10^{19}/\text{cm}^3$
      for germanium, $N_v = 5.45 \times 10^{18}/\text{cm}^3$
                  $N_c = 9.92 \times 10^{18}/\text{cm}^3$
8-5. ionization probability $= 0.9976$
8-8. for silicon, $\lambda_{\max} = 1.12 \times 10^4$ Å
      for germanium, $\lambda_{\max} = 1.86 \times 10^4$ Å
8-11. $\sigma = 5.04 \times 10^{-4}$ mhos/m
      $N_d = 4.16 \times 10^{14}/\text{cm}^3$
      $N_a = 3.04 \times 10^{14}/\text{cm}^3$
8-12. for silicon, $L_h = 1.225 \times 10^{-2}$ cm
                $L_e = 1.94 \times 10^{-2}$ cm
      for germanium, $L_h = 0.218$ cm
                  $L_e = 0.314$ cm

8-13. $\mathcal{E} < 6.67$ V/cm
8-15. $n = 8.57 \times 10^{22}/\text{cm}^3$
$\mu = 49$ cm²/V-sec
8-16. $n = 8.34 \times 10^{14}/\text{cm}^3$
$\mu = 750$ cm²/V-sec
$\tau = 2.35 \times 10^{-13}$ sec
8-18. $\left(\dfrac{\partial \rho}{\rho}\right) = \dfrac{-E_g}{2\kappa T}\left(\dfrac{\partial T}{T}\right)$
8-19. $v_D = 10^6$ cm/sec
$\mu = 330$ cm²/V-sec

Chapter 9

9-2. $\phi_0 = -0.748$ V
9-5. $w_n = 7.28 \times 10^{-5}$ cm
$\mathcal{E}_{0\,\text{max}} = 1.15 \times 10^4$ V/cm
9-6. $\nabla p = 5.64 \times 10^{22}/\text{cm}^4$
$J_h = 4.41 \times 10^5$ A/cm²
9-7. $J_h = 4.55 \times 10^3$ A/cm²
9-8. $\delta n_p = \delta p_n = 5.12 \times 10^{16}/\text{cm}^3$
for majority carriers, 51.2%
9-11. $L_e = 1.06 \times 10^{-2}$ cm
9-12. $I_s = 4.88$ μA
9-13. for silicon, 0.163/°K
for germanium, 0.103/°K
9-14. $r_0 = 5.13 \times 10^3$ Ω
9-15. $r = 25$ Ω
9-18. $\phi_0 = -1.01$ V
$w_n = w_p = 8.1 \times 10^{-7}$ cm
$\mathcal{E}_{0\,\text{max}} = 1.24 \times 10^6$ V/cm
9-20. $I_{sc} = 3.9 \times 10^{-4}$ A
$I_{sc} = 80\,I_s$
$V_{oc} = 0.11$ V

Chapter 10

10-4. $L_{hB} = 5 \times 10^{-3}$ cm
$\alpha = 0.979$
10-5. $h_{ib} \cong 28$ Ω
$h_{rb} \cong 0.1$
$h_{fb} \cong -1.0$
$h_{ob} \cong 10$ μmhos
10-7. $h_{fe} \cong 46.6$

Answers to Selected Problems

10-8. $h_{ie} \cong 375\ \Omega$
       $h_{re} \cong 0.25$
       $h_{fe} \cong 100$
       $h_{oe} \cong 600\ \mu\text{mhos}$

Chapter 11

11-2. $V_{GP} \cong -4.4\ \text{V}$
       $V_{DP} \cong 8\ \text{V}$
11-4. $g_m = 2400\ \mu\text{mhos}$
       $r_d \cong 36 \times 10^3\ \Omega$
11-6. $g_m = 1500\ \mu\text{mhos}$
       $r_d \cong 13.3 \times 10^3\ \Omega$
11-7. $g_m = 2200\ \mu\text{mhos}$
       $r_d \cong 20 \times 10^3\ \Omega$

Chapter 12

12-1. $C = 3.45 \times 10^3\ \mu\mu\text{F/cm}^2$
12-2. $R_s = 8 \times 10^4\ \Omega$
12-3. $C = 9.3 \times 10^{-8}\ \text{F/cm}^2$
       $C = 0.599\ \mu\mu\text{F}$

Index

Numbers in parentheses reference a problem.

A

Abrupt junction, 257
Absorption, optical (see Hole-electron pairs, optical generation)
Acceptors:
  dopant elements, 214
  energy states, 214
Accumulation:
  limited space-charge, Gunn effect, 242
  MOS capacitor, 358
  MOSFET, depletion-type, 362
Ambipolar diffusion (see Diffusion, ambipolar)
Amplifier:
  analysis, 9, 13 (1-7)
  current, ideal, 5
  three-terminal representation, 5
  transistor, 10, 310
  tunnel diode, 300
  two-port representation, 5
Avalanche breakdown:
  IMPATT diode, 301
  p-n junction, 282
Average values:
  classical particles in equilibrium:
    energy, 39 (2-1)
    r.m.s. velocity, 39 (2-1)
    speed, 39 (2-4)
  from distribution function, 23
  from macroscopic description, 18

B

Band theory:
  degeneracies, 171
  single electron theory, 176
  state splitting, 171
Base, junction transistor, 311
Basis vector, 170
Bilateral switch, 329
Bipolar, 343
Bipolar junction transistor (see Junction transistor)
Bloch wave function, 177
Boltzmann statistics (see Maxwell-Boltzmann distribution)
Bonding:
  interaction forces, 161
  types, 169
Bose-Einstein probability, 162
Boundary conditions in quantum mechanics, 130
Bragg reflection, 124
Bravais lattice, 170
Breakdown:
  avalanche, 282, 301
  p-n junction, reverse biased, 281
  temperature dependence, 284
  voltage, 283
  Zener, 283, 296
Brillouin zone, 182

Brownian motion, 57
Built-in field (see Contact potential)
Buried layer, 384 (see also Junction transistor)

C

Capacitance:
  abrupt junction, 290-93
  depletion layer, 290
  diffusion, 292
  linearly graded junction, 292
  MOS, 353-59
Cathode ray tube, 90
Channel:
  MOSFET, depletion-type, 360
Charged-coupled devices, 367
Chemical potential, 162
Circuit models (see specific devices)
Circuit symbols (see specific devices)
Channel:
  JFET, 343
  MOSFET, depletion-type, 360
Charge-coupled devices, 367
Collector, junction transistor, 311
Collision:
  cross section:
    hard spheres interaction, 59
    relation to mean free path, 62
  interaction potential:
    hard spheres, 59
    realistic, 65
  mean collision time, 57, 64
  mean free path, 59, 63
  probability, 61
Collision integral, 54
Common-base configuration (see Junction transistor)
Common-emitter configuration (see Junction transistor)
Common-source configuration (see Junction field-effect transistor)
Conduction band:
  definition, 175
  effective density of states, 220
  electron density, 222-24, 253 (8-5)
Conduction current (see Ohm's law)
Conductivity:
  definition, 45
  mean particle model, 67
  negative differential (see Negative differential conductivity)
  selected materials (table), 46
  semiconductors, 229
Constant source diffusion, 374
Contact potential:
  between metals, 204

Contact potential (cont.)
  difference in work functions, 206
  metal-semiconductor junctions, 243
  plasma sheath, 106
  p-n junctions, 258, 262
Continuity equation:
  charge, 50
  particles, 47
Cross over, integrated circuit, 383
Crystal, 169
Current density, electrical:
  electrons in a band, 185
  holes, 216
  Ohm's law, 45
  relation to electric current, 43
  relation to particle current density, 44
Current density, particle:
  definition, 44, 48
  diffusion, 47
  relation to distribution function, 44
  relation to electric current density, 44
Current-voltage characteristic (see I-V characteristic)
Curvature:
  effective mass, energy band, 186
  second derivative, 131
Cutoff biasing, junction transistor, 317
Cutoff frequency, junction transistor:
  common-base, 327
  common-emitter, 330
Cyclotron frequency, 189
Cyclotron resonance, 189

D

d.c. operating point (see Load line analysis)
de Broglie's hypothesis (see Wave-particle duality)
Debye screening length (see Screening length)
Defects (see Imperfections)
Degeneracy:
  comparison of statistics, 164
  eigenstates, 150
  semiconductors, 220, 297
Density of states:
  semiconductors, effective, 220
  square well potential, 156
  metals, 192
Depletion approximation:
  MOS capacitor, 354
  p-n junction, 263
Depletion layer:
  MOS capacitor, 354, 355
  p-n junction, 258
Depletion layer capacitance, 290

Depletion mode:
  MOS capacitor, 355
  MOSFET, depletion-type, 360
Dielectric isolation, 380
Diffusion:
  ambipolar:
    coefficient, 112
    definition, 110
    extrinsic semiconductors, 233
  device fabrication:
    constant source, 374
    limited source, 374
  Fick's law, 47
  transport coefficient, 42
Diffusion capacitance, 292
Diffusion coefficient:
  impurities into silicon (graph), 80
  limited statistical model, 73
Diffusion current, 48
Diffusion equation:
  minority carriers:
    base of junction transistor, 319
    p-n junction, 277
  time-dependent solution, 86 (3-11)
Diffusion length, 277
Diode:
  ideal characteristic, 3, 287
  IMPATT, 301
  light-emitting, 305
  photo, 96, 303
  p-n junction (see p-n junction, diode)
  p-n-p-n, 332
  thermionic, 96
  tunnel, 296
  vacuum tube, with space charge, 97
Direct gap, 188
Distribution function:
  classical:
    derivation, 26
    in potential field, 38
    (see also Maxwell-Boltzmann distribution)
  definition:
    energy space, 20
    phase space, 21
  description of macrostate, 17
  probability, 22
  properties, 22-24
  quantum mechanical, 160 (see also Fermi-Dirac probability and Bose-Einstein probability)
  relation to spatial density, 24
Donors:
  dopant elements, 214
  energy states, 213, 214
Drain:
  JFET, 343
  MOSFET, 360
Drift velocity, 66

E

E vs. k diagram:
  germanium (graph), 188
  reduced zone plot, 184
  simplified Kronig-Penney model, 183
  symmetries, 182
Ebers-Moll model (see Junction transistor)
Effective density of states (see Density of states, effective)
Effective mass:
  anisotropy, 188
  definition, 186
  of different states, 188
  holes, 217
  holes and electrons (table), 221
  measurement, 189
Effective momentum, 186
Einstein's relation, 76, 233
Electroluminescence, 306
Electron beam:
  bunching, 93
  energy, 89
  space-charge effects, 89
  velocity distribution, 89
Electron gun, 88
Electron-hole pairs (see Hole-electron pairs)
Emission, electron:
  field, 201
  photo, 196
  secondary, 197
  thermal, 198
Emitter, junction transistor, 311
Energy Band:
  definition, 174
  from Kronig-Penney model, 180
  also see Band theory
Epitaxial growth, 375
Equilibrium:
  local thermodynamic, 74
  steady state, 42
  thermodynamic, 24, 28, 158, 162
Exclusion principle, 152
Extrinsic semiconductor (see Semiconductors, n-type and p-type)

F

Fermi-Dirac probability:
  derivation, 157
  properties, 160
Fermi energy:
  chemical potential, 162
  insulators, 175
  intrinsic semiconductors, 215, 221
  metals, 175, 192
  n-type semiconductors, 215, 224
  p-type semiconductors, 216, 225

Fermi energy (cont.)
 partition function, 159
 relation to density and temperature, 161
 semiconductors, 175, 208
 semimetals, 175
 thermodynamic equilibrium, 162
Fermi level (see Fermi energy)
Fermions, 160
Field-effect transistor (see Junction field-effect transistor and Metal-oxide-semiconductor field-effect transistor)
Fick's law, diffusion, 47
Field emission, 201
Fixed source diffusion, 374
Forward-blocking state (see p-n-p-n diode)
Forward-conducting state (see p-n-p-n diode)
Four-layer devices (see p-n-p-n diode)
Free particle:
 energy, 133
 wave function, 133
 wave number, 133

G

Gate:
 JFET, 343
 MOSFET, 360
 SCR, 337
Generation, charge carrier:
 optical, 225
 thermal, 218
Graded junction, 257
Group velocity:
 electrons in solids, 184
 energy propagation, 121
 wave packet, 142
Gunn effect: 238-43
 domains, 240
 limited space-charge accumulation, 242
 transit time, 242

H

Hall coefficient, 236
Hall effect, 234-37
Haynes-Shockley experiment, 233
Heisenberg uncertainty principle (see Uncertainty principle)
Hole-electron pairs:
 optical generation:
  direct transition, 225
  indirect transition, 226
 recombination, 226
 thermal generation, 218

Holes:
 current model, 216
 effective mass, 217
 generation, 216, 218, 225
 recombination, 217, 226
Holding current, SCR, 338
Hybrid representations (see specific devices)
Hybrid-π equivalent circuit, 331
Hydrogen atom:
 eigenenergies, 150
 Schrödinger's equation, 149
 quantum numbers, 150

I

IC (see Integrated circuit)
IMPATT diode, 301-3
Imperfections:
 effect on band structure, 212
 essential, 210
 interstitial, 212
 nonessential, 210
 replacement, 212
 vacancy, 212
Impurity compensation, 253 (8-8)
Incremental equivalent circuit model:
 high frequency a.c., 11
 low frequency a.c., 9
 also see specific devices
Indirect gap, 189
Indistinguishability of quantum particles, 154
Inductors in integrated circuits, 391
Initial conditions, 15
Injection, low level, 271
Insulated-gate field-effect transistor (IGFET) (see Metal-oxide-semiconductor field-effect transistor)
Integrated circuit (IC), 11, 371-96
 fabrication, 373-78
 hybrid, 396
 interconnection, 382
 isolation:
  air beam-lead, 382
  biased junction, 379
  dielectric, 380
  monolithic: 383-92
 capacitors, 390
 diodes, 387
 distributed networks, 391
 inductors, 391
 resistors, 388
 transistors, 384
 substrate, 371
 thick-film, 393, 396
 thin-film, 393-96

Index 411

Integrated circuit (IC) *(cont.)*
  monolithic *(cont.)*
    capacitors, 395
    distributed networks, 395
    resistors, 394
    transistors, 393
Interconnection (*see* Integrated circuit)
Inversion mode:
  MOS capacitor, 355
  MOSFET, enhancement-type, 364
Ion implantation, 376
Isolation (*see* Integrated circuit)
I-V characteristic:
  amplifier, ideal, 5
  bilateral switch, 340
  diode, ideal, 3
  p-n junction, 276-81
  p-n-p-n diode, 332
  rectifying metal-semiconductor junction, 243
  resistor, 2, 6
  SCR, 338
  switch, ideal, 3
  thermionic diode, 203
  tunnel diode, 299
  (*see also* Terminal characteristic)

J

JFET (*see* Junction field-effect transistor)
Junction, metal-semiconductor (*see* Metal-semiconductor junction)
Junction, p-n (*see* p-n junction)
Junction field-effect transistor (JFET), 343-52
  capacitive effects, 352
  channel, 343
  circuit symbol, 344
  configuration, physical, 343
  common-source configuration, 344
  drain, 343
  drain current, 347-50
  drain resistance, 351
  equivalent circuits, 351
  gates, 343
  noise, 352
  output characteristic, 345
  pinch-off, 346, 348, 350
  resistance, voltage-controlled, 349
  saturation, 350
  source, 343
  transconductance, 351
Junction isolation, 379
Junction laser, 306
Junction transistor, bipolar, 311-32
  alpha, 317
  base transport factor, 318, 321

Junction transistor, bipolar *(cont.)*
  bipolar, 312
  buried collector, 385
  circuit symbols, 314
  collector efficiency, 318
  common-base configuration, 323-27
    alpha cutoff, 327
    characteristics, 324
    forward current transfer ratio, 317
    hybrid incremental equivalent circuit, 325
    hybrid parameters, 325, 326
    hybrid representation, 323
  common-emitter configuration, 327-31
    characteristics, 328
    cutoff frequency, 330
    hybrid parameters, 328, 330
    hybrid-π equivalent circuit, 331
    hybrid representation, 327
  configuration, physical, 311
  cutoff biasing, 317
  Ebers-Moll model, 313
  emitter efficiency, 317, 321
  integrated forms, 384
  inverse biasing, 317
  lateral, 385
  minority carriers in base, 319
  noise, 331
  normal biasing, 312, 317
  saturation biasing, 317
  sign conventions, 314
  two-diode model, 313

K

Kinetic energy, 19, 39 (2-1)
Kinetic equation:
  continuity in phase space, 51-55
  distribution function, 41
  iterative solution, 70, 75
  relaxation approximation, 56
  role of collisions, 55 (*see also* Collision integral)
Klystron, 92
Kronig-Penney model, 178-84
  band structure, 180
  limiting case, 180

L

Laser, 152, 306
  p-n junction, 306
  p-n multijunction, 307
  stimulated emission, 153, 306
Lateral transistor, 385

Lattice:
 definition, 170
 symmetries, 170
Lifetime, holes and electrons, 226
Light-emitting diode, 305
Light-sensitive diode (see Photodiode)
Limited source diffusion, 374
Limited space-charge accumulation, 242
Load line analysis:
 d.c. operating point, 8
 low frequency a.c., 9
 low frequency a.c., large amplitude, 13 (1-7)
 non-linear d.c. circuits, 7
Low level injection, 271
Luminescence, 227

M

Macrostate:
 definition, 16
 equilibrium, 16, 26
 steady state, 42
Magnetron, 93
Majority carrier, 223
Mass, effective (see Effective mass)
Maxwell-Boltzmann distribution:
 comparison with normal distribution, 35
 energy space, 35
 phase space, 32
 speed space, 39 (2-4)
Mean collision time:
 definition, 57
 relation to cross section, 64
Mean free path:
 definition, 59
 relation to cross section, 63
Metal:
 band structure, 192
 density of allowed states, 192
 electron mobility, 195
 work function, 192
Metal-oxide-semiconductor (MOS) capacitor, 353-59
 accumulation mode, 358
 capacitance, 355, 359
 configuration, physical, 353
 depletion layer, 354, 355, 357
 depletion mode, 355
 inversion mode, 355
  depletion layer, 357
  turn-on voltage, 357
 sign convention, 353, 355
Metal-oxide-semiconductor field-effect transistor (MOSFET):
 circuit representations, 366
 depletion-type, 360-63

(MOSFET) (cont.)
 accumulation mode, 362
 circuit symbol, 360
 configuration, physical, 360
 depletion mode, 360
 drain characteristic, 365
 enhancement-type, 363-66
  circuit symbol, 364
  configuration, physical, 363
  drain characteristic, 365
  inversion mode, 364
  turn-on voltage, 364
Metal-semiconductor junction, 243-51
 ohmic contact, 243
 rectifying contact:
  I-V characteristic, 243
  Schottky barrier, 249
 realization:
  ohmic contact, 250
  rectifying contact, 244, 249
Microstate, 16
Minority carrier, 223
Minority carrier diffusion equation, 277, 319
Mobility:
 definition, 46
 electrons and holes in semiconductors (table), 230
 electrons in materials (table), 46
 electrons in metals, 195
 limited statistical model, 69-72
 mean particle model, 67-69
 temperature dependence, 230
Monolithic integrated circuit (see Integrated circuit)
MOS capacitor, 353-59
MOSFET (see Metal-oxide-semiconductor field-effect transistor)
Multilayer metalization, 382
Multipliers, undetermined, method of, 29, 159

N

Negative differential conductivity:
 Gunn effect, 239
 IMPATT diode, 301
 tunnel diode, 296, 299
Newton's laws of motion, 15
Noise:
 bipolar junction transistor, 331
 corpuscular, 83
 current, 83, 84
 fluctuations, 81
 JFET, 352
 p-n junction, 284-87
 shot, 84

Index

Noise *(cont.)*
  thermal, 82
  voltage, 83
Normal biasing, junction transistor, 312, 317
Normal distribution, 35
  mean value, 35
  standard deviation, 35
Normalization condition, 130
np product, 261
n-type semiconductor *(see* Semiconductor, n-type)

O

Ohmic contact, 243, 250
Ohm's law, 45
Ohms per square, 389
Operating point *(see* Load line analysis)
Optical absorption *(see* Hole-electron pairs, optical generation)
Optical generation *(see* Hole-electron pairs, optical generation)
Optical spectra, 152
Oxidization of silicon, 377

P

Partition function, 30
Pauli exclusion principle *(see* Exclusion principle)
Phase space:
  continuity *(see* Kinetic equation)
  definition, 21
  distribution function, 32
Phonons, 211, 232
Photoconductivity, 225
Photodiode, 303
Photoemission:
  Einstein's theory, 121
  energy diagram, 196
Photolithography, 377
Photomultiplier tube, 198
Photoresist, 377
Photovoltaic cell, 304
Pinch-off, JFET:
  drain, 346, 350
  gate, 346, 348
Planck's constant, 123
Plasma:
  contact potential, 103-5
  definition, 101
  electron saturation current, 110
  features, 102
  floating potential, 108
  ion saturation current, 109

Plasma *(cont.)*
  potential, 110
  sheath, 103-5, 107-10
Plasma frequency, 115
p-n junction, 255-307
  abrupt, 257
  arrow convention, 287
  biased, 267-76
    carrier densities, 270, 271, 274
    depletion layer, 269
    forward bias, 269
    forward bias current, 271
    low level injection, 271
    net potential drop, 269
    reverse bias, 269
    reverse bias current, 274
    sign convention, 267
  breakdown:
    avalanche, 282
    temperature dependence, 284
    voltage, 283
    Zener, 283
  contact potential, 258
  current, 278
  depletion layer, 258
  depletion layer capacitance, 290
  diffusion capacitance, 292
  diode, 287-96
    capacitive effects, 290-94
    circuit symbol, 287
    equivalent circuit, 288
    switching times, 294
  equilibrium:
    carrier densities, 261
    contact potential, 262, 266
    depletion approximation, 263
    depletion layer, 265, 266
    electric field, 265
  fabrication, 256
  graded, 257
  I-V characteristic, 276-81
  minority carriers:
    continuity equation, 277
    current densities, 278
    densities, 271
  noise, 284-87
    equivalent resistance, 285
    shot noise, 285
  saturation current, 279, 280
  space-charge layer, 258
  switching times, 294
p-n-p-n diode, 332-37
  configuration, physical, 333
  circuit symbol, 333
  forward blocking state, 334
  forward conducting state, 334
  I-V characteristic, 333
  reverse blocking state, 334

Index

Plasma frequency *(cont.)*
  minority carriers *(cont.)*
    switching, 335
    two-transistor model, 336
    also see Silicon controlled rectifier
pn product, 261
Primitive cell, 170
Pseudoclassical description, 185-87 *(see also* Effective mass)
p-type semiconductor *(see* Semiconductor, *p*-type)
Pump frequency, 153
Punch-through, 321

Q

Quantum mechanics, 119-64
  Davisson-Germer experiment, 125
  quantization, 145, 155
  statistics, 154-64
    Bose-Einstein probability, 162
    Fermi-Dirac probability, 157-62
  tunneling, 136
  wave equation, 126-29 *(see also* Schrödinger's equation)
  wave function *(see* Wave function)
  wave packet *(see* Wave packet)
  wave-particle duality *(see* Wave-particle duality)

R

Radiation:
  atomic spectra, 152
  stimulated emission, 153, 306
Radiation detectors, 237
Radiative recombination, 227
Read diode *(see* IMPATT diode)
Recombination, 226-28
  nonradiative, 227
  radiative, 227, 305
Recombination centers, 214
Relaxation approximation, 56
Relaxation time, 56
Representative points in phase space, 20
Resistor:
  integrated forms, 388, 394
  I-V characteristic, 2, 6
  models, 6
  MOS, 390
  voltage-controlled, JFET, 349
Reverse blocking state *(see p-n-p-n* diode)
Reverse current *(see p-n* junction)
Richardson-Dushman equation, 200

S

Saturation current:
  JFET, 350
  p-n junction, 279, 280
Scatter-limited velocity, 232
Schottky barrier, 249 *(see also* Metal-semiconductor junction, rectifying)
Schottky effect, 201
Schrödinger's equation, 126-29
  derivation, 126
  separation of variables, 128
  stationary state, 129
  time dependence, 129
SCR *(see* Silicon controlled rectifier)
Screening length, 105
Secondary emission, 197
Semiconductors, 208-37
  conductivity, 229
  intrinsic:
    carrier densities, 221
    Fermi energy, 221
  n-type:
    carrier densities, 222-4, 253 (8-5)
    donor impurities, 214
    energy diagram, 224
    Fermi energy, 215
    work function, 243
  p-type:
    acceptor impurities, 214
    carrier densities, 253 (8-5)
    energy diagram, 224
    Fermi energy, 216
    work function, 244
Semiconductor controlled rectifier *(see* Silicon controlled rectifier)
Semimetals, 229
Separation of variables, technique of, 128
Sheet resistance, 389
Shockley diode *(see p-n-p-n* diode)
Shockley-Haynes experiment, 233
Sign conventions, see specific devices
Silicon controlled rectifier (SCR), 337-39
  configuration, physical, 337
  circuit symbol, 338
  gate, 337
  holding current, 338
  I-V characteristic, 338
  triggering, 337
Solar cell, 304
Source:
  JFET, 343
  MOSFET, 360
Space-charge capacitance *(see* Depletion layer capacitance)
Space-charge layer *(see* Depletion layer)

Index

Space-charge limited current, 97-101
Spectra, optical, 152
Spin:
 exclusion principle, 152
 quantization, 150
Sputtering, 251, 377
Square well potential:
 one-dimensional, 144-47
 quantization, 145
 uncertainty principle, 147
 wave functions, 145, 147
 three-dimensional, 147-48
 density of states, 156
 eigensolutions, 148
Stimulated emission, 153, 306
Stirling's approximation, 28
Storage capacitance (see Diffusion capacitance)
Switch, idealized, 3
Switching:
 bilateral p-n-p-n diode, 339
 p-n junction diode, 294
 p-n-p-n diode, 332
 Schottky barrier diode, 295
 SCR, 337
 transistors, 310
 tunnel diode, 300
Symbol, circuit (see specific devices)

T

Temperature, thermodynamic, 30, 159
Terminal characteristic:
 JFET, 345
 junction transistor, 324, 328
 MOSFET, 363, 365
 (see also I-V characteristic)
Thermal emission, 198-201
 applied field, 202
 cathode materials (table), 201
 current density, 200
Thermistor, 238
Thermodynamic equilibrium, 24, 28, 158, 162
Thick-film integrated circuit, 393, 396
Thin-film deposition, 376
Thin-film integrated circuit, 393-96
Thin-film transistor, 393
Transistor:
 junction, bipolar, 311-32
 JFET, 343-52
 MOSFET, 359-67
 (see also Integrated circuits)
Transitions:
 direct, 225, 305

Transitions (cont.)
 indirect, 226, 305
Traps, 214
Traveling wave tube, 95
Triggering, SCR, 337
Tunnel diode: 296-301
 amplifier, 300
 bistable switch, 300
 I-V characteristic, 299
 negative differential conductivity, 299
Tunneling, 136, 296
Turn-on voltage, MOS:
 inversion, 357
 MOSFET, enhancement-type, 364
Two-diode model, junction transistor, 313
Two-transistor model, p-n-p-n diode, 336

U

Uncertainty principle, 140
 square well, 147
 wave packet, 140
Unipolar, 343
Unit cell, 170

V

Valence band:
 definition, 175
 effective density of states, 220
 hole density, 225, 253 (8-5)
Variables, separation of, technique, 128
Velocity:
 drift, 66
 group, 121, 142, 184
 relation to mobility, 46
 r.m.s., 39 (2-1), 72
 scatter-limited, 232
 thermal (see Velocity, r.m.s.)
 wave packet, 142
Voltage, turn-on, MOS:
 inversion, 357
 MOSFET, enhancement-type, 364
Voltage-controlled resistance, 349

W

Wave equation (see Schrödinger's equation)
Wave function:
 boundary conditions, 130
 free particle, 133
 normalization condition, 130
 physical interpretation, 129

Wave function *(cont.)*
  probability density, 129
  stationary state, 129
  time dependence, 129
Wave mechanics *(see* Quantum mechanics)
Wave motion, 120
Wave packet, 137-42
  amplitude spectrum, 139
  Fourier transform pair, 139
  group velocity, 142
  motion, 141
  superposition of plane waves, 138
  uncertainty principle, 140
Wave-particle duality, 123-24
  Davisson-Germer experiment, 125, 165
    (5-4)

Wave-particle duality *(cont.)*
  de Broglie hypothesis, 124
Work function:
  comparison of experimental data, 206
    (7-3)
  metals, 192
  n-type semiconductors, 243
  p-type semiconductors, 244

Z

Zener breakdown:
  p-n junction, 283
  tunnel diode, 296